International Review of

A Survey of

Cytology Cell Biology

VOLUME 161

International Review of Cytology

A Survey of Cell Biology

Edited by

Kwang W. Jeon
Department of Zoology
The University of Tennessee
Knoxville, Tennessee

Jonathan Jarvik
Department of Biological Sciences
Carnegie Mellon University
Pittsburgh, Pennsylvania

VOLUME 161

ACADEMIC PRESS
San Diego   New York   Boston   London   Sydney   Tokyo   Toronto

Academic Press, Inc.
A Division of Harcourt Brace & Company
525 B Street, Suite 1900, San Diego, California 92101-4495

United Kingdom Edition published by
Academic Press Limited
24-28 Oval Road, London NW1 7DX

International Standard Serial Number: 0074-7696

International Standard Book Number: 0-12-364564-6

PRINTED IN THE UNITED STATES OF AMERICA
95   96   97   98   99   00   EB   9   8   7   6   5   4   3   2   1

CONTENTS

Analyzing Renal Glomeruli with the New Stereology

John F. Bertram

Membrane Mechanisms and Intracellular Signalling in Cell Volume Regulation

Else K. Hoffmann and Philip B. Dunham

Bacterial Stimulators of Macrophages

Sunna Hauschildt and Bernhard Kleine

CONTRIBUTORS

Numbers in parentheses indicate the pages on which the authors' contributions begin.

John F. Bertram (111), *Department of Anatomy and Cell Biology, University of Melbourne, Parkville, Victoria 3052, Australia*

Philip B. Dunham (173), *Biological Research Laboratories, Syracuse University, Syracuse, New York 13244*

Sunna Hauschildt (263), *Institut für Immunobiologie der Universität, D-79104 Freiburg, Germany*

Else K. Hoffmann (173), *Biological Department, August Krogh Institute, University of Copenhagen, DK-2100 Copenhagen O, Denmark*

Shigeyuki Kawano (49), *Department of Biological Sciences, Graduate School of Science, University of Tokyo, Hongo, Tokyo 113, Japan*

Michael Klagsbrun (1), *Department of Surgery, Harvard Medical School and Children's Hospital, Boston, Massachusetts 02115*

Bernhard Kleine (263), *Institut für Immunobiologie der Universität, D-79104 Freiburg, Germany*

Tsuneyoshi Kuroiwa (49), *Department of Biological Sciences, Graduate School of Science, University of Tokyo, Hongo, Tokyo 113, Japan*

Marsha A. Moses (1), *Department of Surgery, Harvard Medical School and Children's Hospital, Boston, Massachusetts 02115*

Yuen Shing (1), *Department of Surgery, Harvard Medical School and Children's Hospital, Boston, Massachusetts 02115*

Hiroyoshi Takano (49), *Department of Biological Sciences, Graduate School of Science, University of Tokyo, Hongo, Tokyo 113, Japan*

The Role of Growth Factors in Vascular Cell Development and Differentiation

Marsha A. Moses, Michael Klagsbrun, and Yuen Shing
Department of Surgery, Harvard Medical School and Children's Hospital,
Boston, Massachusetts 02115

The control of vascular growth and differentiation is a complex system of activity and interaction between positive and negative modulators of these processes. A number of important stimulators and inhibitors of both smooth muscle cells and endothelial cells have now been purified and biochemically characterized. Imbalances in the activity of these factors can result in serious pathologies. In this chapter, we briefly discuss the biology of blood vessel development and growth, review the current literature which describes these stimulators and inhibitors, and discuss current therapeutic strategies designed around these growth modulators.

KEY WORDS: Endothelial cells, Smooth muscle cells, Vascular cell inhibitors, Tumor angiogenesis, Atherosclerosis.

I. Introduction

Normal vascular growth and development is the result of a series of complex cellular and molecular mechanisms which act in concert to guarantee a properly functioning vascular system. Deregulated control of vascular growth can result in a number of serious pathologies. In this chapter, we review the current literature describing the factors, both stimulatory and inhibitory, which can influence the function of the normal blood vessel, and discuss the therapeutic potential of some of these factors, some of which are already being tested in the clinic.

II. Blood Vessel Development

In the mature vasculature, large blood vessels are composed of three distinct layers: the intima, the media, and the adventitia. These layers can be distinguished from each other on the basis of unique structural and functional characteristics. The intima is composed of a single layer of endothelial cells (ECs) which remains in direct contact with circulating blood, the subendothelial membrane, and the internal elastic lamina. Smooth muscle cells (SMCs) interspersed with elastic fibers comprise the media layer. The adventitia is composed of small capillaries and connective tissue (Dodge and D'Amore, 1992). It has been demonstrated that ECs of the intima and SMCs of the media interact by way of myoendothelial junctions. As revealed by ultrastructural studies, these specialized membrane contacts are made through fenestrae in the internal elastic lamina (Simionescu and Simionescu, 1983). For an extensive review of the developmental biology of vascular smooth muscle cells, see Schwartz et al. (1986).

In the microvasculature, the vessels are composed of two types of cells: the EC and the microvascular mural cell, the pericyte. The lumen of a mature capillary and postcapillary venule is formed by an EC joining with another EC or with itself. The pericytes are intimately associated with the ECs, forming a layer of cells which covers the abluminal EC surface to varying degrees, depending on the site of the vascular bed (Sims, 1986). In this anatomical context, there are frequent sites of contact between the EC and the pericyte (Orlidge and D'Amore, 1987).

It has been suggested that pericytes may be the microvascular analog of SMCs in the large blood vessels (Rouget, 1879; Zimmerman, 1923; Meyrick et al., 1981; Rhodin and Fujita, 1989). It has also been suggested that pericytes that have differentiated in response to certain physiological or pathological stimuli may be the precursors of SMCs (Meyrick et al., 1981; Rhodin and Fujita, 1989). Pericytes and SMCs do share certain structural and functional characteristics, such as the presence of the contractile proteins, muscle actin (Herman and D'Amore, 1985) and muscle myosin, and tropomyosin (Joyce et al., 1985); an inhibited proliferatory response to heparin (Castellot et al., 1982; Orlidge and D'Amore, 1986); and the ability of both cell types to modulate EC growth (Orlidge and D'Amore, 1987). There are, however, important differences between these two cell types. For example, pericytes contain nonmuscle isoactin localized along stress fibers (Orlidge and D'Amore, 1987); are never thicker than a single layer; and are surrounded by a continuous basement membrane that surrounds the capillary (Crocker et al., 1970; Castellot et al., 1982; Davies et al., 1985).

The interactions between EC and pericytes appear to regulate capillary growth. Pericytes have been shown to inhibit EC proliferation in vitro, an inhibition which is dependent upon cell–cell contact (Orlidge and D'Amore, 1987). This inhibitory process is mediated by the activation of latent transforming growth factor-β (TGF-β) produced by both cell types (Antonelli-Orlidge et al., 1989; Sato and Rifkin, 1989). On the other hand, in the absence of contact, ECs can stimulate the growth of cocultured vascular pericytes (Yamagishi *et al.*, 1993). Furthermore, it appears that this proliferative activity is mediated by endothelin 1 (Yamagishi *et al.*, 1993). It has been speculated that endothelial dysfunction, such as the deficiency of endothelin 1, could slow down or halt pericyte growth, which may in turn allow endothelium to grow unchecked.

There is a marked heterogeneity between large vessel and microvascular endothelium. Differences have been observed not only in terms of their morphological heterogeneity but also in such functional characteristics as the expression of certain surface antigens and responses to growth factors, among others (Gerritsen, 1987; Turner *et al.*, 1987; Fajardo, 1989).

It is well recognized that extracellular matrix (ECM) plays a pivotal role in the angiogenic process. Both acidic fibroblast growth factor (aFGF) and basic fibroblast growth factor (bFGF) have been shown *in vitro* and *in vivo* to bind tightly to ECM (Vlodavsky *et al.*, 1987; Weiner and Swain, 1989; Gonzalez *et al.*, 1990), presumably by their high affinity for heparin-like molecules. Basic FGF is an essential component of ECM for the stimulation of cell proliferation and differentiation (Rogelj *et al.*, 1989). However, neovascularization is also dependent on the precise regulation of the synthesis and degradation of the ECM (Banda *et al.*, 1987; Ingber and Folkman, 1989). Numerous studies have been reported on the effects of matrix composition and organization on endothelial cell replication (Ingber *et al.*, 1987; Ingber, 1990; Ingber and Folkman, 1989; Madri and Marx, 1992). Cultured endothelial cells under permissive conditions form tubes with patent lumina, a process referred to as angiogenesis *in vitro* (Foldman and Haudenchild, 1980; Ingber and Folkman, 1989). It has recently been proposed that type I collagen and SPARC (secreted protein that is acidic and rich in cysteine) are specifically related to the tube formation displayed by endothelial cells *in vitro* (Iruela-Arispe *et al.*, 1991b). Moreover, it has been suggested that the mechanochemical properties of ECM are able to translate the forces of cellular traction into templates that direct the formation of the cellular network *in vitro* (Vernon *et al.*, 1992).

Vascular endothelial growth factor (VEGF), when added to microvascular endothelial cells grown on the surface of three-dimensional collagen gels, induces the cells to invade the underlying matrix and to form capillary-like tubules (Pepper *et al.*, 1992). Although its activity is only half as potent as bFGF, a potent synergism between VEGF and bFGF is observed (Goto *et al.*, 1993). This synergism may play an important role in the control of angiogenesis *in vivo*. In addition to the requirement of growth factors for capillary tube formation, it has been reported that synthesis of certain type of oligosaccharides by the endothelial cells also appears to be required (Nguyen *et al.*, 1992).

An accumulating body of evidence suggests that in addition to the modulating proteins discussed earlier, certain proteolytic enzymes play a key role in controlling vascular growth. These include heparanase, plasminogen activator, and certain matrix metalloproteinases. In studies examining the role of heparanase in neovascularization, Vlodavsky and co-workers (1988) demonstrated that bFGF is sequestered in the ECM bound to heparan sulfate and is released in an active form when the ECM-heparan sulfate is degraded by heparanase (Vlodavsky *et al.*, 1988). Recent studies have evaluated the direct effect of heparanase on the angiogenic process. Heparanase I (heparin lyase, EC 4.2.2.7) and heparinase III (heparan-sulfate lyase, EC 4.2.2.8) but not heparinase II (no EC number) have been shown to inhibit both growth factor-stimulated capillary EC proliferation *in vitro* and angiogenesis *in vivo* (Sasisekharan *et al.*, 1994). The authors have suggested that the differences in the effects of the three enzymes on neovascularization may be due to the different substrate specificities of the enzymes, which in itself would influence the availability of the specific heparin fragments that modulate the heparin-binding cytokines involved in angiogenesis.

The serine proteases have also been shown to play a role in the process of neovascularization. The subendothelial ECM contains tissue type- and urokinase type-plasminogen activators which can participate in cell invasion, tissue remodelling, and angiogenesis. Plasminogen activators (PA) convert plasminogen to plasmin, which can activate other matrix-degrading proteases such as collagenase (Saksela *et al.*, 1987). Capillary ECs have been shown to produce both PA and collagenase (Gross *et al.*, 1982), enzymes that have been demonstrated to be required for the degradation of the basal lamina and interstitial stroma by ECs during neovascularization (Gross *et al.*, 1983). Basic FGF induces increased production of PA by ECs as well as increasing the number of PA receptors on the cell surface (Pepper *et al.*, 1987). Montesano and co-workers (1983, 1990) and Pepper and co-workers (1990) have shown that the inherent serine proteinase/proteinase-inhibitor balance can significantly affect the morphology of a capillary tubule so that the inhibition of serine protease activity results in the formation of solid endothelial cords rather than capillary tubes (Pepper *et al.*, 1990).

During neovascularization, ECs must breach the investing membrane of the parent venule, and migrate through the surrounding matrix in response to an angiogenic stimulus, eventually forming a capillary sprout. A key enzyme family whose activity is required in these processes is the matrix metalloproteinase (MMP) family. It is important that the angiogenic protein bFGF has been shown to increase the synthesis of the MMP collagenase by ECs (Rifkin *et al.*, 1990). Blocking MMP activity can block the process of neovascularization. For example, Mignatti and co-workers (1989) reported that TIMP-1, the tissue inhibitor of metalloproteinases, type-1, inhibits the migration of ECs on a human amniotic membrane. Cartilage-derived inhibitor (CDI), a MMP inhibitor, inhibits angiogenesis *in vivo* and *in vitro* (Moses *et al.*, 1990). This MMP inhibitor blocks FGF-stimulated capillary EC proliferation and migration, as well as capillary tube formation *in vitro* (Moses and Langer, 1991). In addition, TIMP-2 has been shown to inhibit FGF-stimulated capillary EC proliferation as well (Murphy *et al.*, 1993). Taken together, these results suggest that shifts in the proteolytic balance, either positively (proenzyme) or negatively (proinhibitor), can have profound effects on vascular growth.

III. Vascular Cell Stimulators

The questions of how normal blood vessels develop and how migration and proliferation of endothelial cells and smooth muscle cells are induced in pathological processes such as tumor-induced angiogenesis and SMC hyperplasia have resulted in the identification, characterization, purification, and cloning of numerous vascular cell growth factors. The best-characterized ones are fibroblast growth factor (Klagsbrun, 1989; Burgess and Maciag, 1989), platelet-derived growth factor (PDGF) (Raines *et al.*, 1990), and vascular endothelial cell growth factor/vascular permeability factor (VEGF/VPF) (Ferrara and Henzel, 1989; Gospodarowicz and Lau, 1989; Connolly *et al.*, 1989; Leung *et al.*, 1989; Plouet *et al.*, 1989). FGF is mitogenic for both EC and SMC. On the other hand, PDGF stimulates SMCs but not ECs, while VEGF is highly specific for EC. More recently, two members of the epidermal growth factor (EGF) family, heparin-binding EGF-like growth factor (HB-EGF) (Higashiyama *et al.*, 1991) and betacellulin (Shing *et al.*, 1993), have been identified as potent mitogens for SMCs but not for ECs.

A. Fibroblast Growth Factors

Fibroblast growth factors are multifunctional polypeptides which play an important role in a variety of processes such as embryonic development, neuronal survival, angiogenesis, wound healing, and tumorigenesis. The FGFs constitute a family of at least nine structurally related polypeptides that include acidic FGF, basic FGF, *int-2*, *hst*/K-FGF, FGF-5, FGF-6, keratinocyte growth factor (KGF) (Klagsbrun, 1989; Burgess and Maciag, 1989; Brem and Klagsbrun, 1993), FGF-8 (Tanaka *et al.*, 1992) and FGF-9 (Miyamoto *et al.*, 1993). A strong affinity for heparin is an important property of most FGF family members. The role of heparin-like molecules may be to stabilize the haprin-binding growth factors by protecting them from denaturation and degradation. FGFs have also been shown to interact with cell surface heparan sulfate proteoglycan (HSPG). Cell surface HSPGs constitute a class of relatively low-affinity binding sites (K_D of 2×10^{-9}, 600,000–1,000,000 sites per cell) for bFGF (Moscatelli, 1987). While bFGF is internalized through its high-affinity receptor, it is also internalized directly through its interaction with cell-surface heparan sulfates (Roghani and Moscatelli, 1992). In addition, the binding of bFGF to cell-surface HSPG is a prerequisite for high-affinity receptor binding (Yayon et al., 1991) and bFGF-induced myoblast differentiation (Rapraeger et al., 1991). More recent studies have also demonstrated that in murine neural precursor cells, the activity of FGF is regulated by HSPG, and this interaction is a prerequisite for the binding of growth factor to the signal-transducing tyrosine kinase receptor (Nurcombe et al., 1993).

FGF receptors are tyrosine kinases. Four homologous receptor proteins encoded by different genes have been identified (Werner et al., 1992). FGF receptors can also bind heparin-like molecules via a specific sequence in the extracellular domain of the FGF tyrosine kinase receptor, suggesting that heparin or heparan sulfate glycosaminoglycan is part of the active FGF binding site of the FGF receptor (Kan et al., 1993). This view is supported by the fact that point mutations of lysine residues in this FGF receptor sequence abrogate both heparin- and FGF-binding activities of the receptor. Taken together, it is likely that activation of signal transduction pathways linked to the FGF receptor may involve a three-way interaction between FGF, tyrosine kinase receptor, and heparan sulfate proteoglycan. Recently, the crystal structure of aFGF complexed to the heparin-like sucrose octasulfate has been determined, which suggests a tight binding between aFGF and the heparin-like molecule (Zhu et al., 1993).

It is well documented that aFGF and bFGF are potent mitogens of ECs and SMCs. Both aFGF and bFGF stimulate angiogenesis in the chick chorioallantoic membrane (CAM), in the rabbit cornea, and in sponges implanted subcutaneously into rats (Klagsbrun, 1989). More recently, it has been demonstrated in a canine experimental myocardial infarct model that bFGF increases the number of arterioles and capillaries in the infarct (Yanagisawa-Miwa et al., 1992). Basic FGF has been shown to enhance myocardial collateral flow in a canine model as well (Unger et al., 1994). Therefore, bFGF might be of therapeutic use in reducing damage to ischemic myocardium by stimulating the development of coronary collateral vessels and blood flow. Basic FGF has also been shown to substantially increase the SMC proliferation in vivo that occurs after infusion of the mitogen into rats subjected to carotid artery injury (Lindner et al., 1990). In addition, antibodies that neutralize bFGF decrease SMC proliferation by 80% following carotid injury (Lindner and Reidy, 1991; Olson et al., 1992). Therefore, both exogenous and endogenous bFGF can stimulate

SMC hyperplasia *in vivo* following injury, suggesting a possible role for this mitogen in mediating the repair of vascular vessels.

EC (Vlodavsky *et al.*, 1987) and SMC (Weich *et al.*, 1990) synthesize both aFGF and bFGF. EC and SMC-derived FGF is mostly cell associated and some of it can be released by injuring the cells *in vitro*. Therefore, one might speculate that, in response to vascular injury and ischemia *in vivo*, endogenous FGF is released by injured EC and/or SMC and is available for stimulating vascular cell proliferation and neovascularization.

B. Vascular Endothelial Growth Factor

Vascular endothelial growth factor, also known as vascular permeability factor, is a specific mitogen for vascular endothelial cells. Molecular characterization of the VEGF/VPF gene and protein has revealed that VEGF/VPF is structurally related to the B chain of platelet-derived growth factor (Keck *et al.*, 1989) and exists in four different glycosylated homodimeric isoforms (VEGF$_{121}$, VEGF$_{165}$, VEGF$_{189}$, and VEGF$_{206}$) (Leung *et al.*, 1989; Houck *et al.*, 1991; Tischer *et al.*, 1991). VEGF$_{121}$ does not bind heparin and is a diffusible mitogen. In contrast, the longer isoforms are bound to heparin-containing extracellular matrix but can be released in a diffusible form by a variety of agents such as heparin and plasmin (Houck *et al.*, 1992; Park *et al.*, 1993).

VEGF/VPF binds to its receptors, which are specifically expressed on vascular endothelial cells. This binding appears to be potentiated by heparin. Indeed, VEGF/VPF's interaction with heparan sulfate proteoglycan is required for its binding to the high-affinity receptor (Gitay-Goren *et al.*, 1992). Three tyrosine kinases have been identified as VEGF receptors. The flt-1 (de Vries *et al.*, 1992) and flk-1/KDR (Terman *et al.*, 1992; Millauer *et al.*, 1993; Quinn *et al.*, 1993) receptors have been shown to bind VEGF with high affinity. The newly cloned flt4 receptor tyrosine kinase gene encodes a protein which shows a high degree of amino acid sequence similarity with the flt1 and flk1/KDR proteins (Galland *et al.*, 1993). All of these three receptors contain an extracellular region composed of seven immunoglobulin-like domains associated with an interrupted intracellular kinase domain. It is now clear that the receptor tyrosine kinases encoded by this flt gene family contribute significantly to regulation of the growth and differentiation of blood vessels (Kaipainen *et al.*, 1993).

VEGF/VPF is a potent angiogenesis factor *in vitro* as well as *in vivo*. When added to microvascular ECs grown on the surface of three-dimensional collagen gels, VEGF/VPF stimulates the cells to invade the underlying matrix and to form capillary-like tubes (Pepper *et al.*, 1992). VEGF/VPF is significantly less potent than bFGF in stimulating angiogenesis *in vitro*. However, when added together, VEGF/VPF and bFGF induced an angiogenic response far greater then additive and with greater rapidity then either EC mitogen alone. Therefore, these two EC mitogens act synergistically, presumably because they activate different high-affinity receptors on EC surfaces.

VEGF/VPF is synthesized and secreted by a number of transformed cell lines. *In situ* hybridization studies demonstrate intense expression of VEGF mRNA in a variety of human tumors. Recently, it has been shown that VEGF/VPF is an important tumor angiogenesis factor (Shweiki *et al.*, 1992; Plate *et al.*, 1992). Direct evidence for the role of VEGF in tumor angiogenesis has been provided by the finding that anti-VEGF monoclonal antibodies have the ability to block the growth of several human tumor cell lines in nude mice, but the antibodies have no inhibitory effect on the growth rate of the tumor cells *in vitro* (Kim *et al.*, 1993). Histologic examination reveals a decreased density of vessels in the antibody-treated tumors compared with controls. Furthermore, VEGF/VPF mRNA was expressed in tumor cells but not in tumor blood vessels.

Cellular production of VEGF/VPF is hypoxia-inducible. It has been shown that VEGF mRNA levels are dramatically increased within a few hours of exposing different cell cultures to hypoxia and return to background when a normal oxygen supply is resumed (Shweiki *et al.*, 1992). Furthermore, it has also been shown that hypoxic conditions can promote cardiac vascular cell growth by inducing VEGF expression (Ladoux and Frelin, 1993). VEGF protein is able to induce collateral vessel growth in models of hind-limb ischemia, suggesting that VEGF has a therapeutic potential for disorders characterized by impaired tissue perfusion, such as obstructive atherosclerosis.

C. Angiogenin

Angiogenin is a 14-kDa polypeptide first purified from conditioned medium of a human adenocarcinoma cell line based on its ability to induce neovascularization in the chick chorioallantoic membrane assay (Fett *et al.*, 1985). It has also been purified from normal plasma (Shapiro *et al.*, 1987; Bond and Vallee, 1988) and milk (Maes *et al.*, 1988). Angiogenin is highly homologous to pancreatic ribonucleases (Strydom *et al.*, 1985) and has a unique ribonucleolytic activity (Shapiro *et al.*, 1986; St. Clair *et al.*, 1987). This enzymic activity is essential for angiogenin to induce neovascularization.

Angiogenin stimulates endothelial cells to produce diacylglycerol (Bicknell and Vallee, 1988) and secrete prostacyclin (Bicknell and Vallee, 1989) by phospholipase activation; binds specifically to calf pulmonary artery endothelial cells (Badet *et al.*, 1989); and supports endothelial and fibroblast cell adhesion (Soncin, 1992). These experimental results indicate that angiogenin might act via a specific cell-membrane receptor. Indeed, in one report it has been demonstrated that the mitogenic activity of angiogenin on bovine brain capillary endothelial cells can be correlated with the occurrence of a specific membrane receptor of 49 kDa on this type of cell (Chamoux *et al.*, 1991). On the other hand, an angiogenin binding protein has recently been purified from a transformed endothelial cell line and identified as a member of the actin family (G. F. Hu *et al.*, 1993). This binding protein is a cell-surface protein with an apparent molecular mass of 42 kDa. It is not clear as yet whether actin is the functional receptor of angiogenin. However, it is postulated that binding of angiogenin to actin molecules could lead to endothelial cell detachment from the ECM followed by migration, proliferation, and differentiation into microvessels, all of which are components of the angiogenic process (G. F. Hu *et al.*, 1993).

In a related report, the roles of His-13 and His-114 in the ribonucleolytic and angiogenic activities of angiogenin have been studied using site-directed mutagenesis. Replacement of either of these residues by alanine significantly decreased the enzymatic activity of angiogenin toward tRNA (app. 10,000-fold) and abolished angiogenic activity on the CAM (Shapiro and Vallee, 1987). Taken together, these studies suggest that ribonuclease inhibitor may participate in the *in vivo* regulation of angiogenin as a stimulator of neovascularization.

D. Platelet-Derived Growth Factor

Platelet-derived growth factor is a potent mitogen that stimulates the migration and proliferation of smooth muscle cells. It is composed of two distinct but homologous peptides, PDGF A ($\sim$17 kDa) and PDGF B ($\sim$14 kDa) (Raines *et al.*, 1990). Dimerization of PDGF A and B to form PDGF AA, BB, or AB is crucial for the activity of PDGF. The PDGF receptor is a 170–180-kDa tyrosine kinase composed of two subunits, α- and β-chain. It also exists in three possible dimeric forms denoted as $\alpha\alpha$, $\beta\beta$, and $\alpha\beta$ receptors. PDGF AA can only bind to the $\alpha\alpha$ receptor; PDGF BB can bind to all three receptors; and PDGF AB can bind to either $\alpha\alpha$ or $\alpha\beta$ receptors. While PDGF has typically not been considered to be a mitogen for EC, it has been demonstrated that PDGF-BB is chemotactic for rat brain capillary ECs *in vitro* and stimulates angiogenesis *in vivo* (Risau *et al.*, 1992). These data suggest that PDGF might be involved in EC migratory rather than proliferative events that accompany angiogenesis. On the other hand, PDGF may act as an indirect angiogenic factor *in vivo* by the induction of bFGF and VEGF gene expression in cells resident near endothelial cells (Brogi *et al.*, 1994). In addition, it has been demonstrated that PDGF might amplify angiogenesis through direct action on endothelially expressed PDGF β-receptors (Battegay et al., 1994).

Both EC and SMC synthesize PDGF (Casscells, 1992). ECs express predominantly PDGF B mRNA while SMCs express predominantly PDGF A mRNA in culture. A number of agents such as angiotensin II, TGF-β, thrombin, and interleukin-1 (IL-1) (Raines et al., 1989) induce PDGF mRNA expression and PDGF secretion by SMCs. A recent study has demonstrated that metabolic labeling of ECs overexpressing PDGF B revealed an inefficient rate of constitutive PDGF release, with the majority of newly synthesized PDGF remaining associated with the extracellular matrix (Soyombo and DiCorleto, 1994). Thrombin treatment, however, led to a dramatic increase in the amount of PDGF released into the medium. It is suggested that modulation of PDGF release by selective cleavage of preformed, matrix-bound precursors may represent a significant mechanism for acute regulation of release of this growth factor independent of the rate of synthesis. On the other hand, in a recent study on the effects of thrombin and PDGF on the induction of immediate-early gene expression and cellular proliferation in smooth muscle cells, it was demonstrated that thrombin is a hypertrophic agent and that PDGF is a proliferative agent (Rothman et al., 1994). PDGF has been speculated to play a role in atherosclerosis (Ross, 1986). However, no cause-effect relationship between the expression of this mitogen and the SMC replication that occurs in this disease has yet been established.

E. EGF-like Growth Factors

Heparin-binding EGF-like growth factor (HB-EGF) is a recently described member of the epidermal growth factor family and was given that name because it has a strong affinity for immobilized heparin (Higashiyama et al., 1991, 1992; Abraham et al., 1993). HB-EGF is synthesized by inflammatory cells, including monocytes and macrophages (Higashiyama et al., 1991), CD4$^+$ T lymphocytes (Blotnick et al., 1994), and eosinophils (Powell et al., 1993); vascular cells, including endothelial cells (Yoshizumi et al., 1992) and smooth muscle cells (Dluz et al., 1993; Klagsbrun and Dluz, 1993; Nakano et al., 1993; Higashiyama et al., 1994); and epithelial cells, including keratinocytes (Marikovsky et al., 1993) and uterine epithelial cells (Das et al., 1994). It is a potent mitogen for fibroblasts, smooth muscle cells, and keratinocytes, but not endothelial cells.

Mitogenic activity is mediated by interactions of HB-EGF with EGF receptor (Higashiyama et al., 1991, 1992), although it is not clear yet exactly which of the EGF receptors, individually or in combination are involved. There are several possible physiological roles for HB-EGF. A role for HB-

EGF in wound healing is suggested by its appearance in wound fluid within 1 day following skin injury (Powell *et al.*, 1993), its synthesis by inflammatory cells (Higashiyama *et al.*, 1991; Blotnick *et al.*, 1994), and its ability to stimulate fibroblast and keratinocyte migration and proliferation, characteristics of granulation tissue formation and reepithelialization, respectively. The induction of HB-EGF mRNA levels in uterine epithelium several hours prior to mouse blastocyst implantation, its localization only at the implantation site, and its ability to stimulate blastocyst EGF receptor phosphorylation *in vitro* suggest that uterine epithelial-derived HB-EGF might act to attach and signal the mouse blastocyst *in utero* (Das *et al.*, 1994; Wang *et al.*, 1994). Overexpression of HB-EGF might have pathological consequences, for example, in inducing smooth muscle cell hyperplasia. In a hypertension model in which rats are exposed to hyperoxia, HB-EGF, a potent smooth muscle cell mitogen, is synthesized by eosinophils associated with hypertrophic smooth muscle cell-like cells (Powell *et al.*, 1993). HB-EGF mRNA is also induced in arteries following interarterial balloon injury (M. Klagsbrun *et al.*, 1994, unpublished data). The potent mitogenic activity of HB-EGF for smooth muscle cells has been used to develop a strategy that kills smooth muscle cells. In these studies, a chimeric protein in which HB-EGF was fused to *Pseudomonas* exotoxin killed smooth muscle cells at concentrations less than 1 ng/ml (Mesri *et al.*, 1994).

HB-EGF gene expression is highly regulated. For example, in smooth muscle cells, HB-EGF levels are induced by phorbol esters; by thrombin, which is generated at the site of vascular injury during atherogenesis; and by other smooth muscle cell growth factors, such as basic fibroblast growth factor platelet-derived growth factor, and by HB-EGF itself (Dluz *et al.*, 1993; Nakano *et al.*, 1993). Induction of HB-EGF mRNA levels is rapid and transient, a property of early-response genes. In monocytes, HB-EGF levels are induced by lysphosphatidylcholine, a principal lipid mediator of lesion formation in atherosclerosis (Nakano *et al.*, 1994). Taken together, it appears that HB-EGF, a potent smooth muscle cell mitogen, is induced by factors associated with vascular injury and disease.

The structure of HB-EGF has been analyzed intensively. It exists in two biologically active forms—as a soluble secreted protein and as a transmembrane cell surface-associated precursor (Higashiyama *et al.*, 1991, 1992; Abraham *et al.*, 1993). The mature secreted form of HB-EGF is a single-chain, 20–22-kDa glycoprotein, 75–86 amino acids in length that is processed from its precursor. It has a strong affinity for immobilized heparin, a property not shared by EGF or transforming growth factor-α (TGF-α). Within the N-terminal domain of HB-EGF, there is a stretch of 21 amino acids that constitutes a heparin-binding domain responsible for binding to cell-surface heparan sulfate proteoglycan, an interaction that modulates growth factor activity (Higashiyama *et al.*, 1993; Thompson *et al.*,

1994). For example, HB-EGF is a potent mitogen for smooth muscle cells, equivalent to PDGF, but 40–50 times more potent than TGF-α or EGF (Higashiyama *et al.*, 1993). Abrogation of HB-EGF-HSPG interactions by treating cells with heparinase or with a synthetic peptide that corresponds to the heparin-binding domain inhibits HB-EGF binding to EGF receptor and bioactivity for smooth muscle cell by about 70–80% (Higashiyama *et al.*, 1993). Heparin itself modulates HB-EGF activity, being a stimulator or a growth suppressor, depending on the identity of the C-terminal amino acid (Cook *et al.*, 1994).

Besides being processed to the secreted form of HB-EGF, the larger HB-EGF precursor also exists in a form that is tethered to the cell via its transmembrane domain (Higashiyama *et al.*, 1991; Ono *et al.*, 1994; Raab *et al.*, 1994). The transmembrane form of HB-EGF appears to be the predominant one in tissues such as epithelia and skeletal muscle. It is biologically active and interacts in a juxtacrine manner with EGF receptor on neighboring cells to promote receptor phosphorylation and adhesion (Raab and Klagsbrun, 1994). For example, in coculture, cells transfected with transmembrane HB-EGF, but not control cells, stimulate phosphorylation of EGF receptor on adjacent A431 cells. In organ culture, these transfected cells adhere to mouse blastocysts, suggesting a possible role for transmembrane HB-EGF in the uterine implantation process.

Transmembrane HB-EGF can be processed to release the mature form by agents such as phorbol esters (Raab *et al.*, 1994). Processing of the transmembrane to the released form appears to be a mechanism of regulating growth factor activity. Another unique biological property of this growth factor is that transmembrane HB-EGF is the receptor for diphtheria toxin (DT) (Iwamoto *et al.*, 1994; Mitamura *et al.*, 1994; Naglich *et al.*, 1992). Cells transfected with, and expressing transmembrane HB-EGF, are DT-sensitive, unlike parental cells, which are DT resistant. Phorbol esters promote the processing of the transmembrane HB-EGF, release HB-EGF into the medium, and cells become DT resistant (Raab *et al.*, 1994). Taken together, these results indicate that HB-EGF is a multifunctional protein that can be a secreted mitogen, a cell-associated transmembrane juxtacrine factor that mediates cell–cell interaction, or a receptor for a nongrowth factor ligand.

Betacellulin is a new member of the EGF family recently isolated from the conditioned media of insulinoma cells (Shing *et al.*, 1993; Sasada *et al.*, 1993; Watanabe *et al.*, 1994). Betacellulin is a glycoprotein processed from a 177-amino acid precursor that is membrane associated to produce an 80-amino acid, mature-secreted, 32-kDa, heavily glycosylated glycoprotein. The C-terminal 50 amino acids of betacellulin are 50, 40, and 32% homologous to TGF-α, HB-EGF, and EGF respectively. Betacellulin is a potent mitogen for SMC, active at 1 ng/ml, and thus is similar in activity to PDGF and HB-EGF. It is also a potent mitogen for retinal pigment epithelial cells but not EC. The mRNA of betacellulin is widely expressed in normal tissues, with substantially higher amounts found in lung, uterus, and kidney. It is also detected in a number of tumor cell lines. The fact that betacellulin is produced by proliferating β cells and is mitogenic for SMC and pigment epithelial cells suggests the possibility that release of betacellulin from β cells could have a role in the vascular complications associated with diabetes.

Both HB-EGF and betacellulin bind to the classical EGF receptor (HER/erbB). However, it is possible that these two EGF-like growth factors might also bind more optimally to other members of the EGF receptor family; these are a group of receptor tyrosine kinases frequently overexpressed in a variety of epithelial tumors. Three other EGF-receptor homologs, HER-2/erbB-2 (the *neu* oncogene) (Coussens *et al.*, 1985; Bargmann *et al.*, 1986), HER-3/erbB3 (Kraus *et al.*, 1989), and HER-4/erbB4 (Plowman *et al.*, 1993) have been identified, but it is not clear whether they will serve as receptors for the HB-EGF or betacellulin ligands.

F. Other EC Growth Factors

Hepatocyte growth factor (also known as scatter factor) is a glycoprotein of 82 kDa, which when reduced, yielded two polypeptide chains of 69 and 34 kDa (Nakamura *et al.*, 1987). Hepatocyte growth factor/scatter factor (HGF/SC) was initially identified as a potent mitogen for epithelial cells (Russell *et al.*, 1984; Stoker and Perryman, 1985). A recent report has demonstrated that HGF/SC stimulates vascular endothelial cell migration, proliferation, and organization into capillary-like tubes *in vitro* (Bussolino *et al.*, 1992). Moreover, it induces angiogenesis *in vivo,* presumably owing to its effects on both epithelial and vascular endothelial cells (Grant *et al.*, 1993).

Placenta growth factor (PLGF), initially isolated from placenta, is a glycosylated dimeric protein which stimulates vascular endothelial cells *in vitro* (Maglione *et al.*, 1991). It is a secreted angiogenic factor composed of 149 amino acids which are significantly related to VEGF/VPF. Similar to VEGF/VPF, PLGF exists in different forms which are probably generated by differential splicing (Hauser and Weich, 1993). In a recent study it was demonstrated that mRNA for PLGF was expressed in 21 of 23 hypervascular renal cell carcinoma tissues (91%), but was not detected in the adjacent normal kidney tissues (Takahashi *et al.*, 1994). This result suggests that PLGF plays a role in tumor angiogenesis in addition to its possible physiological roles during placental development and differentiation.

Platelet-derived endothelial cell growth factor (PD-ECGF), initially isolated from platelets, is a 45-kDa, single-chain polypeptide (Miyazono *et al.*,
1987). It lacks a signal peptide and is not secreted. It stimulates EC DNA
synthesis and migration in culture and angiogenesis *in vivo* (Ishikawa *et
al.*, 1989). PD-ECGF is produced by certain normal and transformed cultured cells (Usuki *et al.*, 1989). Analysis of PD-ECGF produced by cultured
cells revealed that it contains nucleotide(s) covalently bound to serine
residues (Usuki *et al.*, 1991). Recently, PD-ECGF was reported to be identical to thymidine phosphorylase (TP) (Sumizawa *et al.*, 1993). The mechanism by which PD-ECGF/TP stimulates angiogenesis remains unclear.
However, it has been reported that the angiogenic activity of PD-ECGF/
TP is probably mediated by 2-deoxy-D-ribose, a dephosphorylated product
derived from 2-deoxy-D-ribose-1-phosphate (Haraguchi *et al.*, 1994). The
function of PD-ECGF/TP *in vivo* is not known, although its EC specificity
suggests roles in angiogenesis.

Macrophages are the major source of angiogenic activity in a healing
wound. Macrophages release tumor necrosis factor-α (TNF-α), which is a
known angiogenic mediator (Leibovich *et al.*, 1987). In addition, interleukin-
8 (IL-8) has recently been demonstrated as a macrophage-derived mediator
of angiogenesis (Koch *et al.*, 1992).

G. Other SMC Growth Factors

Vasoconstrictor peptides such as angiotensin II play an important role in
smooth muscle cell hypertrophy. Angiotensin II induces delayed mitogenesis in smooth muscle cells by increasing the expression of specific endogenous growth factors, including TGF-β1 and PDGF A-chain (Weber *et al.*,
1994). It is suggested that enhanced endogenous growth factor expression
may represent the direct mechanism by which angiogentsin II promotes
smooth muscle cell growth in some vascular hyperproliferative diseases.

IGFs are potent smooth muscle mitogens. They are secreted by a number
of cell types *in vitro* and *in vivo*, including smooth muscle cells and macrophages. It has been shown that regulation of the expression of insulin-like
growth factor I (IGF I) by competence growth factors such as PDGF may
play an important role in controlling the growth of vascular smooth muscle
cells (Delafontaine *et al.*, 1991).

Interleukin-1 induces bFGF gene expression in vascular smooth muscle
cells (Gay and Winkles, 1991). Immunoprecipitation analysis indicates that
IL-1 stimulated cells also express an increased amount of bFGF protein,
which is a potent mitogen for vascular smooth muscle cells.

Lipoprotein (a) and its constituent, apolipoprotein (a), stimulate the
proliferation of human smooth muscle cells (Grainger *et al.*, 1993). This

effect results from its ability to inhibit the generation of activated TGF-β, which is an endogenous inhibitor of smooth muscle cells (Kojima *et al.*, 1991).

IV. Bifunctional Vascular Growth Factors

A. Transforming Growth Factor-β

TGF-β, a homodimer of M_r 25,000, is secreted by a wide range of cell types in a latent form which, as discussed earlier, can be activated in culture by acidification, alkalization, or protease digestion. As described earlier, it is a stimulator of neovascularization *in vivo* and a potent inducer of capillary tube formation *in vitro* (Madri *et al.*, 1988). However, it is also a potent (less than 1 ng/ml) inhibitor of capillary and aortic EC proliferation and migration *in vitro* (Muller *et al.*, 1987, Muthukkaruppan and Auerbach, 1979). This paradoxical effect has been explained in a number of different ways. To the extent that angiogenesis is a two-stage process composed of EC proliferation and EC differentiation, it has been suggested that TGF-β facilitates the differentiation of EC, for example, into capillary tubes (Madri *et al.*, 1988) by inhibiting EC migration and proliferation (Klagsbrun and D'Amore, 1991). TGF-β might also exert an indirect effect on neovascularization by attracting growth factor-producing cells to a site where they can produce growth. In its activated form, TGF-β becomes capable of binding to its receptor and, depending on its concentration, can either stimulate or inhibit SMC proliferation, depending on SMC density (Majack *et al.*, 1990). At both high and low concentrations, TGF-β has been demonstrated to promote an increase in SMC endogenous PDGF production (Battegay *et al.*, 1991). However, at high TGF-β concentrations, expression of the PDGF receptor is downregulated, preventing the response to PDGF (Battegay *et al.*, 1991). Stimulation of SMC proliferation occurs at low TGF-β concentrations since the endogenous PDGF binds to available receptors. Kojima and co-workers (1991) have recently demonstrated that TGF-β can block the migration of SMC as well.

As noted earlier, the inhibitory process by which pericytes inhibit EC proliferation *in vitro* (Orlidge and D'Amore, 1987) is mediated by the activation of latent TGF-β produced by both cell types (Antonelli-Orlidge *et al.*, 1989; Sato and Rifkin, 1989). The activation of latent TGF-β appears to be a self-regulating system (Y. R. Sato *et al.*, 1990). The activation of TGF-β stimulates the production of plasminogen activator inhibitor-1, which subsequently blocks the activation of the protease required for conversion of latent TGF-β to biologically active TGF-β.

B. Tumor Necrosis Factor-α

TNF-α (cachectin), a 17-kDa polypeptide, is a multifunctional cytokine which plays a role in inflammation, hematopoiesis, angiogenesis, and other physiological events. First detected in activated macrophages, TNF-α has now been described in resident mouse peritoneal mast cells as well (Gordon and Galli, 1990). This protein has been reported to be an inhibitor of capillary EC and smooth muscle cell proliferation. Both basal and FGF-stimulated growth were inhibited with a half-maximal concentration of 0.5-1.0 ng/ml TNF-α (Frater-Schroder et al., 1987). Paradoxically, this protein was shown to be angiogenic in vivo in this same study as well as in others (D. E. Hu et al., 1993). It has also been shown to increase transcription of the HB-EGF gene in SMCs (Yoshizumi et al., 1992).

TNF-α does not suppress EC proliferation in vivo nor did it inhibit angiogenesis in vivo but rather it stimulated the neovascularization response. It has been suggested that, in light of the inflammatory response observed after treatment with TNF-α in these studies, the angiogenic activity of this protein may be an indirect result of leukocyte migration into the test site (Frater-Schroder et al., 1987). It has also been suggested that the angiogenic effect may be a function of the molecule's ability to attract macrophages and mast cells to a site where they might, in turn, release angiogenic stimulants (Leibovich et al., 1987; Klagsbrun and D'Amore, 1991). Further studies have shown that TNF-α stimulates chemotaxis of endothelial cells and induces capillary tube formation in vitro (Leibovich et al., 1987).

V. Vascular Cell Inhibitors

The physiological role and therapeutic value of vascular cell inhibitors have only recently been appreciated. It is now widely recognized that normal vascular development and function is the result of a delicate balance between stimulators and inhibitors of vascular cell processes, and an imbalance in favor of either of these modulators can have profound pathological ramifications. Many of the negative modulators of the microvascular system have been extensively reviewed elsewhere (Folkman, 1993; Auerbach and Auerbach, 1994). In this chapter, we review the best characterized of these inhibitors.

As is the case for inhibitors of endothelial cell proliferation, regulation of SMC growth may have broad therapeutic application. There is an extensive list of factors and compounds which, under certain experimental conditions, have been shown to have inhibitory effects on SMC growth. The mechanism by which these factors exert their inhibitory effects varies widely. Given the limited scope of this chapter, a detailed pharmacological approach to the study of inhibitors of SMC growth would be inappropriate. We have reviewed those factors and compounds which have been the most thoroughly studied to date, including certain pharmacological agents, endocrine hormones, and intracellular modulators of proliferation. For more detailed discussion, see a recent review by Jackson and Schwartz (1992).

A. Endothelial Cell Inhibitors

1. Platelet Factor 4

Platelet factor 4 (PF4) is a 30-kDa tetrameric protein which has a strong affinity for heparin and is released during platelet aggregation. The crystal structure of recombinant human PF4 (rPF4) has recently been reported (Zhang et al., 1994). Originally tested for potential antiangiogenic activity because of its strong affinity for heparin, it was shown to be a potent inhibitor of neovascularization (Taylor and Folkman, 1982). More recently, rPF4 has been shown to be an inhibitor of endothelial cell proliferation and angiogenesis in vivo using the chick chorioallantoic membrane assay (Maione et al., 1990). PF4 has also been shown to inhibit capillary tube formation in vitro in a study of the autocrine role of bFGF on capillary tube formation in vitro (Sato et al., 1991). In this report, it was suggested that PF4's mechanism of action might be through blocking the binding of bFGF to its receptor on capillary ECs.

In early studies, the antiangiogenic activity of recombinant PF4 was associated with its carboxy-terminal, heparin-binding regions since the inhibitory activity could be abrogated when heparin was tested in combination with PF4 (Maione et al., 1990). More recently, however, it has been demonstrated that an analog of human PF4 which lacks affinity for heparin retains potent angiostatic activity. This analog was actually inhibitory at lower concentrations than rPF4 in the CAM system and its inhibitory effects were not abrogated by the presence of heparin (Maione et al., 1991). In addition to its antiangiogenic property, PF4 has other biological activities (Maione et al., 1990) including chemotactic activity for neutrophils, monocytes, and fibroblasts; inhibition of bone resorption; prevention of immunosuppression; inhibition of collagenase, and others. The mechanism by which PF4 inhibits neovascularization, however, remains unknown.

Recombinant human PF4 has also been demonstrated to suppress the growth of B16-F10 murine and HCT 116 human colon carcinoma in semi-syngeneic CByB6F1/J female athymic nude mice when administered intralesionally but not intraperitoneally or intravenously. The growth of these two transformed cell lines in culture was not inhibited by rPF4 at levels (50 μg/ml) that significantly inhibited normal endothelial cell proliferation. The nonheparin-binding analog of rPF4 discussed earlier also retained the ability to inhibit the growth of these tumors as well (Maione *et al.*, 1990). Since the inhibition of the growth of these two tumors by PF4 does not appear to be a function of the inhibition of tumor cell growth, the mechanism by which tumor growth was inhibited might be the inhibition of neovascularization, lymphokine-activated killer cell activation, and/or the induction of other cytokines (Sharpe *et al.*, 1990).

2. Cartilage-Derived Inhibitor

Avascular tissues such as cartilage have been studied as potential sources of angiogenesis inhibitors. Pieces of native cartilage were shown to inhibit neovascularization in the rabbit corneal pocket assay (Brem and Folkman, 1975). Later, partially purified cartilage extracts were shown to inhibit angiogenesis in two different animal models (Langer *et al.*, 1976, 1980). However, until recently, no single cartilage-derived molecule had been shown to do the same. A cartilage-derived polypeptide (27,650 Da) has now been demonstrated to be a potent inhibitor of FGF-stimulated capillary EC proliferation and migration at nanomolar concentrations (Moses *et al.*, 1990). CDI is also an inhibitor of both capillary tube and sprout formation *in vitro* as well. In a series of *in vivo* studies, CDI was shown to inhibit embryonic neovascularization in the chick CAM assay at picomole concentrations (Moses *et al.*, 1990). CDI also inhibits tumor-induced angiogenesis *in vivo* when tested in the rabbit corneal pocket assay using V2 carcinoma as the angiogenic stimulus (Moses and Langer, 1991). Studies have shown that chondrocytes secrete an angiogenic inhibitor (ChDI) which shares the same biological activities as its tissue-derived counterpart (Moses *et al.*, 1992).

CDI is also a potent inhibitor of matrix metalloproteinases. The data described earlier represent the first demonstration that a metalloproteinase inhibitor with no other known biological activity inhibits angiogenesis. Since this discovery, it has been reported that TIMP-2 inhibits FGF-stimulated EC proliferation (Murphy *et al.*, 1993). The mechanism by which these inhibitors exert a negative effect on EC proliferation may not simply be a function of the inhibition of the matrix metalloproteinases, since not all of the metalloproteinase inhibitors inhibit EC growth. For example, TIMP-1 actually stimulates the proliferation of endothelial cells (Hayakawa *et al.*, 1992; Moses and Langer, 1991, unpublished results).

3. GP140/Thrombospondin

Thrombospondin (TSP), a major constituent of platelet alpha granules, is a 450-kDa, trimeric, adhesive glycoprotein. It is also produced by a large number of mesenchymal and epithelial cells (Bornstein, 1992) and is incorporated into their cell matrices. Its role in neovascularization was observed with a report that a 140-kDa protein (GP 140) homologous in sequence and function to the C-terminal of human thrombospondin (160 kDa) inhibited angiogenesis *in vivo* and capillary EC migration *in vitro* (Rastinejad *et al.*, 1989; Good *et al.*, 1990). A number of reports in the literature have attempted to clarify TSP's effect on the process of angiogenesis *in vitro*. TSP inhibits bFGF-stimulated migration (Rastinejad *et al.*, 1989; Taraboletti *et al.*, 1992) of capillary and aortic endothelial cells.

Interestingly, a gradient of TSP has been shown to induce EC chemotaxis (Taraboletti *et al.*, 1992). Since this induction of EC chemotaxis was inhibited by heparin, it was suggested that the heparin-binding region of the molecule is responsible for the induction of chemotaxis (Taraboletti *et al.*, 1992). TSP also inhibits the proliferation of large and small vessel EC in a cell-specific manner since TSP is a stimulator of vascular smooth muscle cells and human foreskin fibroblasts (Bagavandoss and Wilks, 1991; Taraboletti *et al.*, 1992). Both TSP and its 140-kDa fragment have been reported to reduce endothelial cell proliferative response to both serum and bFGF (Taraboletti *et al.*, 1992). These same authors have also demonstrated that TSP induces the adhesion and spreading of capillary and aortic endothelial cells (Taraboletti *et al.*, 1992). In addition, TSP interferes with the spontaneous formation of cords by endothelial cells *in vitro* (Iruela-Arispe *et al.*, 1991a). GP 140 was demonstrated to inhibit tumor-induced angiogenesis in a nude mice model (Zajchowski *et al.*, 1990). Since the onset of angiogenesis and tumorigenesis is concomitant with the loss of GP 140 activity, it was suggested that TSP might act as a functional tumor suppressor (Rastinejad *et al.*, 1989).

More recent studies have focused on determining which of the structural motifs present in thrombospondin are responsible for its antiangiogenic activity. Using peptides and truncated molecules, the angiogenesis-inhibiting activity was localized to two domains of the molecule—the procollagen homology region and the properdin-like type 1 repeats. A small peptide consisting of residues 303–309 of thrombospondin demonstrated antiangiogenic activity *in vivo* and *in vitro* (Tolsma *et al.*, 1993). In addition, a number of small peptides derived from two of the three type 1 repeats present in the intact molecule exhibited antiangiogenic activity. These results implicate at least two different structural domains of the large parent molecule in the regulation of neovascularization.

A study of the role of the tumor suppressor gene p53 in the switch to the angiogenic phenotype during malignant conversion was recently conducted using fibroblasts cultured from Li-Fraumeni patients (Dameron *et al.*, 1994). Having inherited one wild-type and one mutant allele of the p53 gene, these patients have an elevated risk of developing sarcomas and other tumors in which the wild-type allele is inactivated. It was found that p53 inhibited the angiogenic phenotype in the fibroblasts by stimulating their thrombospondin production. This study suggests that thrombospondin, or bioactive peptides derived from it, may be useful in the delay or prevention of tumors in carriers of only one wild-type p53 allele (Dameron *et al.*, 1994).

4. Interferon-α and β

Interferons have been studied with respect to their antiviral, antitumor, and immune system effects (Jasmin, 1991) as well as with respect to the vascular system. Interferon-α has been shown to inhibit EC proliferation and tube formation *in vitro*. Interferon α-2a is being used in clinical trial as a treatment for patients with life-threatening hemangioendotheliomas, tumors which are characterized as having unlimited blood vessel proliferation (Orchard *et al.*, 1989). In 18 of 20 patients treated with daily subcutaneous injections of interferon α-2a (up to 3 million units per square meter of body surface area), the hemangiomas regressed by 50% or more after an average of 7.8 months of treatment. No long-term toxicity has been observed (Ezekowitz *et al.*, 1992). More recently it has been shown that alpha and beta interferon downregulate both the transcription and production of bFGF in human renal cancer cells (Singh *et al.*, 1994). This downregulation of bFGF might serve as a mechanism by which interferon may have its antiangiogenic effect. Interferon γ inhibits growth-factor stimulated EC proliferation (Friesel *et al.*, 1987) and can also inhibit cord formation by capillary EC *in vitro* (Y. Sato *et al.*, 1990).

5. 16K Prolactin

In early studies of the formation of prolactin (PRL)-secreting tumors in rats, it was observed that direct arterialization of the tumors had developed (Elias and Weiner, 1984). To determine the factors involved in the vascularization of the pituitary, intact PRL (23 kDa) was tested for its effect on the basal and bFGF-stimulated proliferation of capillary EC. It had no effect (Ferrara *et al.*, 1991). It was also known that a 16-kDa N-terminal fragment of PRL (16K PRL) originally found in the rat pituitary was mitogenic in the mammary gland (Mittra, 1980; Clapp *et al.*, 1988). This fragment is enzymatically cleaved from the intact rat PRL along with an 8-kDa fragment. The fragments are then released following reduction of an internal sulfide bond (Compton and Witorsch, 1984; Wong *et al.*, 1986; Clapp, 1987). When these fragments were tested for their effect on EC proliferation, the 16-kDa fragment of prolactin inhibited both basal and bFGF-stimulated levels of capillary EC proliferation (Ferrara *et al.*, 1991). On the other hand, parental PRL or cleaved but nonreduced PRL had no effect, even when tested at concentrations 100-fold higher than 16K PRL.

16K PRL is an inhibitor of capillary tube formation *in vitro* and also inhibits embryonic neovascularization *in vivo* when tested in the chick CAM (Clapp *et al.*, 1993). To date, the mechanism by which 16K PRL inhibits angiogenesis is unknown. The mechanism does not appear to be through interaction with known PRL receptors nor with FGF receptors. However, Clapp and Weiner (1992) have demonstrated the presence of a specific, high-affinity, saturable binding site for 16K PRL on capillary EC which may mediate 16K PRL's inhibition of capillary EC proliferation.

6. Placental Ribonuclease Inhibitor

Since angiogenin is detected in normal plasma (Bond and Vallee, 1988), it has been suggested that its angiogenic activity must be tightly regulated (Badet *et al.*, 1990). One modulator of angiogenin's activity is human placental ribonuclease inhibitor (PRI), a 50-kDa protein which has been shown to abolish both the *in vivo* angiogenic and ribonucleolytic activities of angiogenin (Shapiro and Vallee, 1987). A recombinant form of PRI, RNAsin, caused a significant reduction in tumor growth when tested in ELVAX-coated sponges implanted subcutaneously underneath an intradermal inoculum of C755 mammary tumor cells. This antitumor effect of RNAsin correlated with its effect on tumor-induced neovascularization (Polakowski and Lewis, 1993). RNAsin also inhibits the angiogenic activity of bFGF.

A detailed analysis of the structure and action of mammalian ribonuclease inhibitor has been recently reported by Lee and Vallee (1993). Studies have also focused on the biochemical analysis of this enzyme/inhibitor complex and the kinetics of inhibition, and have determined that the placental ribonuclease inhibitor exhibits a competitive mode of inhibition (F. S. Lee *et al.*, 1989).

7. Angiogenic Modulator 1470

A synthetic analog of fumagillin, angiogenic modulator 1470 (AGM-1470; MW 401.89) has been shown to inhibit capillary EC proliferation and migration *in vitro* in nanomolar concentrations and neovascularization *in vivo*. This fungal-derived inhibitor has also been shown to suppress the growth of a wide variety of solid tumors (Ingber *et al.*, 1990). Some of these tumors include Lewis lung carcinoma, colon adenocarcinoma, and fibrosarcoma (Brem and Folkman, 1993), and VX-2 carcinoma (Kamei *et al.*, 1993) as well as the growth of several human nerve sheath tumors (Takamiya *et al.*, 1993). *In vitro* studies reveal that AGM-1470 has an inhibitory effect on the growth of eight human cultured cell lines from choriocarcinoma (Yanase *et al.*, 1993). Recently this angiogenesis inhibitor was also shown to suppress collagen-induced arthritis in experimental animals (Peacock *et al.*, 1992). It has been reported that at doses of 1–1000 ng/ml, AGM-1470 selectively inhibited the capillary-like formation of endothelial cells in the rat blood vessel organ culture assay (Kusaka *et al.*, 1991). Current studies are focusing on determining the mechanism of action of this inhibitor.

8. Thalidomide

Oral administration of the potent teratogen thalidomide has been shown to inhibit angiogenesis induced by bFGF in the rabbit corneal pocket assay (D'Amato *et al.*, 1994). This result was obtained using a teratogenic dose of thalidomide (200 mg/kg), leading the authors to postulate that the limb defects that have been observed with thalidomide treatment were secondary to angiogenesis inhibition in the developing fetal limb bud.

The mechanism by which thalidomide inhibits neovascularization is currently under study but it is known that it does not inhibit the proliferation of capillary EC. Since the antiangiogenic effect of thalidomide is observed only upon systemic administration, it is suggested that thalidomide metabolites are responsible for its ability to inhibit angiogenesis. Current studies are focused on the identification of the active thalidomide metabolites and their mechanism of action.

9. Angiostatin

Recently, a protein sharing a greater than 98% sequence homology to a 39-kDa internal fragment of plasminogen was shown to inhibit angiogenesis *in vivo* and to mediate the suppression of metastases by a Lewis lung carcinoma (O'Reilly *et al.*, 1994). This protein, angiostatin, has an N-terminal beginning at amino acid 98 of plasminogen but interestingly, whole plasminogen had no inhibitory effect. Angiostatin was purified from the serum and urine of tumor-bearing mice and specifically inhibits endothelial cell proliferation ($IC_{50} = 14.2$ ng) in a dose-dependent fashion. It had no effect on Lewis lung tumor cell growth or on the proliferation of nonendothelial cells. It is a potent inhibitor of angiogenesis in the chick CAM (O'Reilly *et al.*, 1994).

This work was an outgrowth of a hypothesis proposed to explain the phenomenon by which the removal of certain tumors can be followed by the growth of distant metastases. This hypothesis suggested that a primary tumor initiates and stimulates its own angiogenesis by generating an excess of angiogenesis stimulator(s) over inhibitor(s). The inhibitor(s), possessing a longer half-life in the circulation than the stimulator, would reach the capillary bed of the secondary tumor in higher quantities than the stimulator, resulting in an inhibition of metastases or secondary tumor growth (O'Reilly et al., 1994).

A variant of Lewis lung carcinoma was studied in which removal of the primary tumor was followed by the rapid growth of neovascularized metastases. In this model, the suppression of metastases observed in the presence of the primary Lewis lung tumor was shown to be mediated by angiostatin generated by the primary tumor (O'Reilly et al., 1994). In addition to its therapeutic potential as an inhibitor of neovascularization, angiostatin may be useful as a long-term therapy to inhibit the growth of metastases.

B. Smooth Muscle Cell Inhibitors

1. Heparin and Heparan Sulfate

Some of the most extensive studies of the inhibition of SMC proliferation are those which have focused on heparin and the structurally related compound, heparan sulfate. SMC proliferation in vitro and in vivo following rat arterial injury is inhibited by both heparan sulfate and soluble heparin (Chamley-Campbell and Campbell, 1981; Castellot et al., 1985, 1989; Herman and Castellot, 1987; Edelman et al., 1990; Rifkin and Moscatelli, 1989; Reilly and McFall, 1991). Low-molecular-weight, noncoagulatory fragments of heparin also inhibit SMC proliferation. SMC migration is also inhibited by heparin (Castellot et al., 1981; Clowes and Clowes, 1986). Although the exact mechanism by which heparin inhibits SMC proliferation remains unknown, a number of hypotheses have been postulated. Soluble heparin could displace heparin-binding growth factors, such as bFGF, PDGF, and HB-EGF heparin-binding EGF (Shing et al., 1984; Heath et al., 1991; Klagsbrun and Baird, 1991). In addition, heparin has been shown to downregulate EGF receptors (Reilly et al., 1987). Incorporation of the SMC mitogen thrombospondin into the extracellular matrix has also been shown to be inhibited by heparin (Majack et al., 1988). Heparin has been shown to alter the types and amounts of collagen made by SMC in vitro (Majack and Bornstein, 1985; Snow et al., 1990; Tan et al., 1991). As reviewed by Casscells (1992), heparin has also been shown to inhibit SMC uptake of thymidine and uridine (Castellot et al., 1985), to bind lipoproteins (Alavi et al., 1989), to bind enzymes involved in DNA synthesis such as topoisomerase II, to block Ca^{2+} release by IP_3 (Hill et al., 1987; Kobayashi et al., 1988), and to inhibit Na^+/H^+ exchange (Zaragoza et al., 1990). Clearly, heparin and its related polysaccharides play a role in the regulation of SMC growth, probably through a combination of these mechanisms.

2. Antihypertensives

a. ACE Inhibitors   Angiotensin is a powerful vasoconstrictor and is used therapeutically to raise blood pressure. Angiotensin-converting enzyme (ACE) is a dipeptidyl carboxypeptidase which catalyzes the formation of the vasoactive octapeptide angiotensin II (Ang II) from its inactive decapeptide precursor, angiotensin I. Angiotensin I is itself cleaved from angiotensinogen by renin (Jackson and Schwartz, 1992). Owing to its profound vasocontrictor effects on the vasculature, the role of the angiotensin system has been studied relative to its effects on SMC growth control. Ang II induces vascular hypertrophy and proliferation in injured and normal vessels *in vivo* (Daemen *et al.*, 1991). It stimulates hypertrophic growth of SMC *in vitro* and growth-related oncogenes (c-fos, c-jun, c-myc) are induced in SMC, as is the synthesis of autocrine growth factors such as PDGF-A, bFGF, and inhibitors such as TFG-β1 after treatment with Ang II (Naftilan *et al.*, 1989a,b; R. Itoh *et al.*, 1991; H. Itoh *et al.*, 1993; Gibbons *et al.*, 1992). The regulation of SMC growth may be a function of the balance between these proliferative and antiproliferative factors (Itoh *et al.*, 1993).

Given that Ang II stimulates SMC hypertrophy, a number of inhibitors of angiotensin-converting enzyme have been studied with respect to their effect on SMC growth. The ACE inhibitors which have been studied most extensively are as follows.

1. Ramipril.   The effect of the ACE inhibitor ramipril on the development of neointima 2 and 14 days after injury to rat aorta with a balloon catheter indicated that, while at early time points there was no significant inhibition of SMC proliferation as a response to injury, ramipril did decrease the amount of neointima formed 14 days after injury (Capron *et al.*, 1991). However, other effects such as inhibition of SMC migration and matrix synthesis should also be taken into account when considering the *in vivo* effects of ramipril (Jackson and Schwartz, 1992).

2. Benazeprilat.   In one study, the ACE inhibitor benazeprilat and the Ang II, AT1-specific receptor antagonist DuP753 were compared for their effects on intimal lesion formation as well as SMC migration and proliferation in Sprague-Dawley rats after carotid balloon injury (Prescott *et al.*, 1991). Both significantly reduced intimal lesion formation after balloon injury. Comparative studies revealed that media SMC proliferation was reduced significantly after injury by the AT1 agonist but not by the ACE inhibitor. Both reagents inhibited SMC migration. Based on these studies, it was concluded that benazeprilat reduced intimal lesion size by inhibiting SMC migration alone without affecting SMC proliferation. When both SMC proliferation and migration were inhibited, a more pronounced reduction in lesion size was obtained after AT1 antagonism (Prescott et al., 1991).

3. Cilazapril. Another ACE inhibitor, cilazapril, suppresses the proliferative response to vascular injury induced by balloon catheterization of rat carotid arteries (Powell et al., 1990). Cilazapril suppresses SMC proliferation in vivo by blocking the conversion of angiotensin I to angiotensin II, thereby suggesting a central role for Ang II in the control of the proliferative response after balloon catheter-induced vascular injury. A clinical trial of cilazapril in postangioplasty restenosis (Multicentre European/American Research Trial) revealed no effect on the minimal lumen diameter or on frequency of clinical events (SCRIP, 1991).

The combination of cilazapril and heparin inhibits intimal thickening in rat carotid arteries injured by balloon catheterization. The combination of both of these inhibitors was more effective than either one alone in inhibiting smooth muscle accumulation after rat carotid ballooning procedures. This effect on growth was reflected in decreased SMC proliferation as well (Clowes et al., 1991).

However, there are serious difficulties in interpretation when ACE inhibitors are used to study SMC growth in vivo because of the complex biochemical pathways associated with these molecules (Jackson and Schwartz, 1992). The effect of ACE inhibitors may be either a direct one as a function of decreasing angiotensin II or it could be an indirect one. For example, in addition to reducing the circulating levels of angiotensin II, ACE inhibitors also increase circulating levels of bradykinin (Mersey et al., 1977), an SMC mitogen in vitro (Paquet et al., 1989). In addition, ACE inhibitors may indirectly reduce circulating levels of factors such as aldosterone as well. In addition, since angiotensin II increases sympathetic tone and increases vasopressin release and the release of catecholamines from the peripheral nervous system, ACE inhibitors may have an indirect effect on these factors (Jackson and Schwartz, 1992).

b. Serotonin   Serotonin (5-hydroxytryptamine; 5HT) initiates DNA synthesis in quiescent SMC cultures and stimulates it in growing cultures (Kavanaugh et al., 1988; Paquet et al., 1989). In addition, serotonin alters cultured bovine pulmonary SMC configuration; i.e., causes dendritic formation (S. L. Lee et al., 1989), a morphological change that is associated with the elevation of intracellular cAMP (Lee et al., 1991).

Ketanserin, a serotonin type 2 receptor antagonist, is also a well-documented antihypertensive drug. It inhibits rat vascular SMC growth, probably through the suppression of DNA replication in S phase (Uehara *et al.*, 1991). This antiproliferative effect could also contribute to the vascular protective effects of ketanserin.

3. Somatostatin

This tetradecapeptide hormone inhibits pituitary release of growth hormone and is found in the hypothalamus and in sympathetic nervous system ganglia. Angiopeptin, an octapeptide analog of somatostatin, has been shown to inhibit DNA synthesis in rat carotid artery explants (Vargas *et al.*, 1989). *In vivo* administration of this compound also inhibited neointimal thickening and DNA synthesis in the rat common carotid artery after air-drying injury (Lundergan *et al.*, 1989) and in balloon-injured rabbit arteries (Asotra *et al.*, 1989; Conte *et al.*, 1989). Since this effect was not mimicked by other somatostatin analogs which also inhibit the release of growth hormone, it has been argued that the inhibitory effect of angiopeptin is not mediated by growth hormone (Vargas *et al.*, 1989).

4. Eicosanoids

Prostaglandins, prostacyclins, thromboxanes, and leukotrienes are members of the eicosanoid family of fatty acid oxidation products. A number of investigators have demonstrated that arterial SMC can produce eicosanoids (Jackson and Schwartz, 1992) as well as being significantly modified by these eicosanoids. Fatty acids and their prostaglandin derivatives have been shown to be inhibitors of aortic SMC. For example, PGE_1 and, to a lesser degree, PGE_2, inhibit SMC proliferation. Their precursor fatty acids also suppressed cell proliferation but to a lesser extent. The accumulation of the precursor fatty acids as cholesteryl esters in fatty streaks may decrease their availability for prostaglandin biosynthesis and lead to SMC proliferation (Huettner *et al.*, 1977). Other prostaglandins such as PGA_1, PGA_2, PGB_1, PGD_2 (Nilsson and Olsson, 1984; Smith *et al.*, 1984; Orekhov *et al.*, 1986), and PGJ_2 have also been shown to inhibit SMC proliferation.

Prostacyclin (PGI_2), in both its authentic form as well as in its analog carbacyclin form, has been shown to inhibit tritiated thymidine incorporation by primary human aortic intimal SMCs (Orekhov *et al.*, 1983, 1986; Akopov *et al.*, 1988). See a recent review of the pharmacology of smooth muscle replication by Jackson and Schwartz (1992) for a detailed presentation of the effects of these pharmacological agents on SMC proliferation.

5. Nitrous Oxide

Nitrous oxide (NO) has also been shown to play a role in the vascular system. It is known that in the blood vessels, NO is released by EC and eventually reaches the SMC, ultimately dilating the blood vessel and lowering blood pressure. Nitric oxide-generating vasodilators such as sodium nitroprusside, S-nitro-N-acetylpenicillamine, and isosorbide dinitrate dose-dependently inhibit serum-induced tritiated thymidine incorporation by rat aortic SMCs (Garg and Hassid, 1989). 8-Bromo-cyclic guanosine monophosphate also mimicked the inhibition of mitogenesis and proliferation (Garg and Hassid, 1989). These results strongly suggest that endogenous nitric oxide may function as an inhibitor of SMC growth by a cyclic guanidine monophosphate (cGMP)-mediated mechanism. Similar results have been reported by Kariya and co-workers (1989) working with rabbit SMCs.

VI. Vascular Disease and Therapeutic Strategies

The importance of growth modulators, both positive and negative, in vascular development and differentiation is most significant when considered in the context of potential therapeutic strategies. In this section we review some of these potential strategies with respect to both SMC and capillary EC.

A. Angiogenic Diseases

1. Stimulation of Angiogenesis

There are a number of pathological conditions involving SMC growth which lend themselves to therapeutic strategies that exploit factors that either stimulate or antagonize SMC mitogenesis. For example, bFGF has been reported to contribute to the growth and maintenance of the vasa vasorum, potentially enhancing recovery from ischemia and injury (Edelman et al., 1992). The use of stimulators of collateral blood vessel formation would also be useful as a strategy to circumvent occluded carotid arteries or to increase blood supply to a damaged heart (D'Amore and Thompson, 1987; Baffour et al., 1992; Takeshita et al., 1994; Unger et al., 1994).

Stimulating the growth of the microvasculature can also be a desirable clinical strategy for treating aberrant wound healing. Endogenous growth factors such as FGF, EGF, PDGF, and TGF-β are known to be released at the normal wound site and their activity has been shown to be required for normal wound repair (Martin et al., 1992). All of these factors have been shown to enhance soft tissue wound healing. Furthermore, EGF and a mixture of PDGFs accelerated epidermal regeneration in cutaneous wounds in clinical trials (Hom and Maisel, 1992).

A current clinical trial is based on the premise that the prevention of ulcer formation or the accelerated healing of ulcers by conventional therapies may be FGF dependent (Folkman *et al.,* 1991). These authors demonstrated that an acid-stable form of bFGF (bFGF-CS23) caused a ninefold increase in angiogenesis in the ulcer bed and significantly accelerated ulcer healing. In this case, bFGF-CS23 could be utilized as a replacement therapy in the treatment of duodenal ulcers (Folkman *et al.,* 1991).

2. Inhibition of Angiogenesis

Under normal conditions, new capillary formation is relatively infrequent and stringently controlled. The capillary endothelium is actually considered to be "quiescent," with an extremely slow turnover rate (thousands of days) (Hobson and Denekamp, 1984). It has been suggested that this fine-tuned control of new vessel formation is a function of a naturally occurring equilibrium between stimulators and inhibitors of microvascular growth that is dominated by inhibitory modulators. When the regulatory controls enforcing this equilibrium fail, a number of pathological conditions can ensue. For example, it is widely accepted that solid tumor growth is dependent upon new capillary formation. Solid tumors rarely grow much beyond 2 mm in diameter without a vascular bed to facilitate nutrient delivery and gas exchange (Folkman, 1971). Furthermore, tumor cells which are able to breach the tumor vasculature find a direct route into the primary circulation through which they can metastasize to a secondary site (Liotta *et al.,* 1991). In terms of another application for the control of solid tumor growth and metastasis, it has long been suggested that potent, biologically compatible inhibitors of neovascularization might be useful as an adjunct to chemotherapy or immunotherapy after surgical removal of a primary tumor to prevent metastasis to a secondary site (Langer and Murray, 1983).

There are other cases in which abnormal proliferation of capillary blood vessels results in serious disease—pulmonary capillary hemangiomatosis. Two patterns of this disease have been described—cavernous hemangiomatosis which results in death from hemorrhage, and capillary hemangiomatosis, which results in death from pulmonary hypertension (White *et al.,* 1989). This disease state is particularly relevant to this chapter since its development is in contrast to the normal program of vascular development in which the two constituents of a vessel wall develop in coordination with each other; i.e., pericytes and ECs in developing capillaries and SMCs and ECs in large blood vessels (Folkman, 1989). Specifically, it has been suggested that a key event in the development of a "cavernous" vessel from a normal capillary is the stimulation of EC proliferation and a suppression of SMC growth (Folkman, 1989).

B. Diseases of Smooth Muscle Cell Hyperplasia

Deregulated SMC proliferation has been correlated with atherosclerosis and hypertension as well as restenosis after angioplasty. For example, although it remains controversial whether SMC growth is a first event in atherosclerotic lesion formation, it has been shown that atherosclerotic lesions are localized to the sites of SMC proliferation where the cells facilitate lesion formation by producing extracellular matrix and by increasing their cell mass (Schwartz et al., 1990; Ross, 1986). It has also been reported that these SMCs accumulate lipid deposits as do the SMC ECM and the macrophages within the plaque. Autopsy after sudden cardiac death has revealed that typical atherosclerotic lesions are composed primarily of SMCs and their ECM (Cliff et al., 1988; Roberts, 1989). Therefore, the use of SMC inhibitors might be therapeutically useful in the treatment of atherosclerosis. Ultimately, it will be the fine-tuned balance between growth stimulators and inhibitors that will determine the vascular growth state in vivo.

The high incidence of restenosis after angioplasty in adults provides a significant incentive for drug discovery. It is now widely accepted that following angioplasty, the lesion that forms is a result of deregulated intimal SMC accumulation (Jackson and Schwartz, 1992). Most in vivo studies have utilized a system of balloon catheter injury to the rat common carotid artery. As a result of this injury, the endothelial lining of the vessel is stripped away and the SMC layer is damaged. The first response observed within the first 4 days following injury is a major increase in SMC proliferation in the injured media (Clowes and Schwartz, 1985; Schwartz, 1985; Schwartz et al., 1985; Schwartz and Reidy, 1987; Clowes et al., 1989; Fingerle et al., 1990; Lindner and Reidy, 1991). The second response to injury is the migration of SMCs from the media into the intima where approximately 50% of them proliferate, forming a thick intimal SMC layer (Clowes and Schwartz, 1985; Schwartz et al., 1990). This phase can continue over weeks or months. During this period SMCs have been shown to produce ECM, which ultimately constitutes 80% of the intima. Interestingly, it has been shown that proliferation stops in areas covered by regenerated endothelium first. These three separate but related events in the development of the pathological lesion formed following angioplasty now comprise the "three-wave model" of SMC replication after this injury (Jackson and Schwartz, 1992) and provide separate but related therapeutic intervention sites.

With respect to uncontrolled vascular growth and hypertension, it has been demonstrated that hypertensive vessel walls (both essential and secondary) are thicker than normal vessel walls (Mulvaney, 1991) and that this thickening has been observed to occur before an elevation in blood pressure is detected (Eccleston-Joyner and Gray, 1988; Lee et al., 1987).

The change in smooth muscle mass has been related to DNA synthesis by SMCs. For example, it has been reported that DNA synthesis precedes secondary hypertension (Loeb et al., 1986). In addition, an increase in either the SMC number or DNA content of each SMC has been demonstrated in a number of studies of vessel walls of hypertensive animals (Mulvaney et al., 1985; Owens et al., 1988). A solid causal relationship between SMC proliferation and hypertension remains to be reported; however, studies aimed at inhibiting SMC proliferation and or vessel wall thickening in vivo may provide this link.

C. Therapeutic Strategies to Control Vascular Disease

1. Antibodies to Growth Factors

An antibody to the potent smooth muscle chemoattractant and mitogen, PDGF, inhibited neointimal smooth muscle accumulation after angioplasty (Ferns et al., 1991). PDGF appears to act largely by stimulating smooth muscle migration rather than proliferation in this model. The anti-PDGF antibody inhibited the chemotaxis of rat carotid smooth muscle cells to purified PDGF as well.

A number of studies have investigated the use of activity-blocking antibodies against various mitogens as a means of inhibiting their effects. In restenosis models, the release of FGF from damaged SMCs and/or their ECM is responsible for the initial rounds of SMC proliferation. Following this event is the release of PDGF from the adherent platelets which itself causes the migration of SMCs from the media into the intima where they proceed to proliferate. Therefore, the use of antibodies to FGF might be a useful approach to the treatment of restenosis (Lindner and Reidy, 1991).

The administration of an antibody to bFGF blocked the SMC proliferation that occurred 48 hr after vascular injury, suggesting that endogenous bFGF is the major mitogen controlling the growth of SMC postinjury (Lindner and Reidy, 1991). These studies were extended in a later report which examined the role of bFGF in intimal smooth muscle cell proliferation after balloon catheter injury. When a neutralizing antibody to bFGF was administered 4 to 5 days after injury, it was found to have no effect on intimal smooth muscle cell proliferation (Olson et al., 1992). Furthermore, the amount of bFGF protein and mRNA levels actually decreased after injury. Taken together, these results suggested that although bFGF is an important mitogen for medial SMC immediately following injury, it is not an important growth factor in the chronic replication of SMC that occurs after balloon injury (Olson et al., 1992). These studies suggest that perhaps a combination of antibodies against both PDGF and FGF, so that anti-FGF antibodies could be used to inhibit the early proliferation of SMCs in the media and anti-PDGF antibodies could be used to inhibit migration of SMCs into the intima, might be therapeutically efficacious.

An immunoneutralizing monoclonal antibody against bFGF powerfully suppressed bFGF-induced EC stimulation and significantly inhibited solid tumor growth (Hori et al., 1991). This series of experiments has provided direct causal evidence that tumor growth is dependent on angiogenesis.

In the case of VEGF/VPF, monoclonal antibodies directed against the mitogen significantly inhibited the growth and reduced the weight of both rhabdomyosarcoma and glioblastoma multiforme in nude mice (Kim et al., 1993). The anti-VEGF antibodies did not inhibit the growth of these tumor cells in vitro; however, the density of blood vessels was decreased in treated tumors relative to controls. This suggests that angiogenesis inhibition is the mechanism by which tumor growth was inhibited by anti-VEGF antibodies.

2. Antisense Oligonucleotides

An antisense strategy has recently been utilized to inhibit intimal arterial SMC accumulation in vivo. Antisense oligonucleotides may represent a new class of therapeutics which, when locally delivered, can inhibit SMC accumulation by inhibiting specific gene products. The reasoning for this experimental strategy is the following. The proto-oncogene c-myb has been shown to be involved in the mitogen-induced proliferation of vascular SMCs (Ross, 1986). Elevated c-myb message levels had been shown to follow the proliferation of SMCs in vitro and this proliferation can be suppressed by antisense oligonucleotide (Brown et al., 1992; Clowes et al., 1983). In the most recent study, local delivery of antisense c-myb oligonucleotide has reportedly suppressed intimal accumulation of carotid SMCs in a rat carotid injury model (Simons et al., 1992).

3. Toxins

Fusion proteins of growth factor-toxin combinations (mitoxins) may prove to be therapeutically useful. For example, with the rationale that some FGF receptors are upregulated after balloon injury, a cytocidal conjugate of bFGF with saporin has been developed (Lappi et al., 1991). Saporin is a plant enzyme which can inactivate ribosomes. The bFGF-saporin conjugate can inhibit SMC protein synthesis while stimulating aortic ECs (Cascells et al., 1993), presumably because ECs have cell surface receptors with a density that is four times less than that of SMC (Biro et al., 1991). For obvious reasons, this cell specificity is a clinically advantageous feature of this reagent. *In vivo* studies using the bFGF-saporin recombinant fusion protein are currently being conducted in the rat carotid artery model (Casscells *et al.,* 1993).

4. Drugs

Strategies focused on shifting the natural balance away from the stimulation of small blood vessels to their inhibition are under study (see Section V,A) as a way to control solid tumor growth and metastasis. For example, clinical trials utilizing the EC inhibitor PF4 in several different cancer populations, including colon carcinoma and Kaposi's sarcoma, are under way. Another angiogenesis inhibitor, AGM-1470, the fumagillin derivative, is also being tested in clinical trials for the treatment of a variety of solid tumors and Kaposi's sarcoma, and recombinant interferon α-2a is being used to effectively treat pulmonary hemangiomatosis in patients.

Control of the abnormal proliferation of SMCs following balloon angioplasty has recently been accomplished using a gene therapy approach in a pig model. Therapy with a viral gene which encodes thymidine kinase (tk) combined with ganciclovir, the anti-viral drug, has been shown to block the deregulated SMC proliferation and significantly decrease artery wall thickening (Ohno *et al.,* 1994). Current studies are focused on the therapeutic efficacy of this approach in humans.

VII. Summary and Conclusion

In the course of this chapter we have reviewed important stimulators and inhibitors of EC and SMC growth in the context of the vascular system within which these cells reside. Although much of this work has been conducted *in vitro,* investigators are now developing the *in vivo* models to complement these studies. As reviewed in Section VI, a number of serious vascular diseases lend themselves to therapies utilizing either vascular cell stimulators or inhibitors. Future work in the area of control of vascular growth will focus not only on the discovery of novel vascular growth modulators but also on the clinical application of some of these factors.

References

Abraham, J. A., Damm, D., Bajardi, A., Miller, J., Klagsbrun, M., and Ezekowitz, R. E. B. (1993). Heparin-binding EGF-like growth factor: Characterization of rat and mouse cDNA clones, protein domain conservation across species and transcript expression in tissues. *Biochem. Biophys. Res. Commun.* **190**, 125–133.

Akopov, S. E., Orekhov, A. N., Tertov, V. V., Khashimov, K. A., Gabrielyan, E. S., and Smirnov, V. N. (1988). Stable analogues of prostacyclin and thromboxane A2 display contradictory influences on atherosclerotic properties of cells cultured from human aorta. *Atherosclerosis (Shannon, Irel.)* **72**, 245–248.

Alavi, M. Z., Richardson, M., and Moore, S. (1989). The in vitro interactions between serum lipoproteins and proteoglycans of the neointima of rabbit aorta after a single balloon catheter injury. *Am. J. Pathol.* **134**, 287–294.

Antonelli-Orlidge, A., Saunders, K. B., Smith, S. R., and D'Amore, P. A. (1989). An activated form of TGF-β is produced by co-cultures of endothelial cells and pericytes. *Proc. Natl. Acad. Sci. U.S.A.* **86**, 4544–4548.

Asotra, S., Foegh, M. L., Conte, J. V., Cai, B. R., and Ramwell, P. W. (1989). Inhibition of 3H thymidine incorporation by angiopeptin in the aorta of rabbits after balloon angioplasty. *Transplant. Proc.* **21**, 3695–3696.

Auerbach, W., and Auerbach, R. (1994). Angiogenesis inhibition: A review. *Pharmacol. Thera.* **63**, 265–311.

Badet, J., Soncin, F., Guitton, J. D., Lamare, O., Cartwright, T., and Barritault, D. (1989). Specific binding of angiogenin to calf pulmonary artery endothelial cells. *Proc. Natl. Acad. Sci. U.S.A.* **86**, 8427–8431.

Badet, J., Soncin, F., Nguyen, T., and Barritault, D. (1990). In vivo and in vitro studies of angiogenin—a potent angiogenic factor. *Blood Coagulation Fibrinol.* **1**, 721–724.

Baffour, R., Berman, J., Garb, J. L., et al. (1992). Enhanced angiogenesis and growth of collaterals by in vivo administration of recombinant basic fibroblast growth factor in a rabbit model of acute lower limb ischemia: Dose-responsive effect of basic fibroblast growth factor. *J. Vasc. Surg.* **16**, 181.

Bagavandoss, P., and Wilks, J. W. (1991). Specific inhibition of endothelial cell proliferation by thrombospondin. *Biochem. Biophys. Res. Commun.* **170**, 867–872.

Banda, M. J., Herron, G. S., Murphy, J., and Werb, Z. (1987). Regulation of metalloproteinase activity by microvascular endothelial cells. *In* "Angiogenesis: Mechanisms and Pathobiology" (D. B. Rifkin and M. Klagsbrun, eds.), pp. 101–109. Cold Spring Harbor Lab., Cold Spring Harbor, NY.

Bargmann, C. I., Hung, M. C., and Weinberg, R. A. (1986). The neu oncogene encodes an epidermal growth factor receptor-related protein. *Nature (London)* **319**, 226–230.

Battegay, E. J., Raines, E. W., Seifert, R. A., Bowen-Pope, D. F., and Ross, R. (1991). TGF-beta induces bimodal proliferation of connective tissue cells via complex control of an autocrine PDGF loop. *Cell (Cambridge, Mass.)* **63**, 515–524.

Battegay, E. J., Rupp, J., Iruela-Arispe, L., Sage, E. H., and Pech, M. (1994). PDGF-BB modulates endothelial proliferation and angiogenesis in vitro via PDGF beta-receptors. *J. Cell Biol.* **125**, 917–928.

Bicknell, R., and Vallee, B. L. (1988). Angiogenin activates endothelial cell phospholipase C. *Proc. Natl. Acad. Sci. U.S.A.* **85**, 5961–5965.

Bicknell, R., and Vallee, B. L. (1989). Angiogenin stimulates endothelial cell prostacyclin secretion by activation of phospholipase A2. *Proc. Natl. Acad. Sci. U.S.A.* **86**, 1573–1577.

Biro, S., Lappi, D. A., Yu, Z. X., Baird, A., and Casscells, W. (1991). Stimulation and inhibition of endothelial and smooth muscle cells by basic fibroblast growth factor-saporin in vitro and in vivo. *Circulation, Suppl.* **84**, 553 (abstr.).

Blotnick, S., Peoples, G. E., Freeman, M. R., Eberlein T. J., and Klagsbrun, M. (1994). T Lymphocytes synthesize and export heparin-binding EGF-like growth factor and basic fibroblast growth factor, mitogens for vascular cells and fibroblasts; Differential production and release by CD4+ and CD8+. *Proc. Natl. Acad. Sci. U.S.A.* **91**, 2890–2894.

Bond, M. D., and Vallee, B. L. (1988). Isolation of bovine angiogenin using a placental ribonuclease inhibitor binding assay. *Biochemistry* **27,** 6282–6287.

Bornstein, P. (1992). Thrombospondins: Structure and regulation of expression. *FASEB J.* **6,** 3290–3299.

Brem, H., and Folkman, H. (1975). Inhibition of tumor angiogenesis mediated by cartilage. *J. Exp. Med.* **141,** 427–439.

Brem, H., and Folkman, J. (1993). Analysis of experimental antiangiogenic therapy. *J. Pediatr. Surg.* **28,** 445–450.

Brem, H., and Klagsbrun, M. (1993). The role of fibroblast growth factors and related onco-genes in tumor growth. *In* "Oncogenes and Tumor Suppressor Genes in Human Malignan-cies" (C. C. Benz and E. T. Liu, eds.), pp. 211–231. Kluwer Academic Publishers, Dordrecht, The Netherlands.

Brogi, E., Wu, T., Namiki, A., and Isner, J. M. (1994). Indirect angiogenic cytokines upregulate VEGF and bFGF gene expression in vascular smooth muscle cells, whereas hypoxia upregu-lates VEGF expression only. *Circulation* **90,** 649–652.

Brown, K. E., Kindy, M. S., and Sonenshein, G. E. (1992). Expression of the c-myb proto-oncogene in bovine vascular smooth muscle cells. *J. Biol. Chem.* **267,** 4625–4630.

Burgess, W. H., and Maciag, T. (1989). The heparin-binding (fibroblast) growth factor family of proteins. *Annu. Rev. Biochem.* **58,** 575–606.

Bussolino, F., Di Renzo, M. F., Ziche, M., Bocchietto, E., Olivero, M., Naldini, L., Gaudino, G., Tamagnone, L., Coffer, A., and Comoglio, P. M. (1992). Hepatocyte growth factor is a potent angiogenic factor which stimulates endothelial cell motility and growth. *J. Cell Biol.* **119,** 629–641.

Capron, L., Heudes, D., Chajara, A., and Bruneval, P. (1991). Effect of ramipril, an inhibitor of angiotensin converting enzyme, on the response of rat thoracic aorta to injury with a balloon catheter. *J. Cardiovasc. Pharmacol.* **18,** 207–211.

Casscells, W. (1992). Smooth muscle cell growth factors. *Prog. Growth Factor Res.* **3,** 177–206.

Casscells, W., Lappi, D. A., and Baird, A. (1993). Molecular atherectomy for restenosis. *Trends Cardiovasc. Med.* **3,** 225–243.

Castellot, J. J., Jr., Addonizio, M. L., Rosenberg, R., and Karnovsky, M. J. (1981). Cultured endothelial cells produce a heparin-like inhibitor of smooth muscle cell growth. *J. Cell Biol.* **90,** 372–379.

Castellot, J. J., Jr., Favreau, L. V., Karnovsky, M. J., and Rosenberg, R. D. (1982). Inhibition of vascular smooth muscle cell growth by endothelial cell-derived heparin. Possible role of a platelet endoglycosidase. *J. Biol. Chem.* **257,** 11256–11260.

Castellot, J. J., Jr., Cochran, D. L., and Karnovsky, M. J. (1985). Effect of heparin on vascular smooth muscle cells. *J. Cell. Physiol.* **124,** 21–28.

Castellot, J. J., Jr., Pukac, L. A., Caleb, B. L., Wright, T. C., Jr., and Karnovsky, M. J. (1989). Heparin selectively inhibits a protein kinase C-dependent mechanism of cell cycle progression in cell aortic smooth muscle cells. *J. Cell Biol.* **109,** 3147–3155.

Chamley-Campbell, J. H., and Campbell, G. R. (1981). What controls smooth muscles' pheno-type? *Atherosclerosis (Shannon, Irel.)* **40,** 347–357.

Chamoux, M., Dehouck, M. P., Fruchart, J. C., Spik, G., Montreuil, J., and Cecchelli, R. (1991). Characterization of angiogenin receptors on bovine brain capillary endothelial cells. *Biochem. Biophys. Res. Commun.* **176,** 833–839.

Clapp, C. (1987). Analysis of the proteolytic cleavage of prolactin by the mammary gland and liver of the rat. Characterization of the cleaved and 16K forms. *Endocrinology (Balti-more)* **121,** 2055–2064.

Clapp, C., and Weiner, R. I. (1992). A specific, high, affinity, saturable binding site for the 16-kilodalton fragment of prolactin on capillary endothelial cells. *Endocrinology (Baltimore)* **130,** 1380–1386.

Clapp, C., Sears, P. S., Russell, D. H., Richards, J., Levay-Young, B. K., and Nicoll, C. S. (1988). Biological and immunological characterization of cleaved and 16K forms of rat prolactin. *Endocrinology (Baltimore)* **122**, 2892–2898.

Clapp, C., Martial, J. A., Guzman, R. C., Rentier-Delure, F., and Weiner, R. I. (1993). The 16-kilodalton N-terminal fragment of human prolactin is a potent inhibitor of angiogenesis. *Endocrinology (Baltimore)* **133**, 1292–1299.

Cliff, W. J., Heathcote, C. R., Moss, N. S., and Reichenbach, D. D. (1988). The coronary arteries in cases of cardiac and noncardiac sudden death. *Am. J. Pathol.* **132**, 319–329.

Clowes, A. W., and Clowes, M. M. (1986). Kinetics of cellular proliferation after arterial injury. IV. Heparin inhibits rat smooth muscle mitogenesis and migration. *Circ. Res.* **58**, 839–845.

Clowes, A. W., and Schwartz, S. M. (1985). Significance of quiescent smooth muscle migration in the injured rat carotid artery. *Circ. Res.* **56**, 139–145.

Clowes, A. W., Reidy, M. A., and Clowes, M. M. (1983). Kinetics of cellular proliferation after arterial injury. I. Smooth muscle growth in the absence of endothelium. *Lab. Invest.* **49**, 327–333.

Clowes, A. W., Clowes, M. M., Fingerle, J., and Reidy, M. A. (1989). Kinetics of cellular proliferation after injury. V. Role of acute distention in the induction of smooth muscle proliferation. *J. Cardiovasc. Pharmacol.* **4**, Suppl. 6, S12–S15.

Clowes, A. W., Clowes, M. M., Vergel, S. C., Muller, R. K., Powell, J. S., Hefti, F., and Baumgartner, H. R. (1991). Heparin and cilazapril together inhibit injury-induced intimal hyperplasia. *Hypertension (Dallas)* **18**, 1165–1169.

Compton, M. M., and Witorsch, R. J. (1984). Proteolytic degradation and modification of rat prolactin by subcellular fractions of the rat ventral prostate gland. *Endocrinology (Baltimore)* **115**, 476–484.

Connolly, D. T., Heuvelman, D. M., Nelson, R., Olander, J. V., Eppley, B. L., Delfino, J. J., Siegel, N. R., Leimgruber, R. M., and Feder, J. (1989). Tumor vascular permeability factor stimulates endothelial cell growth and angiogenesis. *J. Clin. Invest.* **84**, 1478–1489.

Conte, J. V., Foegh, M. L., Calcagno, D., Wallace, R. B., and Ramwell, P. W. (1989). Peptide inhibition of myotintimal proliferation following angioplasty in rabbits. *Transplant. Proc.* **21**, 3686–3688.

Cook, P. W., Damm, D., Garrick, B. L., Wood, K. M., Karkaria, C. E., Higashiyama, S., Klagsbrun, M., and Abraham, J. (1994). Carboxy-terminal truncation of leucine$_{76}$ converts heparin-binding EGF-like growth factor (HB-EGF) to a heparin suppressible growth factor. *J. Cell. Physiol.* (in press).

Coussens, L., Yang, F. T. L., Liao, Y. C., Chen, E., Gray, A., McGrath, J., Seeburg, P. H., Libermann, T. A., Schlessinger, J., Francke, U., Levinson, A., and Ullrich, S. A. (1985). Tyrosine kinase receptor with extensive homology to EGF receptor shares chromosomal location with neu oncogene. *Science* **230**, 1132–1139.

Crocker, D. J., Murad, T. M., and Greer, J. C. (1970). Role of the pericyte in wound healing. An ultrastructural study. *Exp. Mol. Pathol.* **13**, 51–65.

Daemen, M. J. A. P., Lombardi, D. M., Bosman, F. T., and Schwartz, S. M. (1991). Angiotensin II induces smooth muscle cell proliferation in the normal and injured rat arterial wall. *Circ. Res.* **68**, 450–456.

D'Amore, P. A., and Thompson, R. W. (1987). Collateralization in peripheral vascular disease. In "Vascular Diseases" (D. Strandness, P. Didsheim, A. Clowes, and J. Waton, eds.), pp. 319–333. Grune and Stratton, Inc., Orlando, Florida.

D'Amato, R. J., Loughnan, M. S., Flynn, E., and Folkman, J. (1994). Thalidomide is an inhibitor of angiogenesis. *Proc. Natl. Acad. Sci. U.S.A.* **91**, 4082–4085.

Dameron, K. M., Volpert, O. V., Tainsky, M. A., and Bouck, N. (1994). Control of angiogenesis in fibroblasts by p53 regulation of thrombospondin-1. *Science* **265**, 1582–1584.

Das, K. S., Wang, X.-N., Paria, B. C., Damm, D., Abraham, J. A., Klagsbrun, M., Andrews, G. K., and Dey, S. K. (1994). Uterine heparin-binding EGF-like growth factor gene is induced temporally by the blastocyst at the site of apposition: A potential ligand for interaction with blastocyst EGF receptor in implantation. *Development (Cambridge, UK)* **120,** 1071–1083.

Davies, P. F., Truskey, G. A., Warren, H. B., O'Connor, S. E., and Eisenhaure, B. H. (1985). Metabolic cooperation between vascular endothelial cells and smooth muscle cells in co-culture: Changes in low density lipoprotein metabolism. *J. Cell Biol.* **101,** 871–879.

Delafontaine, P., Lou, H., and Alexander, R. W. (1991). Regulation of insulin-like growth factor I messenger RNA levels in vascular smooth muscle cells. *Hypertension (Dallas)* **18,** 742–747.

de Vries, C., Escobedo, J. A, Ueno, H., Houck, K., Ferrara, N., and Williams, L. T. (1992). The fms-like tyrosine kinase, a receptor for vascular endothelial growth factor. *Science* **255,** 989–991.

Dluz, S. M., Higashiyama, S., Damm, D., Abraham, J. A., and Klagsbrun, M. (1993). Heparin-binding EGF like growth factor expression in cultured fetal human vascular smooth muscle cells: Induction of mRNA levels and secretion of active mitogen. *J. Biol. Chem.* **268,** 18330–18334.

Dodge, A. B., and D'Amore, P. A. (1992). Cell-cell interactions in diabetic angiopathy. *Diabetes Care* **9,** 1168–1180.

Eccleston-Joyner, C. A., and Gray, S. D. (1988). Arterial hypertrophy in the fetal and neonatal spontaneously hypertensive rat. *Hypertension (Dallas)* **12,** 513–518.

Edelman, E. R., Adams, D. H., and Karnovsky, M. J. (1990). Effect of controlled adventitial heparin delivery on smooth muscle cell proliferation following endothelial injury. *Proc. Natl. Acad. Sci. U.S.A.* **82,** 3773–3777.

Edelman, R. R., Nugent, M. A., Smith, L. T., et al. (1992). Basic fibroblast growth factor enhances the coupling of intimal hyperplasia and proliferation of vasa vasorum in injured rat arteries. *J. Clin. Invest.* **89,** 465.

Elias, K., and Weiner, R. (1984). Direct arterial vascularization of estrogen-induced prolactin-secreting anterior pituitary tumors. *Proc. Natl. Acad. Sci. U.S.A.* **81,** 4549–4553.

Ezekowitz, R. A., Mulliken, J. B., and Folkman, J. (1992). Interferon alfa-2 therapy for life-threatening hemangiomas of infancy. *N. Engl. J. Med.* **326,** 1456–1463.

Fajardo, L. F. (1989). The complexity of endothelial cells. *Am. J. Clin. Pathol.* **92,** 241–250.

Ferns, G. A. A., Raines, E. W., Sprugel, K. H., Motani, A. S., Ross, R., and Reidy, M. (1991). Anti-PDGF antibody significantly inhibits neointimal smooth muscle accumulation after angioplasty. *Science* **253,** 1129–1132.

Ferrara, N., and Henzel, W. J. (1989). Pituitary follicular cells secrete a novel heparin-binding growth factor specific for vascular endothelial cells. *Biochem. Biophys. Res. Commun.* **161,** 851–858.

Ferrara, N., Clapp, C., and Weiner, R. (1991). The 16K fragment of prolactin specifically inhibits basal or fibroblast growth factor stimulated growth of capillary endothelial cells. *Endocrinology (Baltimore)* **129,** 896–900.

Fett, J. W., Strydom, D. J., Lobb, R. R., Alderman, E. M., Bethune, J. L., Riordan, J. F., and Vallee, B. L. (1985). Isolation and characterization of angiogenin, an angiogenic protein from human carcinoma cells. *Biochemistry* **24,** 5480–5486.

Fingerle, J., Au, Y., Clowes, A. W., and Reidy, M. A. (1990). Intimal lesion formation in rat arteries after endothelial denudation in absence of medial injury. *Arteriosclerosis (Dallas)* **10,** 1082–1087.

Folkman, J. (1971). Tumor angiogenesis: Therapeutic implications. *N. Engl. J. Med.* **285,** 1182–1186.

Folkman, J. (1989). Successful treatment of an angiogenic disease [editorial]. *N. Engl. J. Med.* **320,** 1211–1212.

Folkman, J. (1993). Tumor angiogenesis. *In* "Cancer Medicine" (J. F. Holland, E. Frei, III, R. C. Bast, D. W. Kufe, D. L. Morton, and R. R. Weichselbaum, eds.), pp. 153–170. Lea & Febiger, Philadelphia.

Folkman, J., and Haudenschild, C. (1980). Angiogenesis in vitro. *Nature (London)* **288**, 551–556.

Folkman, J., Szabo, S., Stovroff, M., McNeil, P., Li, W., and Shing, Y. (1991). Duodenal ulcer: Discovery of a new mechanism and development of angiogenic therapy that accelerates healing. *Ann. Surg.* **214**, 414–425.

Frater-Schroder, M., Risau, W., Hallman, R., Gautschi, R., and Bohlen, P. (1987). Tumor necrosis factor type alpha, a potent inhibitor of endothelial cell growth in vitro. *Proc. Natl. Acad. Sci. U.S.A.* **84**, 5277–5281.

Friesel, R., Komoriya, A., and Maciag, T. (1987). Inhibition of endothelial cell proliferation by gamma-interferon. *J. Cell Biol.* **104**, 689–696.

Galland, F., Karamysheva, A., Pebusque, M. J., Borg, J. P., Rottapel, R., Dubreuil, P., Rosnet, O., and Birnbaum, D. (1993). The FLT4 gene encodes a transmembrane tyrosine kinase related to the vascular endothelial growth factor receptor. *Oncogene* **8**, 1233–1240.

Garg, U. C., and Hassid, A. (1989). Nitric oxide-generating vasodilators and 8 bromo cyclic guanosine monophosphate inhibit mitogenesis and proliferation of cultured rat vascular smooth muscle cells. *J. Clin. Invest.* **83**, 1774–1777.

Gay, C. G., and Winkles, J. A. (1991). Interleukin 1 regulates heparin-binding growth factor 2 gene expression in vascular smooth muscle cells. *Proc. Natl. Acad. Sci. U.S.A.* **88**, 296–300.

Gerritsen, M. E. (1987). Functional heterogeneity of vascular endothelial cells. *Biochem. Pharmacol.* **36**, 2701–2711.

Gibbons, G. H., Pratt, R. E., and Dzau, V. J. (1992). Vascular smooth muscle cell hypertrophy vs. hyperplasia. Autocrine transforming growth factor-beta 1 expression determines growth response to angiotensin II. *J. Clin. Invest.* **90**, 456–461.

Gitay-Goren, H., Soker, S., Vlodavsky, I., and Neufeld, G. (1992). The binding of vascular endothelial growth factor to its receptors is dependent on cell surface-associated heparin-like molecules. *J. Biol. Chem.* **267**, 6093–6098.

Gonzalez, A. M., Buscaglia, M., Ong, M., and Baird, A. (1990). Distribution of basic fibroblast growth factor in the 18-day rat fetus: Localization in the basement membranes of diverse tissues. *J. Cell Biol.* **110**, 753–765.

Good, D. J., Polverini, P. J., Rastinejad, F., LeBeau, M. M., Lemons, R. S., Frazier, W. A., and Bouck, N. P. (1990). A tumor suppressor-dependent inhibitor of angiogenesis is immunologically and functionally indistinguishable from a fragment of thrombospondin. *Proc. Natl. Acad. Sci. U.S.A.* **87**, 6624–6628.

Gordon, J. R., and Galli, S. J. (1990). Mast cells as a source of both preformed and immunologically inducible TNF-alpha/cachectin. *Nature (London)* **346**, 274–276.

Gospodarowicz, D., and Lau, K. (1989). Pituitary follicular cells secrete both vascular endothelial growth factor and follistatin. *Biochem. Biophys. Res. Commun.* **165**, 292–298.

Goto, F., Goto, K., Weindel, K., and Folkman, J. (1993). Synergistic effects of vascular endothelial growth factor and basic fibroblast growth factor on the proliferation and cord formation of bovine capillary endothelial cells within collagen gells. *Lab. Invest.* **69**, 508–517.

Grainger, D. J., Kirschenlohr, H. L., Metcalfe, J. C., Weissberg, P. L., Wade, D. P., and Lawn, R. M. (1993). Proliferation of human smooth muscle cells promoted by lipoprotein(a). *Science* **260**, 1655–1658.

Grant, D. S., Kleinman, H. K., Goldberg, I. D., Bhargava, M. M., Nickoloff, B. J., Kinsella, J. L., Polverini, P., and Rosen, E. M. (1993). Scatter factor induces blood vessel formation in vivo. *Proc. Natl. Acad. Sci. U.S.A.* **90**, 1937–1941.

Gross, J. L., Moscatelli, D., Jaffe, E. A., and Rifkin, D. B. (1982)., Plasminogen activator and collagenase production by cultured capillary endothelial cells. *J. Cell Biol.* **95**, 974–981.

Gross, J. L., Moscatelli, D., and Rifkin, D. B. (1983). Increased capillary endothelial cell protease activity in response to angiogenic stimuli in vitro. *Proc. Natl. Acad. Sci. U.S.A.* **80**, 2623–2627.

Haraguchi, M., Miyadera, K., Uemura, K., Sumizawa, T., Furukawa, T., Yamada, K., Akiyama, S., and Yamada, Y. (1994). Angiogenic activity of enzymes [letter] *Nature (London)* **368**, 198.

Hauser, S., and Weich, H. A. (1993). A heparin-binding form of placenta growth factor (PlGF-2) is expressed in human umbilical vein endothelial cells and in placenta. *Growth Factors* **9**, 259–268.

Hayakawa, T., Yamashita, K., Tanzawa, K., Uchijima, E., and Iwata, K. (1992). Growth-promoting activity of tissue inhibitor of metalloproteinases-1 (TIMP-1) for a wide range of cells. *FEBS Lett.* **298**, 29–32.

Heath, W. F., Cantrell, A. S., Mayme, N. G., and Jaskunas, S. R. (1991). Mutations in the heparin-binding domains of human basic fibroblast growth factor alter its biological activity. *Biochemistry* **30**, 5608–5615.

Herman, I. M., and D'Amore, P. A. (1985). Microvascular pericytes contain muscle and nonmuscle actins. *J. Cell Biol.* **101**, 43–52.

Herman, I. M., and Castellot, J. J., Jr. (1987). Regulation of vascular smooth muscle cell growth by endothelial-synthesized extracellular matrices. *Arteriosclerosis (Dallas)* **7**, 463–469.

Higashiyama, S., Abraham, J. A., Miller, J., Fiddes, J. C., and Klagsbrun, M. (1991). A heparin-binding growth factor secreted by macrophage-like cells that is related to EGF. *Science* **251**, 936–939.

Higashiyama, S., Lau, K., Besner, G., Abraham, J. A., and Klagsbrun, M. (1992). Structure of heparin-binding EGF-like growth factor: Multiple forms, primary structure and glycosylation of the mature protein. *J. Biol. Chem.* **267**, 6205–6212.

Higashiyama, S., Abraham, J., and Klagsbrun, M. (1993). Heparin-Binding EGF-like growth factor (HB-EGF) stimulation of smooth muscle cell migration; Dependence on interactions with heparan sulfate. *J. Cell Biol.* **122**, 933–940.

Higashiyama, S., Abraham, J. A., and Klagsbrun, M. (1994). Heparin-binding EGF-like growth factor (HB-EGF) synthesis by smooth muscle cells. *Hormone Res.* **42**, 9–13.

Hill, T. D., Berggren, P. O., and Boynton, A. L. (1987). Heparin inhibits inositol trisphosphate-induced calcium release from permeabilized rat liver cells. *Biochem. Biophys. Res. Commun.* **149**, 897–901.

Hobson, B., and Denekamp, J. (1984). Endothelial proliferation in tumors and normal tissues: Continuous labelling studies. *Br. J. Cancer* **49**, 405–413.

Hom, D. B., and Maisel, R. H. (1992). Angiogenic growth factors: Their effects and potential in soft tissue wound healing. *Ann. Otol.* **101**, 349–354.

Hori, A., Sasada, R., Matsutani, E., Naito, K., Sakura Y., Fujita, T., and Kozai, Y. (1991). Suppression of solid tumor growth by immunoneutralizing monoclonal antibody against basic fibroblast growth factor. *Cancer Res.* **51**, 6180–6184.

Houck, K. A., Ferrara, N., Winer, J., Cachianes, G., Li, B., and Leung, D. W. (1991). The vascular endothelial growth factor family: Identification of a fourth molecular species and characterization of alternative splicing of RNA. *Mol. Endocrinol.* **5**, 1806–1814.

Houck, K. A., Leung, D. W., Rowland, A. M., Winer, J., and Ferrara, N. (1992). Dual regulation of vascular endothelial growth factor bioavailability by genetic and proteolytic mechanisms. *J. Biol. Chem.* **267**, 26031–26037.

Hu, D. E., Hori, Y., and Fan, T. P. (1993). Interleukin-8 stimulates angiogenesis in rats. *Inflammation* **17**, 135–143.

Hu, G. F., Strydom, D. J., Fett, J. W., Riordan, J. F., and Vallee, B. L. (1993). Actin is a binding protein for angiogenin. *Proc. Natl. Acad. Sci. U.S.A.* **90**, 1217–1221.

Huettner, J. J., Gwebu, E. T., Panganamala, R. V., Milo, G. E., Cornwell, D. G., Sharma, H. M., and Geer, J. C. (1977). Fatty acids and their prostaglandin derivatives: Inhibitors of proliferation in aortic smooth muscle cells. *Science* **197**, 289–291.

Ingber, D. E. (1990). Fibronectin controls capillary endothelial cell growth by modulating cell shape. *Proc. Natl. Acad. Sci. U.S.A.* **87**, 3579–3583.

Ingber, D. E., and Folkman, J. (1989). Mechanochemical switching between growth and differentiation during fibroblast growth factor-stimulated angiogenesis in vitro: Role of extracellular matrix. *J. Cell Biol.* **109**, 317–330.

Ingber, D. E., Madri, J. A., and Folkman, J. (1987). Endothelial growth factors and extracellular matrix regulate DNA synthesis through modulation of cell and nuclear shape. *In Vitro* **23**, 387–394.

Ingber, D. E., Fujita, T., Kishimoto, S., Sudo, K., Kanamaru, K., Brem, H., and Folkman, J. (1990). Synthetic analogues of fumagillin that inhibit angiogenesis and suppress tumour growth. *Nature (London)* **348**, 555–557.

Iruela-Arispe, M. L., Hasselaar, P., and Sage, H. (1991a). Differential expression of extracellular proteins is correlated with angiogenesis in vitro. *Lab. Invest.* **64**, 174–186.

Iruela-Arispe, M. L., Bornstein, P., and Sage, H. (1991b). Thrombospondin exerts an antiangiogenic effect on cord formation by endothelial cells in vitro. *Proc. Natl. Acad. Sci. U.S.A.* **88**, 5026–5030.

Ishikawa, F., Miyazono, K., Hellman, U., Drexler, H., Wernstedt, C., Hagiwara, K., Usuki, K., Takaku, F., Risau, W., and Heldin, C. H. (1989). Identification of angiogenic activity and the cloning and expression of platelet-derived endothelial cell growth factor. *Nature (London)* **338**, 557–561.

Itoh, H., Mukoyama, M., Pratt, R. E., Gibbons, G. H., and Dzau, V. J. (1993). Multiple autocrine growth factors modulate vascular smooth muscle cell growth response to angiotensin II. *J. Clin. Invest.* **91**, 2268–2274.

Itoh, H., Pratt, R. E., and Dzau, V. J. (1991). Interactions of atrial natriuretic polypeptide and angiotensin II on protooncogene expression and vascular cell growth. *Biochem. Biophys. Res. Commun.* **176**, 1601–1609.

Iwamoto, R., Higashiyama, S., Mitamura, T., Taniguchi, N., Klagsbrun, M., and Mekada, E. (1994). Heparin-binding EGF-like growth factor, which acts as the diphtheria toxin receptor, forms a complex with membrane protein DRAP27/CD9, which upregulates functional receptors and diphtheria toxin sensitivity. *EMBO J.* **13**, 2322–2330.

Jackson, C. L., and Schwartz, S. M. (1992). Pharmacology of smooth muscle cell replication. *Hypertension (Dallas)* **20**, 713–736.

Jasmin, C. (1991). Human interferons: Biological and clinical promises and premises. *Schweiz. Med. Wochenschr.* **121**, 463–466.

Joyce, N. C., Haire, M. F., and Palade, G. (1985). Contractile proteins in pericytes. I. Immunoperoxidase localization of tropomysin. *J. Cell Biol.* **100**, 1379–1386.

Kaipainen, A., Korhonen, J., Pajusola, K., Aprelikova, O., Persico, M. G., Terman, B. I., and Alitalo, K. (1993). The related FLT4, FLT1, and KDR receptor tyrosine kinases show distinct expression patterns in human fetal endothelial cells. *J. Exp. Med.* **178**, 2077–2088.

Kamei, S., Okada, H., Inoue, Y., Yoshioka, T., Ogawa, Y., and Toguchi, H. (1993). Antitumor effects of angiogenesis inhibitor TNP-470 in rabbits bearing VX-2 carcinoma by arterial administration of microspheres and oil solution. *J. Pharmacol. Exp. Ther.* **264**, 469–474.

Kan, M., Wang, F., Xu, J., Crabb, J. W., Hou, J., and McKeehan, W. L. (1993). An essential heparin-binding domain in the fibroblast growth factor receptor kinase. *Science* **259**, 1918–1921.

Kariya, K., Kawahara, Y., Araki, S., Fukuzaki, H., and Takai, Y. (1989). Antiproliferative action of cyclic GMP-elevating vasodilators in cultured rabbit aortic smooth muscle cells. *Atherosclerosis (Shannon, Irel.)* **80**, 143–147.

Kavanaugh, W. M., Williams, L. T., Ives, H. E., and Coughlin, S. R. (1988). Serotonin-induced deoxyribonucleic acid synthesis in vascular smooth muscle cells involves a novel, pertussis toxin-sensitive pathway. *Mol. Endocrinol.* **2**, 599–605.

Keck, P. J., Hauser, S. D., Krivi, G., Sanzo, K., Warren, T., Feder, J., and Connolly, D. T. (1989). Vascular permeability factor, an endothelial cell mitogen related to PDGF. *Science* **246**, 1309–1312.

Kim, D. K., Li, B., Winer, J., Armanini, M., Gillett, N., Phillips, H. S., and Ferrara, N. (1993). Inhibition of vascular endothelial growth factor-induced angiogenesis suppresses tumour growth in vivo. *Nature (London)* **362**, 841–844.

Klagsbrun, M. (1989). The fibroblast growth factor family: Structural properties and biological properties. *Prog. Growth Factor Res.* **1**, 207–235.

Klagsbrun, M., and Baird, A. (1991). A dual receptor system is required for basic fibroblast growth factor activity. *Cell* (*Cambridge, Mass.*) **67**, 229–231.

Klagsbrun, M., and D'Amore, P. A. (1991). Regulators of angiogenesis. *Annu. Rev. Physiol.* **53**, 217–258.

Klagsbrun, M., and Dluz, S. (1993). Smooth muscle cell and endothelial growth factors. *Trends Cardiovasc. Med.* **3**, 213–217.

Kobayashi, S., Somlyo, A. V., and Somlyo, A. P. (1988). Heparin inhibits the inositol 1, 4, 5-trisphosphate-dependent, but not the independent, calcium release induced by guanine nucleotide in vascular smooth muscle. *Biochem. Biophys. Res. Commun.* **153**, 625–631.

Koch, A. E., Polverini, P. J., Kunkel, S. L., Harlow, L. A., DiPietro, L. A., Elner, V. M., Elner, S. G., and Strieter, R. M. (1992). Interleukin-8 as a macrophage-derived mediator of angiogenesis. *Science* **258**, 1798–1801.

Kojima, S., Harpel, P. C., and Rifkin, D. B. (1991). Lipoprotein (a) inhibits the generation of transforming growth factor beta: An endogenous inhibitor of smooth muscle cell migration. *J. Cell Biol.* **113**, 1439–1445.

Kraus, M. H., Issing, W., Miki, T., Popescu, N. C., and Aaronson, S. A. (1989). Isolation and characterization of ERBB3, a thrid member of the ERB/epidermal growth factor receptor family: Evidence for overexpression in a subset of human mammary tumors. *Proc. Natl. Acad. Sci. U.S.A.* **86**, 9193–9197.

Kusaka, M., Sudo, K., Fujita, T., Marui, S., Itoh, F., Ingber, D., and Folkman, J. (1991). Potent anti-angiogenic action of AGM-1470: Comparison to the fumagillin parent. *Biochem. Biophys. Res. Commun.* **174**, 1070–1076.

Ladoux, A., and Frelin, C. (1993). Hypoxia is a strong inducer of vascular endothelial growth factor mRNA expression in the heart. *Biochem. Biophys. Res. Commun.* **195**, 1005–1010.

Langer, R., and Murray, J. (1983). Angiogenesis inhibitors and their delivery systems. *Appl. Biochem. Biotechnol.* **8**, 9–24.

Langer, R., Brem, H., Falterman, K., Klein, M., and Folkman, J. (1976). Isolation of a cartilage factor that inhibits tumor neovascularization. *Science* **193**, 70–72.

Langer, R., Conn, H., Vacanti, J., Haudenschild, C., and Folkman, J. (1980). Control of tumor growth in animals by infusion of angiogenesis inhibitor. *Proc. Natl. Acad. Sci. U.S.A.* **77**, 4331–4335.

Lappi, D. A., Maher, P. A., Martineau, D., and Baird, A. (1991). The basic fibroblast growth factor-saporin mitotoxin acts through the basic fibroblast growth factor receptor. *J. Cell. Physiol.* **147**, 17–26.

Lee, F. S., and Vallee, B. L. (1993). Structure and action of mammalian ribonuclease (angiogenin) inhibitor. *Proc. Natl. Acad. Sci. U.S.A.* **44**, 1–30.

Lee, F. S., Shapiro, R., and Vallee, B. L. (1989). Tight-binding inhibition of angiogenin and ribonuclease A by placental ribonuclease inhibitor. *Biochemistry* **28**, 220–230.

Lee, R. M., Triggle, C. R., Cheung, D. W., and Coughlin, M. D. (1987). Structural and functional consequence of neonatal sympathectomy on the blood vessels of spontaneously hypertensive rats. *Hypertension* (*Dallas*) **10**, 328–338.

Lee, S. L., Dunn, J., Yu, F. S., and Fanburg, B. L. (1989). Serotonin uptake and configurational change of bovine pulmonary artery smooth muscle cells in culture. *J. Cell. Physiol.* **132**, 178–182.

Lee, S. L., Wang, W. W., Moore, B. J., and Fanburg, B. L. (1991). Dual effect of serotonin on growth of bovine pulmonary artery smooth muscle cells in culture. *Circ. Res.* **68**, 1362–1368.

Leibovich, S. J., Polverini, P. J., Shepard, H. M., Wiseman, D. M., Shively, V., and Nuseir, N. (1987). Macrophage-induced angiogenesis is mediated by tumour necrosis factor-alpha. *Nature* (*London*) **329**, 630–632.

Leung, D. W., Cachianes, G., Kuang, W. J., Goeddel, D. V., and Ferrara, N. (1989). Vascular endothelial growth factor is a secreted angiogenic mitogen. *Science* **246**, 1306–1309.

Lindner, V., and Reidy, M. A. (1991). Proliferation of smooth muscle cells after vascular injury is inhibited by an antibody against basic fibroblast growth factor. *Proc. Natl. Acad. Sci. U.S.A.* **88**, 3739–3743.

Lindner, V., Lappi, D. A., Baird, A., Majack, R. A., and Reidy, M. A. (1990). Role of basic fibroblast growth factor in vascular lesion formation. *Circ. Res.* **68,** 106–113.

Liotta, L. A., Steeg, P. S., and Stetler-Stevensen, W. G. (1991). Cancer metastasis and angiogenesis: An imbalance of positive and negative regulation. *Cell (Cambridge, Mass.)* **64,** 327–336.

Loeb, A. L., Mandel, G., Straw, J. A., and Bean, B. L. (1986). Increased aortic DNA synthesis precedes renal hypertension in rats. An obligatory step? *Hypertension (Dallas)* **8,** 754–761.

Lundergan, C., Foegh, M. I., Vargas, R., Eufemio, M., Bormes, G. W., Kot, P. A., and Ramwell, P. W. (1989). Inhibition of myointimal proliferation of the rat carotid artery by the peptides, angiopeptin and BIM 23034. *Atherosclerosis (Shannon, Irel.)* **80,** 49–55.

Madri, J. A., and Marx, M. (1992). Matrix composition, organization, and soluble factors: Modulators of microvascular cell differentiatin in vitro. *Kidney Int.* **41,** 560–565.

Madri, J. A., Pratt, B. M., and Tucker, A. M. (1988). Phenotypic modulation of endothelial cell by transforming growth factor-beta depends upon the composition and organization of the extracellular matrix. *J. Cell Biol.* **106,** 1375–1384.

Maes, P., Damart, D., Rommens, C., Montreuil, J., Spik, G., and Tartar, A. (1988). The complete amino acid sequence of bovine milk angiogenin. *FEBS Lett.* **241,** 41–45.

Maglione, D., Guerriero, V., Viglietto, G., Delli-Bovi, P., and Persico, M. G. (1991). Isolation of a human placenta cDNA coding for a protein related to the vascular permeability factor. *Proc. Natl. Acad. Sci. U.S.A.* **88,** 9267–9271.

Maione, T. E., Grey, G. S., Petro, J., Hunt, A. J., Donner, A. L., Bauer, S. I., Carson, H. F., and Sharpe, R. J. (1990). Inhibition of angiogenesis by recombinant human platelet factor-4 and related peptides. *Science* **247,** 77–79.

Maione, T. E., Gray, G. S., Hunt, A. J., and Sharpe, R. J. (1991). Inhibition of tumor growth in mice by an analogue of platelet factor 4 that lacks affinity for heparin and retains potent angiostatic activity. *Cancer Res.* **51,** 2077–2083.

Majack, R. A., and Bornstein, P. (1985). Heparin regulates the collagen phenotype of vascular smooth muscle cells: Induced synthesis of an Mr 60,000 collagen. *J. Cell Biol.* **100,** 613–619.

Majack, R. A., Goodman, L. V., and Dixit, V. M. (1988). Cell surface thrombospondin is functionally essential for vascular smooth muscle cell proliferation. *J. Cell Biol.* **106,** 415–422.

Majack, R. A., Majesky, M. W., and Goodman, L. V. (1990). Role of PDGF-A expression in the control of vascular smooth muscle cell growth by transforming growth factor-beta. *J. Cell Biol.* **111,** 239–247.

Marikovsky, M., Breuing, K., Liu, P.-Y., Eriksson, E., Higashiyama, S., Farber, P., Abraham, J., and Klagsbrun, M. (1993). Appearance of heparin-binding-EGF (HB-EGF) mRNA in wound fluid as a response to injury. *Proc. Natl. Acad. Sci. U.S.A.* **90,** 3889–3893.

Martin, P., Hopkinson-Woolley, J., and McClusky, J. (1992). Growth factors and cutaneous wound repair. *Prog. Growth Factor Res.* **4,** 25–44.

Mersey, J. H., Williams, G. H., Hollenberg, N. K., and Dluhy, R. G. (1977). Relationship between aldosterone and bradykinin. *Circ. Res.* **40/41,** Suppl. I, I-84–I-88.

Mesri, E. A., Ono, M., Kreitman, R. J., Klagsbrun, M., and Pastan, I. (1994). The heparin-binding domain of heparin-binding EGF-like growth factor can target pseudomonas exotoxin to kill cell exclusively through heparan sulfate proteoglycans. *J. Cell Sci.* **107,** 2559–2608.

Meyrick, B., Fujiwara, K., and Reid, L. (1981). Smooth muscle myosin in precursor and mature smooth muscle cells in normal pulmonary arteries and the effect of hypoxia. *Exp. Lung Res.* **2,** 303–313.

Mignatti, P., Tsuboi, R., Robbins, E., and Rifkin, D. B. (1989). In vitro angiogenesis on the human amniotic membrane: Requirement for basic fibroblast growth factor-induced proteinases. *J. Cell Biol.* **108,** 671–682.

Millauer, B., Wizigmann-Voos, S., Schnurch, H., Martinez, R., Moller, N. P., Risau, W., and Ullrich, A. (1993). High affinity VEGF binding and developmental expression suggest Flk-1 as a major regulator of vasculogenesis and angiogenesis. *Cell (Cambridge, Mass.)* **72,** 835–846.

Mitamura, T., Higashiyama, S., Taniguchi, N., Klagsbrun, M., and Mekada, E. (1995). Diphtheria toxin binds to the EGF-like domain of human heparin-binding EGF-like growth factor/diphtheria toxin receptor and inhibits specifically its mitogenic activity. *J. Biol. Chem.* **270**, 1015–1019.

Mittra, I. (1980). A novel "cleaved prolactin" in the rat pituitary. II. In vivo mammary mitogenic activity of its N-terminal 16K moiety. *Biochem. Biophys. Res. Commun.* **95**, 1760–1767.

Miyamoto, M., Naruo, K., Seko, C., Matsumoto, S., Konodo, T., and Kurokawa, T. (1993). Molecular cloning of a novel cytokine cDNA encoding the ninth member of the fibroblast growth factor family, which has a unique secretion property. *Mol. Cell. Biol.* **13**, 4251–4259.

Miyazono, K., Okabe, T., Urabe, A., Takaku, F., and Heldin, C. H. (1987). Purification and properties of an endothelial cell growth factor from human platelets. *J. Biol. Chem.* **262**, 4098–4103.

Montesano, R., and Pepper, M. S. (1990). Proteolytic balance and capillary morphogenesis. *Cell Differ. Dev.* **32**, 319–328,

Montesano, R., Orci, L., and Vassalli, P. (1983). In vitro rapid organization of endothelial cells into capillary-like networks is promoted by collagen matrices. *J. Cell Biol.* **97**, 1648–1652.

Montesano, R., Pepper, M. S., Mohle-Steinlen, U., Risau, W., Wagner, E. F., and Orci, L. (1990). Increased proteolytic activity is responsible for the aberrant morphogenetic behavior of endothelial cells expressing the middle T oncogene. *Cell (Cambridge, Mass.)* **62**, 435–445.

Moscatelli, D. (1987). High and low affinity binding sites for basic fibroblast growth factor on cultured cells: Absence of a role for low affinity binding in the stimulation of plasminogen activator production by bovine capillary endothelial cells. *J. Cell. Physiol.* **131**, 123–130.

Moses, M. A., and Langer, R. (1991). Angiogenesis inhibitors. *Bio/Technology* **9**, 630–634.

Moses, M. A., Sudhalter, J., and Langer, R. (1990). Identification of an inhibitor of neovascularization from cartilage. *Science* **248**, 1408–1410.

Moses, M. A., Sudhalter, J., and Langer, R. (1992). Isolation and characterization of an inhibitor of neovascularization from chondrocytes. *J. Cell Biol.* **119**, 475–482.

Muller, G., Behrens, J., Nussbaumer, U., Bohlen, P., and Birchmeier, W. (1987). Inhibitory action of transforming growth factor-β on endothelial cells. *Proc. Natl. Acad. Sci. U.S.A.* **84**, 5600–5604.

Mulvaney, M. J. (1991). Are vascular abnormalities a primary cause or secondary consequence of hypertension? *Hypertension (Dallas)* **18**, Suppl. 3, I-52–I-57.

Mulvaney, M. J., Baandrup, U., and Gundersen, H. L. G. (1985). Evidence for hyperplasia in mesenteric resistance vessels of spontaneously hypertensive rats using a three-dimensional dissector. *Circ. Res.* **57**, 794–800.

Murphy, A. N., Unsworth, E. J., and Stetler-Stevenson, W. G. (1993). Tissue inhibitor of metalloproteinases-2 inhibits bFGF-induced human microvascular endothelial cell proliferation. *J. Cell. Physiol.* **157**, 351–358.

Muthukkaruppan, V., and Auerbach, R. (1979). Angiogenesis in the mouse cornea. *Science* **205**, 1416–1417.

Naftilan, A. J., Pratt, R. E., Eldridge, C. S., Lin, H. L., and Dzau, V. J. (1989a). Angiotensin II induces c-fos expression in smooth muscle via transcriptional control. *Hypertension (Dallas)* **13**, 706–711.

Naftilan, A. J., Pratt, R. E., and Dzau, V. J. (1989b). Induction of platelet-derived growth factor A-chain and c-myc gene expressions by angiotensin II in cultured rat vascular smooth muscle cells. *J. Clin. Invest.* **83**, 1419–1424.

Naglich, J. G., Metherall, J. E., Russell, D. W., and Eidels, L. (1992). Expression cloning a diphtheria toxin receptor: Identity with a heparin-binding EGF-like growth factor precursor. *Cell (Cambridge, Mass.)* **69**, 1051–1061.

Nakamura, T., Nawa, K., Ichihara, A., Kaise, N., and Nishino, T. (1987). Purification and subunit structure of hepatocyte growth factor from rat platelets. *FEBS Lett.* **224**, 311–316.

Nakano, T., Raines, E., Abraham, J., Wenzel, F. G., Higashiyama, S., Klagsbrun, M., and Ross, R. (1993). Glucocorticoid inhibits thrombin-induced expression of PDGF-A chain and HG-EGF in human aortic smooth muscle cells. *J. Biol. Chem.* **268**, 22941–22947.

Nakano, T., Raines, E. W., Abraham, J. A., Klagsbrun, M., and Ross, R. (1994). Lysophophatidylcholine upregulates the level of heparin-binding EGF-like growth factor mRNA in human monocytes. *Proc. Natl. Acad. Sci. U.S.A.* **91**, 1069–1073.

Nguyen, M., Folkman, J., and Bischoff, J. (1992). 1-Deoxymannojirimycin inhibits capillary tube formation in vitro: Analysis of N-linked oligosaccharides in bovine capillary endothelial cells. *J. Biol. Chem.* **267**, 26157–26165.

Nilsson, J., and Olsson, A. G. (1984). Prostaglandin E1 inhibits DNA synthesis in arterial smooth muscle cells stimulated with platelet-derived growth factor. *Atherosclerosis (Shannon, Irel.)* **53**, 77–82.

Nurcombe, V., Ford, M. D., Wildschutt, J. A., and Bartlett, P. F. (1993). Developmental regulation of neural response to FGF-1 and FGF-2 by heparan sulfate proteoglycan. *Science* **260**, 103–106.

Ohno, T., Gordon, D., San, H., Pompili, V. J., Imperiale, M. J., Nabel, G. J., and Nabel, E. G. (1994). Gene therapy for vascular smooth muscle cell proliferation after arterial injury. *Science* **264**, 781–784.

Olson, N. E., Chao, S., Lindner, V., and Reidy, M. A. (1992). Intimal smooth muscle cell proliferation after balloon catheter injury. The role of basic fibroblast growth factor. *Am. J. Pathol.* **140**, 1017–1023.

Ono, M., Raab, G., Lau, K., Abraham, J. A., and Klagsbrun, M. (1994). Purification and characterization of transmembrane forms of heparin-binding EGF-like growth factor. *J. Biol. Chem.* **269**, 31315–31321.

Orchard, P. J., Smith, C. M., Woods, W. G., Day, D. L., Dehner, L. P., and Shapiro, R. (1989). Treatment of haemangioednotheliomas with alpha interferon [letter]. *Lancet* **2**, 565–567.

O'Reilly, M. S., Holmgren, L., Shing, Y., Chen, C., Rosenthal, R. A., Moses, M., Lane, W. S., Cao, Y., Sage, E. H., and Folkman, J. (1994). Angiostatin: A novel angiogenesis inhibitor that mediates the suppression of metastases by a Lewis lung carcinoma. *Cell (Cambridge, Mass.)* **79**, 315–328.

Orekhov, A. N., Tertov, V. V., and Smirnov, V. N. (1983). Prostacyclin analogues as antiatherosclerotic drugs [letter]. *Lancet* **2**, 521.

Orekhov, A. N., Tertov, V. V., Kudryashov, S. A., Khashimov, K. H. A., and Smirnov, V. N. (1986). Primary culture of human aortic intima cells as a model for testing antiatherosclerotic drugs. Effects of cyclic AMP, prostaglandins, calcium antagonists, antioxidants, and lipid-lowering agents. *Atherosclerosis (Shannon, Irel.)* **60**, 101–110.

Orlidge, A., and D'Amore, P. A. (1986). Cell specific effects of glycosaminoglycans on the attachment and proliferation of vascular wall components. *Microvasc. Res.* **31**, 41–53.

Orlidge, A., and D'Amore, P. A. (1987). Inhibition of capillary endothelial cell growth by pericytes and smooth muscle cells. *J. Cell Biol.* **105**, 1455–1462.

Owens, G. K., Schwartz, S. M., and McCanna, M. (1988). Evaluation of medial hypertrophy in resistance vessels of spontaneously hypertensive rats. *Hypertension (Dallas)* **11**, 198–207.

Paquet, J.-L., Baudouin-Legros, M., Marche, P., and Meyer, P. (1989). Enhanced proliferating activity of cultured smooth muscle cells from SHR. *Am. J. Hypertens.* **2**, 108–110.

Park, J. E., Keller, G. A., and Ferrara, N. (1993). The vascular endothelial growth factor (VEGF) isoforms: Differential deposition into the subepithelial extracellular matrix and bioactivity of extracellular matrixbound VEGF. *Mol. Biol. Cell* **4**, 1317–1326.

Peacock, D. J., Banquerigo, M. L., and Brahn, E. (1992). Angiogenesis inhibition suppresses collagen arthritis. *J. Exp. Med.* **175**, 1135–1138.

Pepper, M. S., Vassalli, J. D., Montesano, R., and Orci, L. (1987). Urokinase-type plasminogen activator is induced in migrating capillary endothelial cells. *J. Cell Biol.* **105**, 2535–2541.

Pepper, M. S., Belin, D., Montesano, R., Orci, L., and Vassalli, J. D. (1990). Transforming growth factor-beta 1 modulates basic fibroblast growth factor-induced proteolytic and angiogenic properties of endothelial cells in vitro. *J. Cell Biol.* **111**, 743–755.

Pepper, M. S., Ferrara, N., Orci, L., and Montesano, R. (1992). Potent synergism between vascular endothelial growth factor and basic fibroblast growth factor in the induction of angiogenesis in vitro. *Biochem. Biophy. Res. Commun.* **189**, 824–831.

Plate, K. H., Breier, G., Weich, H. A., and Risau, W. (1992). Vascular endothelial growth factor is a potential tumour angiogenesis factor in human gliomas in vivo. *Nature* (*London*) **359,** 845–848.

Plouet, J., Schilling, J., and Gospodarowicz, D. (1989). Isolation and characterization of a newly identified endothelial cell mitogen produced by AtT-20 cells. *EMBO J.* **8,** 3801–3806.

Plowman, G. D., Culouscou, J. M., Shitney, G. S., Green, J. M., Carlton, G. W., Foy, L., Neubauer, M. G., and Shoyab, M. (1993). Ligand-specific activation of HER4/p180erbB4, a fourth member of the epidermal growth factor receptor family. *Proc. Natl. Acad. Sci. U.S.A.* **90,** 1746–1750.

Polakowski, I. J., and Lewis, M. K. (1993). A ribonuclease inhibitor expresses anti-angiogenic properties and leads to reduced tumor growth in mice. *Am. J. Pathol.* **143,** 507–517.

Powell, J. S., Muller, R. K., Rouge, M., Kuhn, H., Hefti, F., and Baumgartner, H. R. (1990). The proliferative response to vascular injury is suppressed by angiotensin-converting enzyme inhibition. *J. Cardiovasc. Pharmacol.* **16,** Suppl., S42–S49.

Powell, P., Klagsbrun, M., Abraham, J. A., and Jones, R. (1993). Eosinophil expressing heparin-binding EGF-like growth factor localizes around microvessels in pulmonary hypertension. *Am. J. Pathol.* **143,** 784–793.

Prescott, M. F., Webb, R. L., and Reidy, M. A. (1991). Angiotensin-converting enzyme inhibitor versus angiotensin II, AT1 receptor antagonist. Effects on smooth muscle cell migration and proliferation after balloon catheter injury. *Am. J. Pathol.* **139,** 1291–1296.

Quinn, T. P., Peters, K. G., De Vries, C., Ferrara, N., and Williams, L. T. (1993). Fetal liver kinase 1 is a receptor for vascular endothelial growth factor and is selectively expressed in vascular endothelium. *Proc. Natl. Acad. Sci. U.S.A.* **90,** 7533–7537.

Raab, G., and Klagsbrun, M. (1994). Submitted for publication.

Raab, G., Higashiyama, S., Hetelekidis, S., Abraham, J. A., Damm, D., Ono, M., and Klagsbrun, M. (1994). Biosynthesis and processing by phorbol ester of the cell surface-associated form of heparin- binding EGF-like growth factor. *Biochem. Biophys. Res. Commun.* 204, 592–597.

Raines, E. W., Dower, S. K., and Ross, R. (1989). Interleukin-1 mitogenic activity for fibroblasts and smooth muscle cells is due to PDGF-AA. *Science* **243,** 393–396.

Raines, E. W., Bowen-Pope, D. F., and Ross, R. (1990). Platelet-derived growth factor. *In* "Peptide Growth Factors and Their Receptors" (M. B. Sporn and A. B. Roberts, eds.), pp. 173–262. Springer-Verlag, Berlin.

Rapraeger, A. C., Krufka, A., and Olwin, B. B. (1991). Requirement of heparen sulfate for bFGF-mediated fibroblast growth and myoblast differentiation. *Science* **252,** 1075–1078.

Rastinejad, F., Polverini, P. J., and Bouck, N. P. (1989). Regulation of the activity of a new inhibitor of angiogenesis by a cancer suppressor gene. *Cell* (*Cambridge, Mass.*) **56,** 345–355.

Reilly, C. F., and McFall, R. C. (1991). Platelet-derived growth factor and transforming growth factor-beta regulate plasminogen activator inhibitor-1 synthesis in vascular smooth muscle cells. *J. Biol. Chem.* **266,** 9419–9427.

Reilly, C. F., Fritze, L. M. S., and Rosenberg, R. D. (1987). Antiproliferative effects of heparin on vascular smooth muscle cells are reversed by epidermal growth factor. *J. Cell. Physiol.* **131,** 149–157.

Rhodin, J. A. G., and Fujita, H. (1989). Capillary growth in the mesentery of normal young rats. Intravital video and electron microscopy. *J. Submicrosc. Cytol. Pathol.* **21,** 1–34.

Rifkin, D. B., and Moscatelli, D. (1989). Recent developments in the cell biology of basic fibroblast growth factor. *J. Cell Biol.* **109,** 1–6.

Rifkin, D. B., Moscatelli, D., Bizik, J., Quarto, N., Blei, F., Dennis, P., Flaumenhaft, R., and Mignatti, P. (1990). Growth factor control of extracellular proteolysis. *Cell. Differ. Dev.* **32,** 313–318.

Risau, W., Drexler, H., Mironov, V. *et al.* (1992). Platelet- derived growth factor is angiogenic in vivo. *Growth Factors* **7,** 261–266.

Roberts, W. C. (1989). Qualitative and quantitative comparison of amounts of narrowing by atherosclerotic plaques in the major epicardial coronary arteries at necropsy in sudden coronary death, transmural acute myocardial infarction, transmural healed myocardial infarction and unstable angina pectoris. *Am. J. Cardiol.* **64,** 324–328.

Rogelj, S., Klagsbrun, M., Atzmon, R., Kurokawa, M., Haimovitz, A., Fuks, Z., and Vlodavsky, I. (1989). Basic fibroblast growth factor is an extracellular matrix component required for supporting the proliferation of vascular endothelial cells and the differentiation of PC12 cells. *J. Cell Biol.* **109,** 823–831.

Roghani, M., and Moscatelli, D. (1992). Basic fibroblast growth factor is internalized through both receptor-mediated and heparan sulfate-mediated mechanisms. *J. Biol. Chem.* **267,** 22156–22162.

Rosenberg, C., van Wijk, R., Zandbergen, J., van Aken, W. G., van Mourik, J. A., and deGroot, P. G. (1985). Cell cycle-dependent inhibition of human vascular smooth muscle cell proliferation by prostaglandin E1. *Exp. Cell Res.* **160,** 117–125.

Ross, R. (1986). The pathogenesis of atherosclerosis: An update. *N. Engl. J. Med.* **314,** 488–500.

Rothman, A., Wolner, B., Button, D., and Taylor, P. (1994). Immediate-early gene expression in response to hypertrophic and proliferative stimuli in pulmonary arterial smooth muscle cells. *J. Biol. Chem.* **269,** 6399–6404.

Rouget, C. (1879). Sur la contractilité des capillaires sanguins. *C. R. Hebd. Seances Acad. Sci.* **88,** 916–918.

Russell, W. E., McGowan, J. A., and Bucher, N. R. L. (1984). Partial characterization of a hepatocyte growth factor from rat platelets. *J. Cell. Physiol.* **119,** 183–192.

Saksela, O., Moscatelli, D., and Rifkin, D. B. (1987). The opposing effects of basic fibroblast growth factor and transforming growth factor beta on the regulation of plasminogen activator activity in capillary endothelial cells. *J. Cell Biol.* **105,** 957–963.

Sasada, R., Ono, Y., Taniyama, Y., Shing, Y., Folkman, J., and Igarashi, K. (1993). Cloning and expression of cDNA encoding human betacellulin, a new member of the EGF family. *Biochem. Biophys. Res. Commun.* **190,** 1173–1179.

Sasisekharan, R., Moses, M. A., Nugent, M. A., Cooney, C. L., and Langer, R. (1994). Heparinase inhibits neovascularization. *Proc. Natl. Acad. Sci. U.S.A.* **91,** 1524–1528.

Sato, Y., and Rifkin, D. B. (1989). Inhibition of endothelial cell movement by pericytes and smooth muscle cells: Activation of a latent transforming growth factor-β1-like molecule by plasmin during co-culture. *J. Cell Biol.* **109,** 309–315.

Sato, Y., Nariuchi, H., Tsuruoka, N., Nishihara, T., Beitz, J. G., Calabresi, P., and Frackelton, A. R. (1990). Actions of TNF and IFM-gamma on angiogenesis in vitro. *J. Invest. Dermatol.* **95,** 85S–89S.

Sato, Y., Shimada, T., and Takaki, R. (1991). Autocrinological role of basic fibroblast growth factor on tube formation of vascular endothelial cells in vitro. *Biochem. Biophys. Res. Commun.* **180,** 1098–1102.

Sato, Y. R., Tsuboi, R., Lyons, R., Moses, H., and Rifkin, D. B. (1990). Characterization of the activation of latent TGF-β by co- cultures of endothelial cells and pericytes or smooth muscle cells: A self regulating system. *J. Cell Biol.* **111,** 757–763.

Schwartz, S. M. (1985). Cellular mechanisms in atherosclerosis: Theories and therapies. *Ann. N.Y. Acad. Sci.* **454,** 320–321.

Schwartz, S. M., and Reidy, M. A. (1987). Common mechanisms of proliferation of smooth muscle in atherosclerosis and hypertension. *Hum. Pathol.* **18,** 240–247.

Schwartz, S. M., Reidy, M. A., and Clowes, A. W. (1985). Kinetics of atherosclerosis: A stem cell model. *Ann. N.Y. Acad. Sci.* **454,** 292–304.

Schwartz, S. M., Campbell, G. R., and Campbell, J. H. (1986). Replication of smooth muscle cells in vascular disease. *Circ. Res.* **58,** 427–444.

Schwartz, S. M., Heimark, R. L., and Majesky, M. W. (1990). Developmental mechanisms underlying pathology of arteries. *Physiol. Rev.* **70,** 1177–1209.

SCRIP. (1991). 1648, 23.

Shapiro, R., and Vallee, B. L. (1987). Human placental ribonuclease inhibitor abolishes both angiogenic and ribonucleolytic activities of angiogenin. *Proc. Natl. Acad. Sci. U.S.A.* **84,** 2238–2241.

Shapiro, R., Riordan, J. F., and Vallee, B. L. (1986). Characteristic ribonucleolytic activity of human angiogenin. *Biochemistry* **25,** 3527–3532.

Shapiro, R., Strydom, D. J., Olson, K. A., and Vallee, B. L. (1987). Isolation of angiogenin from normal human plasma. *Biochemistry* **26**, 5141–5146.

Sharpe, R. J., Byers, H. R., Scott, C. F., Bauer, S. I., and Maione, T. E. (1990). Growth inhibition of murine melanoma and human colon carcinoma by recombinant human platelet factor 4. *J. Natl. Cancer Inst.* **82**, 848–853.

Shing, Y., Folkman, J., Sullivan, R., Butterfield, C., Murray, J., and Klagsbrun, M. (1984). Heparin affinity: Purification of a tumor-derived capillary endothelial cell growth factor. *Science* **223**, 1296–1299.

Shing, Y., Christofori, D., Hanahan, D., Ono, Y., Sasada, R., Igarashi, K., and Folkman, J. (1993). Betacellulin: A novel mitogen from pancreatic β tumor cells. *Science* **259**, 1604–1607.

Shweiki, D., Itin, A., Soffer, D., and Keshet, E. (1992). Vascular endothelial growth factor induced by hypoxia may mediate hypoxia-initiated angiogenesis. *Nature (London)* **359**, 843–845.

Simionescu, N., and Simionescu, M. (1983). *In* "Histology, Cell and Tissue Biology" (L. S. Weiss, ed.), pp. 373–433. Elsevier, New York.

Simons, M., Edelman, E. R., Dekeyser, J., Langer, R., and Rosenberg, R. D. (1992). Antisense c-myb oligonucleotides inhibit intimal arterial smooth muscle cell accumulation *in vivo. Nature (London)* **359**, 67–70.

Sims, D. E. (1986). The pericyte—a review. *Tissue Cell* **18**, 153–174.

Singh, R. K., Bucana, C. D., Gutman, M., Fan, D., Wilson, M. R., and Fidler, I. J. (1994). Organ site-dependent expression of basic fibroblast growth factor in human renal cell carcinoma cells. *Am. J. Pathol.* **145**, 365–374.

Smith, D. L., Willis, A. L., and Mahmud, I. (1984). Eicosanoid effects on cell proliferation in vitro: Relevance to atherosclerosis. *Prostaglandins Leukotriene Med.* **16**, 1–10.

Snow, A. D., Bolender, R. P., Wight, T. N., and Clowes, A. W. (1990). Heparin modulates the composition of the extracellular matrix domain surrounding arterial smooth muscle cells. *Am. J. Pathol.* **137**, 313–330.

Soncin, F. (1992). Angiogenin supports endothelial and fibroblast cell adhesion. *Proc. Natl. Acad. Sci. U.S.A.* **89**, 2232–2236.

Soyombo, A. A., and DiCorleto, P. E. (1994). Stable expression of human platelet-derived growth factor B chain by bovine aortic endothelial cells. Matrix association and selective proteolytic cleavage by thrombin. *J. Biol. Chem.* **269**, 17734–17740.

St. Clair, D. K., Rybak, S. M., Riordan, J. F., and Vallee, B. L. (1987). Angiogenin abolishes cell-free protein synthesis by specific ribonucleolytic inactivation of ribosomes. *Proc. Natl. Acad. Sci. U.S.A.* **84**, 8330–8334.

Stoker, M., and Perryman, M. (1985). An epithelial scatter factor released by embryo fibroblasts. *J. Cell Sci.* **77**, 209–223.

Strydom, D. J., Fett, J. W., Lobb, R. R., Alderman, E. M., Bethune, J. L., Riordan, J. F., and Vallee, B. L. (1985). Amino acid sequence of human tumor derived angiogenin. *Biochemistry* **24**, 5486–5494.

Sumizawa, T., Furukawa, T., Haraguchi, M., Yoshimura, A., Takeyasu, A., Ishizawa, M., Yamada, Y., and Akiyama, S. (1993). Thymidine phosphorylase activity associated with platelet-derived endothelial cell growth factor. *J. Biochem. (Tokyo)* **114**, 9–14.

Takahashi, A., Sasaki, H., Kim, S. J., Tobisu, K., Kakizoe, T., Tsukamoto, T., Kumamoto, Y., Sugimura, T., and Terada, M. (1994). Markedly increased amounts of messenger RNAs for vascular endothelial growth factor and placenta growth factor in renal cell carcinoma associated with angiogenesis. *Cancer Res.* **54**, 4233–4237.

Takamiya, Y., Friedlander, R. M., Brem, H., Malick, A., and Martuza, R. L. (1993). Inhibition of angiogenesis and growth of human nerve-sheath tumors by AGM-1470. *J. Neurosurg.* **78**, 470–476.

Takeshita, S., Zheng, L. P., Brogi, E., Kearney, M., Pu, L.-Q., Bunting, S., Ferrara, N., Symes, J. F., and Isner, J. M. (1994). Therapeutic Angiogenesis: A single intraarterial bolus of vascular endothelial growth factor augments revascularization in a rabbit ischemic hind limb model. *J. Clin. Invest.* **93**, 662–670.

Tan, E. M. L., Dodge, G. R., Sorger, T., Kovalsky, I., Unger, G. A., Yang, L., Levine, E. M., and Iozzo, R. V. (1991). Modulation of extracellular matrix gene expression by heparin and endothelial cell growth factor in human smooth muscle cells. *Lab. Invest.* **64,** 474–482.

Tanaka, A., Miyamoto, K., Minamino, N., Takeda, M., Sato, B., Matsuo, H., and Matsumoto, K. (1992). Cloning and characterization of an androgen-induced growth factor essential for the androgen-dependent growth of mouse mammary carcinoma cells. *Proc. Natl. Acad. Sci. U.S.A.* **89,** 8928–8932.

Taraboletti, G., Belotti, D., and Giavazzi, R. (1992). Thrombospondin modulates basic fibroblast growth factor activities on endothelial cells. *EXS* **61,** 210–213.

Taylor, S., and Folkman, J. (1982). Protamine is an inhibitor of angiogenesis. *Nature (London)* **297,** 307–312.

Terman, B. I., Dougher-Vermazen, M., Carrion, M. E., Dimitrov, D., Armellino, D. C., Gospodarowicz, D., and Bohlen, P. (1992). Identification of the KDR tyrosine kinase as a receptor for vascular endothelial cell growth factor. *Biochem. Biophys. Res. Commun.* **187,** 1579–1586.

Thompson, S. A., Higashiyama, S., Wood, K., Pollitt, N. S., Damm, D., McEnroe, G., Garrick, B., Ashton, N., Lau, K., Hancock, N., Klagsbrun, M., and Abraham, J. A. (1994). Characterization of sequences within heparin-binding EGF-like growth factor that mediate interaction with heparin. *J. Biol. Chem.* **269,** 2541–2549.

Tischer, E., Mitchell, R., Hartman, T., Silva, M., Gospodarowicz, D., Fiddes, J. C., and Abraham, J. A. (1991). The human gene for vascular endothelial growth factor. Multiple protein forms are encoded through alternative exon splicing. *J. Biol. Chem.* **266,** 11947–11954.

Tolsma, S. S., Volpert, O. V., Good, D. J., Frazier, W. A., Polverini, P. J., and Bouck, N. (1993). Peptides derived from two separate domains of the matrix protein thrombospondin-1 have antiangiogenic activity. *J. Cell Biol.* **122,** 497–511.

Turner, R. R., Beckstead, J. H., Warnke, T. A., and Wood, G. S. (1987). Endothelial cell phenotypic diversity. *Am. J. Pathol.* **87,** 569–575.

Uehara, Y., Nagata, T., Matsuoka, H., Numabe, A., Hirawa, N., Takada, S., Ishimitsu, T., Yagi, S., and Sugimoto, T. (1991). Antiproliferative effects of the serotinin type 2 receptor antagonist, ketanserin, on smooth muscle cell growth in rats. *J. Cardiovasc. Pharmacol.* **17,** S154–S156.

Unger, E. F., Banai, S., Shou, M., Lazarous, D. F., Jaklitsch, M. T., Scheinowitz, M., Correa, R., Klingbeil, C., and Epstein, S. E. (1994). Basic fibroblast growth factor enhances myocardial collateral flow in a canine model. *Am. J. Physiol.* **266**(4, Pt. 2), H1588–H1595.

Usuki, K., Heldin, N., Miyazono, K., Ishikawa, F., Takaku, F., Westermark, B., and Heldin, C. (1989). Production of platelet-derived endothelial cell growth factor by normal and transformed human cells in culture. *Proc. Natl. Acad. Sci. U.S.A.* **86,** 7427–7431.

Usuki, K., Miyazono, K., and Heldin, C. H. (1991). Covalent linkage between nucleotides and platelet-derived endothelial cell growth factor. *J. Biol. Chem.* **266,** 20525–20531.

Vargas, R., Bormes, G. W., Wroblewska, B., Foegh, M. L., Kot, P. A., and Ramwell, P. W. (1989). Angiopeptin inhibits thymidine incorporation in rat carotid artery in vitro. *Transplant. Proc.* **21,** 3702–3703.

Vernon, R. B., Angello, J. C., Iruela-Arispe, L., Lane, T. F., and Sage, E. H. (1992). Reorganization of basement membrane matrices by cellular traction promotes the formation of cellular networks in vitro. *Lab. Invest.* **66,** 536–547.

Vlodavsky, I., Folkman, J., Sullivan, R., Fridman, R., Ishai-Michaeli, R., Sasse, J., and Klagsbrun, M. (1987). Endothelial cell-derived basic fibroblast growth factor: Synthesis and deposition into subendothelial extracellular matrix. *Proc. Natl. Acad. Sci. U.S.A.* **84,** 2292–2296.

Vlodavsky, I., Michaeli, R. I., Bar-Ner, M., Fridman, R., Horowitz, A. T., Fuks, Z., and Brian, S. (1988). Involvement of heparanase in tumor metastasis and angiogenesis. *Isr. J. Med. Sci.* **24,** 464–470.

Wang, X.-N., Das, S. K., Abraham, J. A., Klagsbrun, M., and Dey, S. K. (1994). Differential regulation of heparin-binding EGF-like growth factor in the adult ovariectomized mouse uterus by progesterone and estrogen. *Endocrinology (Baltimore)* **135,** 1264–1271.

Watanabe, T., Shintani, A., Nakata, M., Shing, Y., Folkman, J., Igarashi, K., and Sasada, R. (1994). Recombinant human betacellulin: Molecular structure, biological activities, and receptor interaction. *J. Biol. Chem.* **269,** 9966–9973.

Weber, H., Taylor, D. S., and Molloy, C. J. (1994). Angiotensin II induces delayed mitogenesis and cellular proliferation in rat aortic smooth muscle cells. Correlation with the expression of specific endogenous growth factors and reversal by suramin. *J. Clin. Invest.* **93,** 788–798.

Weich, H. E., Iberg, N., Klagsbrun, M., and Folkman, J. (1990). Expression of acidic and basic fibroblast growth factors in human and bovine vascular smooth muscle cells. *Growth Factors* **2,** 313–320.

Weiner, H. L., and Swain, J. L. (1989). Acidic fibroblast growth factor mRNA is expressed by cardiac myocytes in culture and the protein is localized to the extracellular matrix. *Proc. Natl. Acad. Sci. U.S.A.* **86,** 2683–2687.

Werner, S., Duan, D. S., DeVries, C., Peter, K. G., Johnson, D. E., and Williams, L. T. (1992). Differential splicing in the extracellular region of fibroblast growth factor receptor 1 generates receptor variants with different ligand-binding specificities. *Mol. Cell Biol.* **12,** 82–88.

White, C. W., Sondheimer, H. M., Crouch, E. C., Wilson, H., and Fan, L. L. (1989). Treatment of pulmonary hemangiomatosis with recombinant interferon alpha-2a. *N. Engl. J. Med.* **320,** 1197–1200.

Wong, V., Compton, M., and Witorsch, R. (1986). Proteolytic modification of rat prolactin by subcellular fractions of the lactating rat mammary gland. *Biochim. Biophys. Acta* **881,** 167–174.

Yamagishi, S., Hsu, C., Kobayashi, K., and Yamamoto, H. (1993). Endothelin 1 mediates endothelial cell-dependent proliferation of vascular pericytes. *Biochem. Biophys. Res. Commun.* **191,** 840–846.

Yanagisawa-Miwa, A., Uchida, Y., Nakamura, T., Tomaru, T., Kido, H., Kamijo, T., Sugimoto, T., Kaji, K., Utsuyama, M., and Kurashima, C., and Ito, H. (1992). Salvage of infarcted myocardium by angiogenic action of basic fibroblast growth factor. *Science* **257,** 1401–1403.

Yanase, T., Tamura, M., Fujita, K., Kodama, S., and Tanaka, K. (1993). Inhibitory effect of angiogenesis inhibitor TNP-470 on tumor growth and metastasis of human cell lines in vitro and in vivo. *Cancer Res.* **53,** 2566–2570.

Yayon, A., Klagsbrun, M., Esko, J. D., Leder, P., and Ornitz, D. M. (1991). Cell surface, heparin-like molecules are required for binding of basic fibroblast growth factor to its high affinity receptor. *Cell (Cambridge, Mass.)* **64,** 841–848.

Yoshizumi, M., Kourembanas, S., Temizer, D. H., Cambria, R. P., Quertermous, T., and Lee, M. E. (1992). Tumor necrosis factor increases transcription of the heparin-binding epidermal growth factor-like growth factor gene in vascular endothelial cells. *J. Biol. Chem.* **267,** 9467–9469.

Zajchowski, D. A., Band, V., Trask, D. K., Kling, D., Connolly, J. L., and Sager, R. (1990). Suppression of tumor-forming ability and related traits in MCF-7 human breast cancer cells by fusion with immortal mammary epithelial cells. *Proc. Natl. Acad. Sci. U.S.A.* **87,** 2314–2318.

Zaragoza, R., Battle-Tracy, K. M., and Owen, N. E. (1990). Heparin inhibits Na^+-H^+ exchange in vascular smooth muscle cells. *Am. J. Physiol.* **258,** 46–53.

Zhang, X., Chen, L., Bancroft, D. P., Lai, C. K., and Maione, T. E. (1994). Crystal structure of recombinant human platelet factor 4. *Biochemistry* **33,** 8361–8366.

Zhu, X., Hsu, B. T., and Rees, D. C. (1993). Structural studies of the binding of the anti-ulcer drug sucrose octasulfate to acidic fibroblast growth factor. *Structure* **1,** 27–34.

Zimmerman, K. (1923). Der Peinere Bau der Blutcapillaren. *Z. Anat. Entwicklungsgesch.* **68,** 29–109.

Sexuality of Mitochondria: Fusion, Recombination, and Plasmids

Shigeyuki Kawano, Hiroyoshi Takano, and Tsuneyoshi Kuroiwa
Department of Biological Sciences, Graduate School of Science, University of Tokyo, Hongo, Tokyo 113, Japan

Mitochondrial fusion, recombination, and mobile genetic elements, which are essential for mitochondrial sexuality, are well established in various organisms. The recombination of mitochondrial DNA (mtDNA) depends upon fusion between parental mitochondria, and between their mtDNA-containing areas (mt-nuclei), to allow pairing between the parental mtDNAs. Such mitochondrial fusion followed by recombination may be called "mitochondrial sex." We have identified a novel mitochondrial plasmid named mF. This plasmid is apparently responsible for promoting mitochondrial fusion and crosses over with mtDNA in successive sexual crosses with mF⁻ strains. Only in mF⁺ strains carrying the mF plasmid did small spherical mitochondria fuse which subsequently underwent fusion between the mt-nuclei that contained the mtDNA derived from individual mitochondria. Several successive mitochondrial divisions followed, accompanied by mt-nuclear divisions. The resulting mitochondria contained recombinant mtDNA with the mF plasmid. Such features remind us also of the bacterial conjugative plasmids such as F plasmid. Therefore, in the final part of this chapter, we discuss the origin of sex and its relationship to the sexuality of mitochondria.
KEY WORDS: *Physarum polycephalum,* Mitochondria, Fusion, Recombination, Plasmid, Sex.

I. Introduction

Mitochondria are semiautonomous organelles with a large variety of functions in cellular metabolism. The question of how mitochondria are formed has fascinated us ever since Altmann (1890) first described these organelles 100 years ago. Initially, mitochondria were regarded as intracellular parasites. In the 1950s, when the resolving power of the electron microscope revealed the multitude of intracellular membranes in eukaryotes, mitochondria were frequently assumed to be just another intracellular membrane system composed of unit membranes. In the early 1960s, the discovery of mitochondrial DNA (mtDNA) led some investigators to propose that mitochondria were self-replicating units. Today, mitochondria are viewed as bona fide organelles that are very much controlled by the nucleus, yet they retain earmarks of their endosymbiotic past. They should harbor not only pieces of genes but also retain unique properties characteristic of ancient mitochondria and aerobic bacteria. One such property is a capacity for fusion and recombination.

The genetic properties of mitochondria have now been precisely described from the results of numerous crosses performed using various combinations of ant^R (resistance to antibiotics) mitochondrial markers of the yeast, Saccharomyces cerevisiae. Dujon (1981) concluded that mitochondrial crosses reveal all the properties of a multiple-copy genetic system, and can best be treated in terms of population genetics at the intracellular level. This model, although deliberately simplified in its formalism, was the first attempt to provide an integrated view of the genetic properties of mitochondrial crosses and gave rise to specific predictions amenable to quantitative experimental verification. One of its premises is random pairing and recombination between the different mtDNA molecules of a cell. Certainly, almost all data of mitochondrial crosses seem to indicate multiple rounds of pairing and recombination of mtDNA molecules as a panmictic pool. Such a situation suggests a cellular mechanism that promotes effective pairing and recombination of mtDNA. It may be simply explained as a result of mitochondrial fusion. Kuroiwa (1982) proposed that mitochondria perform "mitochondrial meiosis" during cell nuclear meiosis, since mitochondrial fusion and the following successive division are observed during meiosis of S. cerevisiae. However, we do not yet fully know how such mitochondrial fusion is regulated and by what kind of induction it occurs, though the past dozen years have witnessed the accumulation of a large body of knowledge on the biogenesis of mitochondria.

There are many reports (Table I), in which mitochondrial fusion is generally thought to be involved in pairing and recombination of mtDNA. In most reports, however, fusion and recombination of mtDNA have been reported independently, and practically no report exists which correlates these two phenomena, even in S. cerevisiae, whose genetic experimental systems have been well established. The simplest approach to demonstrating the relationship between these two phenomena, fusion and recombination, is to isolate a mutant that promotes or suppresses mitochondrial fusion. Recently, separate strains of the true slime mold Physarum polycephalum, one showing mitochondrial fusion and the other not, were isolated, and these mitochondria were the first to show evidence of a relationship between fusion and recombination (Kawano *et al.,* 1991a). Furthermore, numerous other organisms display circumstantial evidence for mitochondrial fusion occurring with recombination of mtDNA. In this chapter, such mitochondrial fusion followed by recombination will be outlined first.

Mitochondrial fusion followed by recombination may be called "mitochondrial sex" or "mitochondrial meiosis" as proposed by Kuroiwa (1982). Historically, the term "sex" was used to describe the type of biparental reproduction exemplified by vertebrate animals. It is only relatively recently that the genetic consequences of sex have been understood along with an appreciation of its ubiquity among not only all eukaryote groups but also prokaryote groups. The term "mitochondrial sex" may also encounter some resistance, since all forms of "true" eukaryotic sex involve cycles of gamete fusion and meiosis with reproduction. However, an alternative view to sex and its origin now exists. Sex could have resulted from the evolution of parasitic DNA sequences that exploit opportunities for horizontal transmission, these opportunities being afforded by cycles of germ cell fusion and fission that allow horizontal spreading of sequences through populations (Hickey and Rose, 1988). Such parasitic elements could, in theory, enhance their own fitness by promoting gamete fusion. In other words, in an attempt to understand the origins of both the transfer of genes from one individual to another and the fusion of gametes, some researchers proposed that a parasitic gene might underlie the process.

Surprisingly, all strains of *P. polycephalum* that display mitochondrial fusion have a unique plasmid, a plasmid that promotes mitochondrial fusion (Kawano *et al.,* 1991a). This mitochondrial fusion-promoting (mF) plasmid is very similar to that proposed for the initial stage in the evolution of sex, as pointed out by Hurst (1991). The existence of an mF plasmid seems to provide direct support for the idea that parasitic, selfish genes are capable of manipulating their hosts (mitochondria) in the manner proposed by Hickey and Rose (1988). Numerous mitochondrial plasmids are well known in protists, fungi, and plants (Table II). They replicate themselves autonomously in the mitochondria independently of mtDNA and may be responsible for the diversity found in the mitochondrial genome. It is possible that such a scenario found in mitochondria reflects the initial stage in the evolution of parasitic sex.

The aim of this chapter is to summarize recent findings on the mitochondrial fusion mediated by the mF plasmid in *P. polycephalum*. The first part of the chapter will thus be concerned with general aspects of mitochondrial fusion and recombination. In the following section, the structure and function of the mF plasmid is presented together with observations of mitochondrial fusion during sporulation and plasmodium formation of *P. polycepha-*

lum. Finally, attention will be drawn to the origin of sex and its relationship to the sexuality of mitochondria.

II. A Morphological Aspect of Mitochondria

A. Giant Mitochondria

Mitochondria generally tend to be considered simply one of the small stationary structures found inside the cell, having an appearance like a drug capsule. This view may partly result from the influence of basic biology textbooks. Inside living cells, however, mitochondrial form is far from static, and harbored within these mobile and plastic organelles is an immense and varied repertoire of biochemical functions.

Since the end of the last century the existence of large filamentous structures in the cell, called "chondriomes," was well known from staining of tissue sections with iron hematoxylin. In the early part of this century, with the supravital demonstration of mitochondrial activity made possible using redox stains such as Janus green B and enzyme-catalyzed reduction of tetrazolium salts to colored and insoluble formazan compounds, some researchers successfully localized viable mitochondria in tissues. However, the insufficient resolving power of microscopes of that time and the granular cytoplasm of most unicellular organisms made the identification of single organelles difficult, leading to a dispute over whether these structures were simply artifacts or real organelles (Cowdry, 1918; Ernster and Schatz, 1981).

The first reliable information on the presence and behavior of mitochondria in living cells was derived from observations of well-spread cells in culture (chick fibroblasts) by Lewis and Lewis (1914). They were among the first to describe mitochondrial dislocation, shape changes, fission, and fusion. After the introduction of phase-contrast microscopy, many studies on mitochondrial morphology and motile activities appeared (Bereiter-Hahn, 1990; Bereiter-Hahn and Voth, 1994). These showed that mitochondrial form changes dynamically not only with the phases of development and differentiation or the rhythm of metabolism, but also with the physiological-chemical (osmotic pressure, ion concentration, etc.) environmental changes outside the cell and also with various experimental procedures. Phase-contrast images provide good lateral resolution and contrast. However, some nonmitochondrial cytoplasmic inclusions might give the same contrast, and in some cases difficulties arise in unambiguously identifying structures as mitochondria. Girbardt (1970) was the first to retrieve single mitochondria in fungal hyphae observed with the phase-contrast microscope and then identify them with the electron microscope.

In electron micrographs of thin sections, profiles of actively respiring mitochondria are rounded or elongated and are regularly distributed in the peripheral cytoplasm. The average cross-sectional diameter of the profiles is generally on the order of 0.3–0.4 μm. However, long filamentous forms, branching profiles, and dumbbell shapes are also typical of certain growth conditions or life-cycle stages (Stevens, 1981). Data such as the position of mitochondria in the cell or the arrangement of their cristae may be obtained from random thin sections. However, knowledge of the entire mitochondrial shape and population within a cell must be obtained by other means.

Since the latter half of the 1950s, a method of reconstructing electron microscopic images of a series of serial thin sections has been utilized for analyzing whole mitochondrial forms. The method is to prepare several consecutive thin sections that collectively cover the entire organelle and use the corresponding electron microscopic images to produce balsam or plastic plates, which can then be used to reconstruct the organelle and reveal its three-dimensional structures. Bang and Bang (1957) first used this method to analyze the mitochondria in human hepatocytes with complex ramification. Osafune (1973) analyzed three-dimensional structures of mitochondria and other cell components based on about 50 serial thin sections of synchronously grown *Euglena gracilis* cells, and found that giant mitochondria with markedly irregular shapes and multiple branching formed during the growth phase of the cell cycle. The matrix of giant mitochondria was very large in volume and no cristae extended into the middle of the matrix. These authors suggested that a giant mitochondrion is formed by fusion of smaller mitochondria, that its newly formed matrix is further enlarged, and that it may continually change its configuration. Spurred by such studies, many investigators analyzed the morphological changes of mitochondria by this serial thin section method and observed that giant mitochondria spread inside the cells of many organisms. Particularly in protists, algae, and fungi, a variety of giant mitochondrial forms have been reported (Table I).

B. Mitochondrial Fusion

A ramified giant mitochondrion resembles the resultant product of mitochondrial fusion. However, when a cell has giant mitochondria, it cannot immediately be assumed that the giant mitochondrion is the product of mitochondrial fusion because, if a cell always has giant mitochondria regardless of cell cycle, life cycle, or growth conditions, the giant mitochondria cannot be the result of fusion. If the giant mitochondria result from fusion, there must be a continuously changing pattern from small to giant mitochondria, and their appearance and disappearance must show a certain periodicity. In some of the reports on giant mitochondria, this periodicity has not been confirmed, so it is not true that all reports about giant mitochondria suggest mitochondrial fusion (Table I).

The study of structural organization of mitochondria in a yeast, *Saccharomyces cerevisiae,* offers a unique opportunity to relate the considerable metabolic, genetic, and biochemical data on this organism to visible morphological entities (Pon and Schatz, 1991). A series of morphological studies on mitochondrial behavior in *S. cerevisiae* were reported first by Yotsuyanagi (1962a,b). He studied the mitochondrial morphology of each phase of the yeast by staining with Janus green, and reported that giant mitochondria occur at the transitional phase between the exponential growth phase and the stationary growth phase. The changes in cellular respiration are paralleled by distinctive changes in mitochondrial morphology.

The presence of a single, highly branched, giant mitochondrion in each cell was first proposed by Hoffman and Avers (1973). Subsequent studies have demonstrated a more complex situation. In general, it has been shown that the giant mitochondrion is not a static entity, but that its form and volume undergo rapid and extensive modifications in accord with changes in the life cycle and physiological state. Examination and measurement of mitochondrial profiles in serial thin sections of 34 whole cells, from several different strains growing vegetatively under a variety of conditions, led to the following basic conclusions (Stevens, 1977): The three-dimensional form varies, with small oval shapes predominant in stationary-phase cells, and some elongated, curved, and branching forms in all states. However, one mitochondrion, or in some rare cases two, is always significantly larger and more complex than the others in the cell. The volume of the giant mitochondrion is clearly a function of the physiological state of the cell: In glucose-repressed cells, it represents only about 3% of the cell volume, whereas in fully respiring cells, it represents 10–12%. These values are in agreement with those given by Grimes *et al.* (1974) and Damsky (1976). In these observations and those on sporulating cells, it appears that the giant mitochondrion exists in a mobile and flexible state within the living yeast cell. It forms a branching network that ramifies throughout the peripheral cytoplasm; its branches constantly fuse together, pinch apart, and change location. The membranes, enzymes, and nucleic acid components are thus being continually redistributed within a "fluid" mitochondrial system.

The most powerful tools for detecting and identifying the periodicity of mitochondrial fusion *in situ* are advanced fluorescence techniques. Fluorochroming endows the organelles with the luminescence. Thus, very fine extensions below the resolving power of a light microscope can be detected because of their fluorescence. In addition, spatial or temporal variations of the fluorescence emission along a single mitochondrion indicate changes

TABLE I

Organisms in Which Giant Mitochondria, Mitochondrial Fusion, Mitochondrial-Nuclear Fusion, and Recombination of mtDNA Are Reported

Organisms	Mitochondria		Mt-nuclei fusion	Recombination	Reference
	Giant	Fusion			
Protists					
Blastocrithidia culicis	+				1
Cryptobia vaginalis	+				2
Trypanosoma brucei	+				3
Trypanosoma cruzi	+				1
Fungi					
Blastocladiella emersonii	+	+			4
Bullera alba	+	+			5
Candida albicans	+	+			6
Candida utilis	+				4, 7
Olpidium brassicae	+				8
Pityrosporum orbicularc	+				9
Physarum polycephalum	+	+	+	+	10–12
Saccharomyces cerevisiae	+	+	+	+	13–18
Schizosaccharomyces pombe	+				19
Algae					
Chlamydomonas reinhardtii	+	+		+	20–25
Chlorella fusca	+				26
Chlorella minutissima	+				27
Chlorococcum infusionum	+	+			28
Chromuliona pusilla	+				29
Eudorina illinoiensis	+				30
Euglena gracilis	+	+			31–38
Friedmannia israelensis	+				39
Hemiselmis rufescens	+				40
Hydrodictyon reticulatum	+	+			41
Pleurochrysis carterae	+				42
Polytoma papillatum	+	+			43–45
Polytomella agilis	+	+			46
Prorocentrum minimum	+				47
Pyramimonas gelidicola	+				48
Rhodella reticulata	+				49
Higher plants					
Pelargonium zonale	+	+	+		50, 51
Zea mays	+				52
Animals					
Xenopus endotherial cells	+	+			53
Rat liver cells	+	+			54
Rat heart cells	+	+			55
Mouse cultured cells (3T6)	+	+			55
Mouse lymphocytes	+	+			56
Mouse ascites tumor cells	+	+			57
Human alcoholic liver cells	+	+			58

(*continues*)

TABLE I *(continued)*

Note: It is well known that mitochondrial fusion occurs during spermatogenesis of mosses, ferns, and animals but such events are omitted from this table. It is also well known that the recombination of mtDNA occurs in the hybrid cells of plants but such events are omitted from this table.

References: (1) Paulin, J. J. (1975). *J. Cell Biol.* **66**, 404–413; (2) Vickerman, K. (1977). *J. Protozool.* **24**, 221–233; (3) Vickerman, K. (1965). *Nature (London)* **208**, 762–766; (4) Bromberg, R. (1974). *Dev. Biol.* **36**, 187–194; (5) Taylor, J. W., and Wells, K. (1979). *Exp. Micol.* **3**, 16–27; (6) Tanaka, K., and Kanbe, T. (1985). *J. Cell Sci.* **73**, 207–220; (7) Keyhani, E. (1980). *J. Cell Sci.* **46**, 289–297; (8) Lange, L., and Olson, L. W. (1976). *Protoplasma* **90**, 33–45; (9) Keddie, F. M., and Barajas, L. (1969). *J. Ultrastruct. Res.* **29**, 260–275; (10) Nishibayashi, S. *et al.* (1987). *Cytologia* **52**, 599–614; (11) Kawano, S. (1991). *Bot. Mag. (Tokyo)* **104**, 97–113; (12) Kawano, S., *et al.* (1991). *Protoplasma* **160**, 167–169; (13) Thomas, D. Y., and Wilkie, D. (1968). *Biochem. Biophys. Res. Commun.* **30**, 368–372; (14) Hoffman, H. P., and Aveers, C. J. (1973). *Science* **181**, 749–751; (15) Davison, M. T., and Garland, P. B. (1977). *J. Gen. Microbiol.* **98**, 147–153; (16) Sando, N., *et al.* (1981). *J. Gen. Appl. Microbiol.* **27**, 511–516; (17) Miyakawa, I. *et al.* (1984). *J. Cell Sci.* **66**, 21–38; (18) Nakagawa, K., *et al.* (1991). *J. Biol. Chem.* **266**, 1977–1984; (19) Nakagawa, K., *et al.* (1992). *EMBO J.* **11**, 2707–2715; (20) Osafune, T., *et al.* (1972). *Plant Cell Physiol.* **13**, 211–227; (21) Grobe, B., and Arnold, C.-G. (1975). *Protoplasma* **86**, 291–294; (22) Grobe, B., and Arnold, C.-G. (1977). *Protoplasma* **93**, 357–361; (23) Blank, R., and Arnold, C. G. (1980). *Protoplasma* **91**, 187–191; (24) Boynton, J. E., *et al. Proc. Natl. Acad. Sci. U.S.A.* **84**, 2391–2395; (25) Remacle, C., *et al.* (1990). *Mol. Gen. Genet.* **223**, 180–184; (26) Atkinson, A. W., Jr., *et al.* (1974). *Protoplasma* **81**, 77–109; (27) Dempsey, G. P., *et al.* (1980). *Phycologia* **19**, 13–19; (28) Chiba, Y., and Ueda, K. (1986). *Phycologia* **25**, 503–509; (29) Manton, I. (1959). *J. Mar. Biol. Assoc. U.K.* **38**, 319–333; (30) Hobbs, M. J. (1971). *Br. Phycol. J.* **6**, 81–103; (31) Calvayrac, R., *et al.* (1972). *Exp. Cell Res.* **71**, 422–432; (32) Osafune, T. (1973). *J. Electron. Microsc.* **22**, 51–61; (33) Osafune, T., *et al.* (1975). *Plant Cell Physiol.* **16**, 313–326; (34) Osafune, T., *et al.* (1975). *J. Electron Microsc.* **24**, 33–39; (35) Osafune, T., *et al.* (1975). *J. Electron Microsc.* **24**, 283–286; (36) Pellegrini, M. (1980). *J. Cell Sci.* **43**, 137–166; (37) Pellegrini, M. (1980). *J. Cell Sci.* **46**, 313–340; (38) Hayashi, Y., and Ueda, K. (1989). *J. Cell Sci.* **93**, 565–570; (39) Melkonian, M., and Berns, B. (1983). *Protoplasma* **144**, 67–84; (40) Santore, J. J., and Greenwood, A. D. (1977). *Arch. Microbiol.* **112**, 207–218; (41) Hatano, K., and Ueda, K. (1988). *Eur. J. Cell Biol.* **47**, 193–197; (42) Beech, P. L., and Wetherbee, R. (1984). *Protoplasma* **123**, 226–229; (43) Gaffal, K. P., and Kreutzer, D. (1977). *Protoplasma* **91**, 167–177; (44) Gaffal, K. P. (1978). *Protoplasma* **94**, 175–191; (45) Gaffal, K. P., and Schnwider, G. J. (1980). *J. Cell Sci.* **46**, 299–312; (46) Burton, M. D., and Moore, J. (1974). *J. Ultrastruct. Res.* **48**, 414–419; (47) Malcolm, S. M., and Wetherbee, R. (1987). *Protoplasma* **138**, 32–36; (48) McFadden, G. I., and Wetherbee, R. (1982). *Protoplasma* **111**, 79–82; (49) Broadwater, S., and Scott, J. (1986). *J. Cell Sci.* **84**, 213–219; (50) Kuroiwa, T., *et al.* (1990). *Protoplasma* **158**, 191–194; (51) Kuroiwa, H., and Kuroiwa, T. (1992). *Protoplasma* **168**, 184–188; (52) Faure, J.-E., *et al.* (1992). *Protoplasma* **171**, 97–103; (53) Bereiter-Hahn, J. (1983). *Publ. Wissenschaftl. Film D 1455* **16**, 1–10; (54) Brandt, J. T., *et al.* (1974). *Biochem. Biophys. Res. Commun.* **59**, 1097–1103; (55) Johnson, L. V., *et al.* (1980). *Proc. Natl. Acad. Sci. U.S.A.* **77**, 990–994; (56) Rancourt, M. W., *et al.* (1975). *J. Ultrastruct. Res.* **51**, 418–424; (57) Koukl, J. F., *et al.* (1977). *J. Ultrastruct. Res.* **61**, 158–165; (58) Inagaki, T., *et al.* (1992). *Hepatology* **15**, 46–53.

within the inner compartment. Most of the fluorochromes used for *in situ* staining of mitochondria are lipophilic cationic dyes. Alkylated nitrogen seems to add to the organelle specificity of dyes. The most widely used group of dyes is the rhodamines, first used by Johannes (1941) to stain mitochondria in plant cells. Improvement of purification methods allowed more detailed studies on the properties of different rhodamines, among which the laser dye rhodamine 123 proved to be least toxic and with the highest specificity for staining mitochondria (Johnson *et al.*, 1980). Other groups with similar properties are the dimethylaminostyryl-pyridinium-iodide derivatives with methyl and ethyl groups (DASPMI and DASPEI) and cyanine dyes (DiOC6, 3,3'-dihexyloxocarbocyanine iodide; DiOC7, 3,3'-dihectaloxocarbocyanine iodide). They have been found to accumulate in mitochondria, making them useful for studying mitochondria *in situ* and isolated in suspension (Chen, 1988, 1989).

On the basis of such improved methods, systematic changes in mitochondrial morphology during the cell cycle and life cycle have been reported for many organisms. In general these changes follow a common series of events: at the onset of interphase the giant mitochondrion consists primarily of one highly reticulated basket-shaped complex that lines the periphery of the cell. During interphase the size of the mitochondrial basket increases, as does the number of additional small mitochondria. During mitosis, the mitochondrial basket is subdivided into several fragments that tend to form clusters, and after cytokinesis the total number of mitochondria is drastically reduced again. Organisms exhibiting such a cycle include *Saccharomyces, Candida, Schizosaccharomyces, Euglena, Chlamydomonas, Chlorella, Chlorococcum, Hydrodiction,* and the dinoflagellate *Prorocentrum minimum* (Table I).

The alternation of mitochondrial networks with fragmentation to individual mitochondria can change the distribution of metabolites and thus create a specific environment for the nucleus. However, this alternation does not seem to be essential for progression through the cycle, because Hayashi and Ueda (1989) found no indication for such a cycle in the mitochondrial organization of *Euglena,* which had previously been reported to exist in this genus. On the other hand, in cultured animal cells, fusion of mitochondria can easily be observed with phase-contrast and fluorescence microscopy. Johnson *et al.,* (1980) demonstrated an interconnecting mitochondrial network in a mouse 3T6 cell using rhodamine 123. Bereiter-Hahn (1983) analyzed fusion of two mitochondria in an amphibian endothelial cell in culture by time-lapse filming. They found that fusion may occur either when two mitochondria collide with their tips and then fuse, or one mitochondrion touches the side of another with its tip, resulting in fusion. The mitochondrial fusion events observed with cultured animal cells seem not to depend on a certain physiological condition; rather, the frequency of such events is related to mitochondrial motility, which does not show cell cycle periodicity.

C. Mitochondrial Nuclear Fusion

There are two types of mitochondrial fusion: One is periodical, depending on the cell cycle and/or the life cycle as observed in protists, algae, and fungi; while the other is accidental, as observed in cultured animal cells. Of course, in the later case, mitochondria may fuse after several efforts, i.e., the tip of one mitochondrion may slide along the side of another or the two may approach each other several times (Johnson et al., 1980; Bereiter-Hahn and Voth, 1994). Electron micrographs show mitochondria which approached each other very closely and behaved as if they were about to fuse though such an event can never be foreseen in fixed cells. The membranes in the contact zone exhibit a high electron density. The connection starts at contact sites between the inner and outer membrane, which are highly specialized zones where hexokinase and creatine kinase are localized (Biermans et al., 1990; Nicolay et al., 1990; Brdiczka, 1991). The structure of contact zones may predispose them for a fusion event, as suggested by Bereiter-Hahn and Voth (1994).

It is important to understand with respect to the genetic requirements whether the mitochondrial fusion in either case is accompanied by redistribution of the mitochondrial nuclei (mt-nuclei; synonymous with mitochondrial nucleoid; Kuroiwa, 1982; Kuroiwa et al., 1994), since this would provide strong morphological support for the considerable genetic data concerning recombination and segregation of mitochondrial genomes in crosses. From serial thin sections, however, it has not been shown that mt-nuclei form a continuous reticulum inside a giant mitochondrion, even in S. cerevisiae. This may be because visualization of mtDNA in situ in thin sections is greatly dependent on the fixation employed and the physiological state of the organism, and is very difficult to observe throughout the mitochondrial fusion event (Stevens, 1981). On the other hand, the introduction of a DNA-specific fluorochrome such as 4'-6-diamidino-2-phenylindole (DAPI) in biological research opened new horizons in the dynamics of mt-nuclei. DAPI was originally used as a highly sensitive and specific fluorescent probe for mtDNA in S. cerevisiae (Williamson and Fennell, 1975) and chloroplast DNA in plants (James and Jope, 1978; Coleman, 1978). The blue-white fluorescence of DAPI is stronger than that obtained with other dyes.

In S. cerevisiae, with the DAPI staining technique, mitochondrial nuclei have been observed as small, discrete, fluorescent spots. They have been observed to appear as a "string of beads" during periods of exponential growth (Williamson and Fennell, 1975, 1979). With the fluorescent dyes DAPI and DASPMI, Sando *et al.,* (1981) and Miyakawa *et al.* (1984) demonstrated that the morphology of yeast mitochondria changes even more dramatically during meiosis and sporulation. In zygotes just after mating, 50–70 mt-nuclei are separated from each other, and each spherical mitochondrion contains only one mt-nucleus. In the later stage of premeiotic DNA synthesis, a single branched giant mitochondrion is formed as a result of complete mitochondrial fusion. All of the mt-nuclei are arranged in an array on a giant mitochondrion and coalesced into a string-like network. Through meiosis I and II, strings of mt-nuclei are observed close to the dividing nuclei. At late meiosis II, a ring of mt-nuclei enclosing each daughter nucleus is formed. In ascospores, discrete small mt-nuclei appear close to each spore as a string of beads. Many mt-nuclei are excluded from the ascospores and remain in the residual cytoplasm of the ascus. Recently, Miyakawa *et al.,* (1994) developed a simple and rapid method for double vital staining of mitochondria and mt-nuclei with DiOC6 and DAPI, and confirmed the relationship between mitochondrial fusion and mt-nuclear fusion in living yeast cells.

The mt-nuclear fusion following mitochondrial fusion or within a giant mitochondrion has been confirmed only in *S. cerevisiae* and *P. polycephalum* (Section IV). In spite of many reports of mitochondrial fusion events and giant mitochondria (Table I), there are no reports that the mt-nuclei fuse in other organisms as drastically as in *S. cerevisiae.* Most of the giant mitochondria reported to date in a variety of organisms were confirmed by serial thin sections. It is almost impossible with electron microscopy to analyze precisely the location, shape, and extent of mt-nuclei and to demonstrate the mt-nuclear fusion event as described by Stevens (1981) and Kuroiwa (1982). This may be one of the reasons for the lack of evidence of mt-nuclei fusion. Moreover, if mitochondria have only a small amount of mtDNA, even by the DAPI staining technique, it is not easy to monitor the behavior of mt-nuclei during mitochondrial fusion events. This also applies to cultured human cells (Satoh and Kuroiwa, 1993). It is more important to understand whether mitochondrial fusion events are accompanied by mt-nuclear fusion than whether they are periodic or accidental events. If it is not accompanied by mt-nuclear fusion, mitochondrial fusion is no better than an accident with respect to genetic requirements.

Studies on mitochondrial fusion have been limited to unicellular or simple organisms such as protists, algae, fungi, and animal cells, including even cultured cells. It is very difficult with conventional microscopy to monitor mitochondrial fusion that accompanies mt-nuclear fusion in cells within tissues and organs of animals and plants. Recently, Kuroiwa *et al.,* (1990, 1991a) developed a new method that enables observation of organelle nuclei in sections prepared in a resin (Technovit 7100) permeable to solutions of

DAPI. This method allows observation of an extremely small amount of organelle DNA (less than 100 kbp) in the cells within tissues and organs, and also enables observation of giant mt-nuclei in the young ovules of *Pelargonium zonale*. The giant mt-nuclei around a nucleus appear in some areas of the young integument. Since the number of giant mt-nuclei per cell is fairly low, it is possible that they are formed not only by preferential mtDNA synthesis but also by the association of small, discrete mt-nuclei. Transformed and giant mitochondria have been observed in the egg cells of some plants: *Zea mays* (Diboll and Larson, 1966), *Pteridium aquilinum* (Tourte, 1975), and *Crepis capillaris* (Kuroiwa, 1982). The three-dimensional reconstruction of isolated egg cell protoplasts of *Z. mays* showed that the giant mitochondria are organized in filamentous and reticulate structures located near the nucleus in the center of the cell (Faure *et al.*, 1992). Kuroiwa and Kuroiwa (1992) demonstrated that such giant mitochondria contain giant mt-nuclei which contain large amounts of DNA (about 4 Mbp) in the Technovit sections of the egg cells of *P. zonale*. Although it is not yet known whether the giant mt-nuclei arose from mt-nuclear fusion or amplification of mtDNA during megagametogenesis, their presence is significant and may enable analysis of the mechanism of recombination and rearrangement of the plant mitochondrial genome as reported elsewhere (Lonsdale *et al.*, 1988; Lonsdale, 1989; Schuster and Brennicke, 1994).

III. A Genetic Aspect of Mitochondria

A. Mitochondrial Genetics

Much has been learned about the rules and mechanisms that govern inheritance of mitochondrial and chloroplast genomes. In many organisms, these organelle genomes are transmitted to progeny predominantly or entirely by only a single parent. This uniparental mode of inheritance has usually been attributed in oogamous species to failure of organelles from the male gamete to enter the egg, or to the presence in the male gamete of comparatively few organelles. Thus, the determination of organelle transmission in such species is dependent upon sexual differentiation. In isogamous species, such as *Saccharomyces cerevisiae*, uniparental inheritance of mitochondrial markers is thought to be due, at least in part, to vegetative segregation; for example, to random partitioning of mitochondria by cell division during sexual development (Birky, 1978; Birky *et al.*, 1982).

Mitochondrial genetics is a major category of transmission genetics. Mitochondrial genes differ at the molecular level from eukaryotic cell nuclear genes in a number of interesting aspects of structure and function. Even more striking are the differences in patterns of inheritance: cell nuclear genes show biparental inheritance while mitochondrial genes most often show uniparental inheritance; cell nuclear alleles segregate only during meiosis with rare exceptions while mitochondrial alleles segregate with high frequency at mitosis as well as meiosis. This phenomenon is called vegetative segregation.

Genes in mitochondria and chloroplasts were first identified by the two unique features that distinguished them from cell nuclear genes: uniparental inheritance and vegetative segregation. A goal of mitochondrial genetics is to explain these two phenomena at the molecular level, in terms of the behavior of mitochondria and mtDNA molecules.

The transmission of mitochondria in isogamous species such as *S. cerevisiae* is most fruitfully dealt with as a problem in population genetics at the intracellular level. Among the parameters and phenomena that must be considered are the input frequencies of alleles in the populations; the mating structure of the populations, e.g., the degree of "panmixis"; random drift of gene frequencies; and the extent of migration of DNA molecules between populations. For reviews of such transmission genetics of mitochondria and chloroplasts, see Sager (1972), Gillham (1978), Dujon (1981), Wilkie (1983), and Birky (1983a,b, 1991). In mitochondrial inheritance, however, additional peculiarities and complexities exist, mainly resulting from some mobile genetic elements and mitochondrial fusion. The situation is thus somewhat complex. In this chapter, primarily recombination and other such mobile genetic elements of mitochondria are reviewed.

B. Recombination

Mitochondria are part of the cell, not distinct organisms, and one cannot score the phenotype of mitochondria but only the phenotype of the cell in which they reside. The first mitochondrial point mutations to be isolated were those of antibiotic resistance (ant^R mutation) in *Saccharomyces cerevisiae* (Linnane *et al.*, 1968; Thomas and Wilkie, 1968), and it was immediately apparent that determination of frequencies of recombination between these mutations would not alone permit construction of a complete genetic map, since rates of recombination can reach a maximum of 20–25% within only 6000 bp or less (Dujon, 1981). Mitochondrial genomes of *S. cerevisiae* are capable of recombination and sorting out. When haploid yeast strains of opposite mating type fuse to form zygotes, mtDNAs from each parent mix and many undergo one or more rounds of pairing. After about 20 cell generations, recombinant and parental genomes are fully segregated among the diploid progeny so that all of the 100 or so mtDNA molecules in any given cell are identical. All cells are therefore homoplasmic with respect to their mitochondrial genomes.

These observations led to the following model of yeast mitochondrial genetics—an analogy between yeast mitochondrial and phage crosses (Dujon, 1981). This model assumes that a fully mixed (panmictic) pool of mtDNA is established rapidly in zygotes and that mtDNA molecules in the zygote undergo repeated rounds of random pairing and recombination. At cell division, the mtDNA molecules are partitioned between mother and bud with the same degree of randomness, so that eventually each cell is homoplasmic for molecules of one genotype or another. Although recombination events continue in the diploid cells, when these cells are homoplasmic or nearly so, all matings will occur between genetically identical molecules and no additional detectable recombinants will be produced. Cell division and vegetative segregation thus limit the number of rounds of effective mating and recombination. Furthermore, all genotypes of molecules replicate at approximately the same rate in all cells. Consequently, the output frequency of a particular allele is equal to the input frequency. Based as it was on random diploid analysis, this model made no provision for differences in the behavior of individual zygotes.

Strausberg and Perlman (1978) investigated segregation of mitochondrial genomes in zygotes by partial pedigree analysis of crosses of S. cerevisiae. Clones derived from first end buds are usually pure (or nearly so) for a parental genotype, reducing the occurrence of detectable recombination of mitochondrial markers in these zygotes. Cells derived from a zygote after removal of the first end bud are predominantly of the other parental genotype. The position of the first bud is important in determining the transmission of mtDNA and the frequency of recombination. Aufderheide and Johnson (1976) have studied the movement of mitochondria from zygotes to buds by phase-contrast microscopy and found little mixing of the zygote cytoplasm before formation of the first bud. This is one reason why the initial end buds are pure for the mitochondrial parental type contributed by the haploid cell which formed that end of the zygote (Wilkie, 1983).

However, zygotes with central buds appearing first may most closely resemble the panmictic situation as Dujon et al. have envisioned it (Dujon, 1981). A first central bud receives cytoplasm from a part of the zygote in which both types of parental mtDNA molecules should be present and also usually yields some recombinant progeny. After formation of the first bud, the cytoplasm apparently mixes more fully so that a panmictic pool is more closely approximated at that time, even for subsequent end buds. Although the behavior of individual mitochondria during zygote formation and budding has not been well studied, the recombination of mtDNA should depend upon fusion between parental mitochondria and between their mt-nuclei to allow pairing between the parental mtDNAs (Sando *et al.*, 1981; Miya-kawa *et al.*, 1984; Tanaka *et al.*, 1985).

C. Mobile Genetic Elements

1. Mobile Intron

Recombination among most loci on the mitochondrial genome of *S. cerevisiae* appears reciprocal at the population level, with parental markers trans-mitted coordinately to the diploid progeny as described earlier. However, a particular phenomenon, called "polarity of recombination," occurs in a specific region (the polar region) of the mitochondrial genome in the imme-diate vicinity of the locus ω. Since the first ant^R mutants (cap^R and ery^R) found localize in this region, this phenomenon has played an important role in elucidating the mechanisms involved in mitochondrial crosses (Bolotin *et al.*, 1971). We now know that the polar region contains the gene for the large rRNA with cap^R and ery^R being point mutations in this gene, which codes for a mobile group I intron encoding endonuclease (Lambowitz, 1989; Scazzocchio, 1989; Dujon, 1989; Lambowitz and Belfort, 1993).

Group I introns form a structural and functional group of introns with widespread but irregular distribution among very diverse organisms and genetic systems. Evidence has now accumulated that several group I introns are mobile genetic elements with properties similar to those originally described for the ω system of *S. cerevisiae:* mobile group I introns encode sequence-specific, double-stranded endonuclease, which recognizes and cleaves intronless genes to insert a copy of the intron by a double-stranded break repair mechanism. Group I introns are of special interest for two reasons. The first reason, shared with group II introns, is that several group I introns are capable of self-splicing *in vitro* (Cech, 1988). The second interesting aspect of group I introns, and so far specific for this group, is their ability to propagate themselves by self-insertion at predetermined positions into intronless sites of the genes. This phenomenon, which origi-nally was discovered with the intron of the mitochondrial large subunit rRNA (LSU) gene of *S. cerevisiae,* has recently been shown to occur with several other group I members. These include mitochondrial, chloroplast, nuclear, and prokaryotic introns of very diverse organisms such as, in addi-tion to yeast, *Chlamydomonas, Neurospora, Physarum, Coprinus,* and bac-teriophage T4 (Dujon *et al.*, 1989; Lambowitz and Belfort, 1993).

Genetic crosses between ω^+ and ω^- strains also revealed that the ω^+ allele itself is transmitted to more than 95% of the progeny, the ω^- allele being almost completely eliminated. At this time, a double-stranded break forms within the intronless LSU gene at a location corresponding to the intron insertion site (Jacquier and Dujon, 1985; Macreadie *et al.*, 1985). This break disappears after a few hours when zygotes age and bud, allowing different mtDNA molecules to segregate out. Zinn and Butow (1985) have determined the kinetics of ω recombination. Within 2–4 hr after ω^+ and ω^- are mixed, a nonparental form of the LSU gene appears that contains the 1.1 kbp intron but not the region flanking the ω^+ parental allele. Although this recombinant is not a major product in the conversion of ω^- to ω^+, it serves as a convenient measure of ω recombination. Since zygotes do not appear until about 2 hr after cells of opposite mating type are mixed, mitochondria from each parent must therefore fuse and the parental mtDNAs become readily available for recombination essentially as soon as zygotes are formed. This conclusion does not conflict with the fluorescent-microscopic observation that mt-nuclei accumulate on a thread-like mitochondrion and disperse into many small ones just after mating (Sando *et al.*, 1981; Miyakawa *et al.*, 1984). If the mt-nucleoids are fully mixed in the zygote cytoplasm, the random arrangement of mt-nucleoids from both parents on a giant mitochondrion would allow significant freedom of access between mtDNAs and extensive recombination of neighboring mtDNAs from ω^- and ω^+.

2. Mitochondrial Plasmid

The size and structural complexity of fungal mitochondrial genomes occupy a middle ground between those of metazoa and higher plants. The organization of genes is not as frugal as that in animal mtDNA nor as loose as that in plant mitochondrial genomes. Although mtDNAs from some fungi match those from vertebrates in the number of encoded genes, this is not so for *S. cerevisiae* (Clark-Walker, 1992). However, the relative poverty of fungal mitochondrial genes is redressed in some species by a richness in structural complexity of intergenic regions and also by the presence of optional introns such as ω. Some fungal mitochondria also carry, in addition to a high-molecular-weight DNA component that represents the main mtDNA, smaller circular and linear DNA molecules such as mitochondrial plasmids. These plasmids replicate autonomously in the mitochondria independently of the mtDNA and may be responsible for the diversity among mitochondrial genomes. It has been demonstrated that some of these plasmids are integrated into mtDNA and cause structural changes in the mtDNA (Samac and Leong, 1989; Lonsdale, 1989; Meinhardt *et al.*, 1990; Griffiths, 1992).

Linear plasmids in eukaryotes have been identified in numerous fungi, protozoa, and plants, and most of them appear to be localized in the mitochondria. The main characteristic feature of these linear mitochondrial plasmids is a terminal inverted repeat (TIR) sequence with a terminal protein covalently linked to the 5' end of the plasmid. A similar terminal structure has been found in adenovirus (Challberg *et al.*, 1980) and in bacteriophage φ29 of *Bacillus subtilis* (Gutierrez *et al.*, 1988). In these linear DNA viruses, TIRs and terminal proteins are important for the recognition of the origin of replication by DNA polymerase or by DNA-binding protein(s), the 3′-hydroxyl priming of DNA polymerase, and subsequent strand displacement during replication of the DNA (Campbell, 1986). Analogously, the TIRs and the terminal proteins on the linear mitochondrial plasmids also function in replication (Samac and Leong, 1989; Meinhardt *et al.*, 1990). The almost complete nucleotide sequences of some linear mitochondrial plasmids have been determined, such as S1 and S2 in maize, pClK1 in *Claviceps purpurea*, pMC3-2 in *Morchella conica, kalilo* in *Neurospora intermedia, maranhar* in *N. crassa*, and pAL2-1 in *Podospora anserina* (Table II). They encode DNA and/or RNA polymerase and a few short open reading frames (ORFs) as follows: DNA polymerase and two short ORFs on S1, RNA polymerase and a short ORF on S2, DNA and RNA polymerases and four short ORFs on pClK1, DNA polymerase and a short ORF on pMC3-2, DNA and RNA polymerases on *kalilo*, DNA and RNA polymerases and three short ORFs on *maranhar*, and DNA and RNA polymerases on pAL2-1.

Circular plasmids in eukaryotes have been also identified in the mitochondria of numerous fungi and plants (Esser *et al.*, 1986; Lonsdale *et al.*, 1988). Extensive work on circular plasmids has been carried out, particularly using *Neurospora*. Based on DNA hybridization studies, most of the circular mitochondrial plasmids of *Neurospora* have been placed in one of three homology groups that are named after the geographical location where the initial isolate was found (Collins *et al.*, 1981; Stohl *et al.*, 1982; Natvig *et al.*, 1984). The circular plasmids studied thus far encode plasmid-specific polymerases (Li and Nargang, 1993). Plasmids of the Mauriceville homology group encode a reverse transcriptase that appears to be involved in replication of the plasmid (Nargang *et al.*, 1984; Kuiper and Lambowitz, 1988). The lone plasmid from the LaBelle homology group contains an open reading frame in which motifs characteristic of reverse transcriptase have been identified, though the lack of some highly conserved residues in the motifs is noted (Pande *et al.*, 1989). The subsequent discovery that the LaBelle plasmid encodes a DNA-dependent DNA polymerase prompted a search for motifs characteristic of DNA-dependent DNA polymerases (Schulte and Lambowitz, 1991). Some similarity to the DNA polymerase family B has been found, but the reverse transcriptase motifs are considered more convincing. ORFs encoding the DNA polymerase family B are also found in some other linear mitochondrial plasmids.

Although extensive studies on sequencing of some mitochondrial plasmids has been performed, the reason for the existence of plasmids is still something of a mystery (Yang and Griffiths, 1993). Most plasmids appear

TABLE II

Mitochondrial Plasmids

Organisms	Name	Size (kbp)	Structure	TIR (bp)	TP	Gene	Reference
Protists							
Paramecium caudatum	Types I, II, III	8.9-1.4	Linear	Yes	Yes		1, 2
P. jenningsi	Type I	6.0, 6.8					2
P. multimicronucleatum	Type I	6.0, 6.8					2
P. polycaryum	Type II	8.2-1.4					2
Fungi							
Agaricus bitorquis	pEM	7.4	Linear	1000	—	D, R-pol	3, 4
	pMPj	3.7	Linear	—	—		3
Ascobolus immersus	pAI2	5.6	Linear	700	Yes	D-pol	5, 6
	pAI3	2.8	—				5
Ascochyta rabiei	—	13.0	Linear	—	—		7
Claviceps purpurea	pClB4	6.7	Linear	—	Yes		8
	pClK1	6.8	Linear	327	Yes	D, R-pol	9
	pClK2	5.5	Linear	—	No		8
	pClK3	1.1	Linear	—	No		8
	pClK9	6.7	Linear	—	Yes		8
	pClT5	7.1	Linear	—	Yes		8
Fusarium merismoides	—	2.1	—				10
	—	1.8	—				10
Fusarium oxysporum							
f. sp. *conglutinans*	pFOXC2	1.9	Linear	—	Yes		11
Fusarium solani	pFSC1	9.2	Linear	—	—		12
f. sp. *cucurbitae*	pFSC2	8.3	Linear	—	—		12
Fusarium sporotrichoides	—	2.1	Linear	—	—		13
Gaeumannomyces graminis	E1	8.4	Linear	—	—		14
	E2	7.2	Linear	—	—		14
Lentinus edodes	pLLE1	11.0	Linear	—	—		15
Morchella conica	—	6.0	Linear	750	—		16, 17
Neurospora crassa	*maranhar*	7.0	Linear	—	Yes		18, 19
	Mauriceville	3.6	Circular			RT	20–22
	Roanoke	5.2	Circular				23
Neurospora intermedia	*kalilo*	8.6	Linear	1366	Yes	D, R-pol	24, 25
	Fuji	5.2	Circular				23
	LaBelle	4.1	Circular			D-pol	26
	Varkud	3.8	Circular			RT	27
Neurospora tetrasperma	Hanalei	5.2	Circular				23
Pleurotus ostreatus	pLPO1	10.0	Linear	—	Yes		28
	pLPO2	9.4	Linear	—	Yes		28
Physarum polycephalum	mF	16.0	Linear	144	Yes	D-pol	29, 30
Podospora anserina	pAL2-1		Linear				31
Tilletia caries	pTCC	7.4	—				32
Tilletia controversa	pTCT	7.4	Linear	—	Yes		32
Tilletia laevis	pTCL	7.4	—				32
Higher plants							
Beta vulgaris	Mc.a	1.6	Circular				33
	Mc.d	1.3	Circular				33
	pO	1.4	Circular				34
	—	7.3	Circular				35
	—	1.4	Circular				35

(*continues*)

TABLE II (continues)

Organisms	Name	Size (kbp)	Structure	TIR (bp)	TP	Gene	Reference
Brassica compestris	—	11.5	Linear	325	Yes		36
Helianthus annuus	P1	1.4	Circular				37
	P2	1.8	—				37
	P3	1.8	—				37
Oryza sativa	B1	2.1	Circular				38
	B2	1.5	Circular				38
	B3	1.5	Circular				39
	B4	1.0	Circular				40
	L1	1.0	Circular				41
	L2	1.4	Circular				41
	L3	1.4	Circular				41
	L4	2.1	Circular				41
Sorghum bicolor	N1	5.8	Linear	—	Yes		42
	N2	5.4	Linear	—	Yes		42
	—	2.3	Circular				43
	—	1.7	Circular				43
	—	1.4	Circular				43
Vicia fuba	—	1.7	Circular				44
	—	1.7	Circular				44
	—	1.5	Circular				44
Zea diploperennis	D1	7.4	Linear	—	Yes		45
	D2	5.4	Linear	—	Yes		45
Zea mays S-cytoplasm	S1	6.4	Linear	208	Yes	DNA pol	46, 47
S-cytoplasm	S2	5.5	Linear	208	Yes	RNA pol	48, 49
RU-cytoplasm	R1	7.5	Linear	187	Yes		50
RU-cytoplasm	R2	5.4	Linear	187	Yes		50
N,S,C-cytoplasm	—	2.3	Linear	170	—	tRNAs	51
N,S,C-cytoplasm	—	1.9	Circular				52
C-cytoplasm	—	1.6	Circular				53
C-cytoplasm	—	1.4	Circular				54

Note: D-pol, DNA polymerase; R-pol, RNA polymerase; RT, reverse transcriptase; —, not decided.

References: (1) Endoh, H., *et al.* (1994). *Curr. Genet.* **27**, 90–94; (2) Tsukii, Y., *et al.* (1994). *Jpn. J. Genet.* **69**, 685–696; (3) Mohan, M., *et al.* (1984). *Curr. Genet.* **8**, 615–619; (4) Robison, M. M., *et al.* (1991). *Curr. Genet.* **19**, 495–502; (5) Meinhardt, F., *et al.* (1986). *Curr. Genet.* **11**, 243–246; (6) Kempken, F., *et al.* (1989). *Mol. Gen. Genet.* **218**, 523–530; (7) Meinhardt, F., *et al.* (1990). *Curr. Genet.* **17**, 89–95; (8) Duvell, A., *et al.* (1988). *Mol. Gen. Genet.* **214**, 128–134; (9) Oeser, B., and Tudzynski, P. (1989). *Mol. Gen. Genet.* **217**, 132–140; (10) Rubidge, T. (1986). *Trans. Br. Mycol. Soc.* **87**, 463–466; (11) Kistler, H. C., and Leong, S. A. (1986). *J. Bacteriol.* **167**, 587–593; (12) Samac, D. A., and Leong, S. A. (1988). *Plasmid* **19**, 57–67; (13) Cullen, D., *et al.* (1985). *J. Cell. Biochem.* **9c**, 169; (14) Honeyman, A. L., and Currier, T. C. (1986). *Appl. Environ. Microbiol.* **52**, 924–929; (15) Katayose, Y., *et al.* (1990). *Nucleic Acids Res.* **18**, 1395–1400; (16) Meinhardt, F., and Esser, K. (1984). *Curr. Genet.* **8**, 15–18; (17) Rohe, M., *et al.* (1991). *Curr. Genet.* **20**, 527–533; (18) Court, D. A., *et al.* (1991). *Curr. Genet.* **19**, 129–137; (19) Court, D. A., and Bertrand, H. (1992). *Curr. Genet.* **22**, 385–397; (20) Nargang, F. E., *et al.* (1984). *Cell (Cambridge, Mass.)* **38**, 441–453; (21) Akins, R. A., *et al.* (1986). *Cell (Cambridge, Mass.)* **47**, 505–516; (22) Kuiper, M. T., *et al.* (1990). *J. Biol. Chem.* **265**, 6936–6943; (23) Taylor, J. W., *et al.* (1985). *Mol. Gen. Genet.* **201**, 161–167; (24)

(continues)

TABLE II (continued)

Griffiths, A. J. F., and Bertrand, H. (1984). *Curr. Genet.* **8,** 387–398; (25) Chan, B. S.-S., *et al.* (1991). *Curr. Genet.* **20,** 225–237; (26) Pande, S., *et al.* (1989). *Nucleic Acids Res.* **17,** 2023–2024; (26) Akins, R. A., *et al.* (1988); *J. Mol. Biol.* **204,** 1–25; (28) Yui, Y., *et al.* (1988). *Biochim. Biophys. Acta* **951,** 53–60; (29) Kawano, S., *et al.* (1991). *Protoplasma* **160,** 167–169; (30) Takano, H., *et al.* (1994). *Curr. Genet.* **26,** 506–511; (31) Hermanns, J., and Osiewacz, H. D. (1992). *Curr. Genet.* **22,** 491–500; (32) Kim, W. K., *et al.* (1990). *Curr. Genet.* **17,** 229–233; (33) Thomas, C. M. (1986). *Nucleic Acids Res.* **14,** 9353–9370; (34) Hansen, B. M., and Marcker, K. A. (1984). *Nucleic Acids Res.* **12,** 4747–4756; (35) Powling, A. (1981). *Mol. Gen. Genet.* **183,** 82–84; (36) Turpen, T. *et al.* (1987). *Mol. Gen. Genet.* **209,** 227–233; (37) Crouzillant, D., *et al.* (1989). *Curr. Genet.* **15,** 283–289; (38) Shikanai, T., *et al.* (1987). *Plant Cell Physiol.* **28,** 1243–1251; (39) Shikanai, T., *et al.* (1989). *Curr. Genet.* **15,** 349–354; (40) Shikanai, T., and Yamada, Y. (1988). *Curr. Genet.* **13,** 441–443; (41) Mignouna, H., *et al.* (1987). *Theor. Appl. Genet.* **74,** 666–669; (42) Chase, C. D., and Pring, D. R. (1986). *Plant Mol. Biol.* **6,** 53–64; (43) Chase, C. D., and Pring, D. R. (1985). *Plant Mol. Biol.* **5,** 303–311; (44) Wahleithner, J. A., and Wolstenholme, D. R. (1987). *Curr. Genet.* **12,** 55–67; (45) Timothy, D. H., *et al.* (1983). *Maydica* **28,** 139–149; (46) Paillard, M., *et al.* (1985). *EMBO J.* **4,** 1125–1128; (47) Kuzmin, E. V., and Levchenko, I. V. (1987). *Nucleic Acids Res.* **15,** 6758; (48) Levings, C. S., III, and Sederoff, R. R. (1983). *Proc. Natl. Acad. Sci. U.S.A.* **80,** 4055–4059; (49) Kuzmin, E. V., *et al.* (1988). *Nucleic Acids Res.* **16,** 4177; (50) Weissinger, A. K., *et al.* (1982). *Proc. Natl. Acad. Sci. U.S.A.* **79,** 1–5; (51) Leon, P., *et al.* (1989). *Nucleic Acids Res.* **17,** 4089–4099; (52) Ludwig, S. R., *et al.* (1985). *Gene* **38,** 131–138; (53) Kemble, R. J., and Bedbrook, J. R. (1980). *Nature* (*London*) **284,** 565–566; (54) Smith, A. G., and Pring, D. R. (1987). *Curr. Genet.* **12,** 617–623.

to be neutral passengers with no obvious detrimental effect on their hosts. In such a sense, they are selfish. Two linear plasmids are unique in that they cause the death of their hosts by inserting into mitochondrial DNA, which disrupts mitochondrial function; these are referred to as senescence plasmids (Griffiths, 1992). Under conditions of prolonged growth at 37°C, the circular plasmids Mauriceville and Varkud occasionally insert themselves into mtDNA, resulting in mitochondrial malfunction and growth irregularities (Akins *et al.,* 1986), but these and other circular plasmids do not confer the strong predisposition to death shown by the linear *kal*DNA and *mar*DNA plasmids. For example, it is known that the senescence in *N. intermedia* and *N. crassa* is induced by the integrations of *kalilo* and *maranhar* plasmids into the mtDNAs, respectively. However, these plasmids encode only DNA and RNA polymerases, and do not encode any genes which cause senescence.

Certainly one of the predominant impressions that has emerged from the present work is that plasmids are the rule in *Neurospora,* and not the exception. The types that we have identified are undoubtedly merely the tip of an iceberg of diversity. The liner plasmids so far described in fungi and plants, despite showing no nucleotide homology, have remarkably similar general structures, which apparently code for the same kinds of proteins related to viral polymerases, and are virtually all mitochondrial in location. Assuming that all linear plasmids will have such properties, is it reasonable to propose that this entire gamut of plasmid types all evolved from one common ancestor that inhabited the original endosymbiont that gave rise to mitochondria? Alternatively, it is possible that only one structure compatible with the selfish DNA lifestyle exists within mitochondria, and that the multitude of types have converged on this form. Thus, linear plasmids are most probably descendent from viral genomes. The demonstration of the existence of virus-like capsids could prove this supposition, as proposed by Meinhardt *et al.* (1990). It is reasonable that bacteriophages are progenitors of the linear plasmids of bacteria. The same may be true for the mitochondrial linear plasmids since it is generally accepted that mitochondria are remnants of endosymbiotic bacteria.

3. Mobile Genetic Elements and Mitochondrial Fusion

Uniparental inheritance of mtDNA is virtually axiomatic in animals, plants, and fungi. Generally, during sexual reproduction in these organisms, only the female parent transmits mitochondria to its progeny, and the maternal lineage of mtDNA is strictly maintained. In contrast, mobile introns, such as ω of *S. cerevisiae* (Zinn and Butow, 1985; Dujon, 1989) and the ω-like intron of *C. smithii* (Boynton *et al.*, 1987; Remacle *et al.*, 1990), promote their insertion into the mtDNA, and as a result, spread through the mitochondrial population. Moreover, many plasmids are known in fungal mitochondria, and some of them may contain ω-like introns (Nargang *et al.*, 1984). Mobile genetic elements such as ω-like introns and plasmids seem to have a scattered distribution within and among species against a background of uniparental mtDNA lineage (Dujon, 1989; May and Taylor, 1989; Collins and Saville, 1990). In particular, ω-like introns are sporadically distributed, not only in mitochondria, but also in chloroplasts, in nuclear genes for ribosomal RNA, and in bacteriophage genes (Dujon *et al.*, 1989; Lambowitz, 1989; Gauthier *et al.*, 1991; Marshall and Lemieux, 1991).

For horizontal transmission of DNA to occur, the genetic material must be routed around at least two hypothetical barriers: transfer barriers that prevent the delivery of genetic information from a donor, and establishment barriers that block inheritance of new molecules (Heinemann and Sprague, 1989; Heinemann, 1991). Some bacterial plasmids and transposons are known to be capable of conjugal transfer among bacterial species and even between bacteria and eukaryotes (Ippen-Ihler and Minkley, 1986; Clewell and Gawron-Burke, 1986; Heinemann, 1991). Conjugation is a reliable method for the transfer and stable maintenance of these mobile genetic elements. Conjugation may also occur in mitochondria. Mitochondrial fusion is also considered to be a basic mechanism for spreading mobile genetic elements through mitochondrial populations within species against a background of uniparental mtDNA transmission (Zinn and Butow, 1985; Boynton *et al.*, 1987; Remacle *et al.*, 1990).

IV. Mitochondrial Fusion in *Physarum polycephalum*

A. Mitochondrial Life Cycle

1. Life Cycle of *Physarum polycephalum*

Recombination between mitochondrial genomes has also been found in the mitochondrial crosses of a few other organisms, the best studies of which have been with the yeast *Saccharomyces cerevisiae* (Table I). In attempts to solve problems between mitochondrial fusion and recombination, the true slime mold, *Physarum polycephalum,* has been found to serve as an extremely useful experimental system because of its controllable life cycle and characteristic mitochondria (Alexopoulos, 1982; Burland *et al.,* 1993). The life history of *P. polycephalum* is that of a typical haplodiplont, with two distinct stages: the diploid syncytial plasmodium and the haploid uninucleate myxamebas (Fig. 1). Furthermore, the complete life history involves resting stages, as represented by spores, spherules, sclerotia, and cysts. The thick-walled, resistant spores hatch to release myxamebas, which act as isogametes; individual isogametes of different mating types pair to form diploid zygotes. The zygotes develop into giant, diploid syncytia called plasmodia, which may reversibly transform themselves into resistant sclerotia (synonymous with spherules) or may undergo a terminal differentiation process that leads to the formation of lobed sporangia as a result of starvation and illumination. Meiosis occurs within the spores.

The mitochondria of *P. polycephalum* also provide a particularly favorable model for the analysis of division, segregation, and fusion of mt-nuclei. They contain about ten times more mtDNA than mitochondria from other sources. The bulk of the mtDNA is usually packed into an electron-dense mt-nucleus together with RNA and proteins. The mtDNA content per mt-nucleus is estimated to be about 32 molecules, each of which is linear, with a molecular weight of 45×10^6 per mitochondrion at the mG1 phase (Kuroiwa, 1982). Since the mtDNA molecules in *P. polycephalum* are homogeneous in terms of their physical structure, the mt-nucleus seems to be polyploid or polytene (Kawano *et al.,* 1982; Suzuki *et al.,* 1982; Takano *et al.,* 1990). This mt-nucleus facilitates observations of the processes of mt-nuclear duplication, division, and fusion, as well as monitoring the behavior of mtDNA by light and electron microscopy. Using *P. polycephalum,*

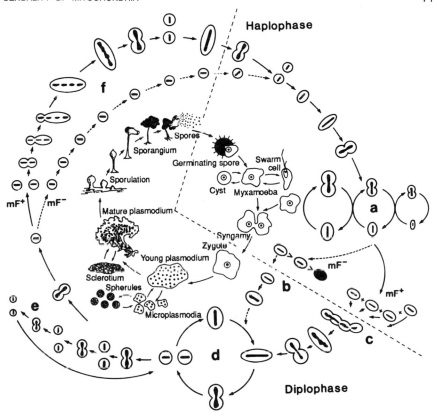

FIG. 1   The life cycle of mitochondria (outer scheme) throughout the life history of *P. polyceph-alum* (inner scheme). Small spherical, oval, and dumbbell-shaped figures in the outer scheme represent mitochondrial bodies and mitochondrial nuclei. (a) Mitochondrial division cycle and stepwise reduction and rise in volume of mitochondria and mtDNA content of mitochondria nonsynchronizing with mtDNA synthesis through the culture of myxamebas. (b) Hierarchical uniparental transmission of mtDNA during the formation of plasmodium by crossing between mF-strains. (c) Mitochondrial fusion accompanying mt-nuclear fusion promoted by the mF plasmid during formation of plasmodium by crossing between mF⁻ and mF⁺ strains. (d) Mitochondrial division cycle synchronizing with mtDNA synthesis in the plasmodium. (e) Stepwise reduction of mitochondria by the division of mitochondria without DNA synthesis during spherulation. (f) Behavior of mitochondria during sporulation, mitochondrial fusion (Mif⁺) carrying the mF plasmid, and mitochondrial fusion-deficient (Mif⁻) strain without the mF plasmid. (Modified from Kawano, 1991; Kuroiwa *et al.*, 1994.) From Kuroiwa, T., Ohta, T., Kuroiwa, H., and Kawano, S. (1994). Molecular and cellular mechanisms of mitochondrial nuclear division and mitochondriokinesis. *Microsc. Res. Technol.* **27,** 220–232. Copyright © 1994 John Wiley and Sons. Reprinted by permission of John Wiley & Sons, Inc.

we have studied various cytoplasmic aspects of mitochondrial biogenesis, following the mitochondrial life cycle throughout its history, as shown in Fig. 1 (Kuroiwa, 1982; Kawano, 1991; Kuroiwa *et al.*, 1994). In particular, in this chapter, we focus on and discuss the behavior of the mitochondrial genome throughout the mitochondrial life cycle, the way that fusion and recombination occur, and how the mitochondrial genome is transmitted to the next generation.

2. Mitochondrial Differentiation during Sclerotization and Sporulation

A model system for studies of mitochondrial differentiation should have a uniform population of cells and controllable cellular differentiation. Furthermore, it must be available in sufficient quantities for isolation and characterization of mitochondria. These criteria are satisfied by the period of transition from plasmodia to sclerotization and sporulation in *P. polycephalum*. Under certain adverse conditions (e.g., desiccation, cold temperature, osmotic pressure, starvation, and "aging"), a plasmodium divides up into small sclerotia (spherules), each containing several cell nuclei (Raub and Aldrich, 1982). Walls are established between these spherules and the whole mass appears cellular. The spherules are resistant structures, and they may play an important role in the survival of the myxomycete. During the very early stage of sclerotization, by starvation and/or aging of the culture, sporulation is induced by regulated illumination. Light is an indispensable requirement for fruiting of pigmented plasmodia (Wheals, 1970).

Microplasmodia differentiate into spherules when cultures are aged for 8–10 days. Rates of respiration in the microplasmodia decrease rapidly with aging to give a 90% decrease in the consumption of oxygen over the course of 9 days. These phenomena were examined by isolating and characterizing mitochondria from microplasmodia and spherules at different stages of spherulation (Kawano *et al.*, 1983). Uptake of oxygen by the isolated mitochondria decreased according to spherulation. Mitochondrial differentiation to an inactive state is characterized by a decrease not only in their dimensions but also in their DNA, RNA, and protein contents. Diminutive mitochondria contain small particle-shaped mt-nuclei. The DNA content, measured by microscopic fluorometry, is about 1.15 or 0.58×10^{-10} g, which corresponds to about 16 or 8 copies of the genome, respectively. Finally, the number of copies of the genome per mitochondrion in the completed spherule equilibrated 4 or 2, namely, 1/8 to 1/16 of the original level in the plasmodium (e.g., 32 copies of the genome per mitochondrion at the mG1 phase). The changes in ploidy of the mt-nucleus during spherulation are due to stepwise reduction in the number of whole mitochondrial genomes as a result of cycles of mitochondrial division without any accompanying mtDNA synthesis.

Recently, a similar phenomenon of the mitochondrial division not being synchronized with mtDNA synthesis has been observed in a culture of myxamebas (Sasaki *et al.*, 1994). A stepwise reduction in the volume and DNA content of organelles has also been observed in plastids of spermatocytes during spermatogenesis in the fern *Pteris vittata* (Kuroiwa *et al.*, 1988), the green algae *Bryopsis maxima* (Kuroiwa and Hori, 1986) and *Chara*

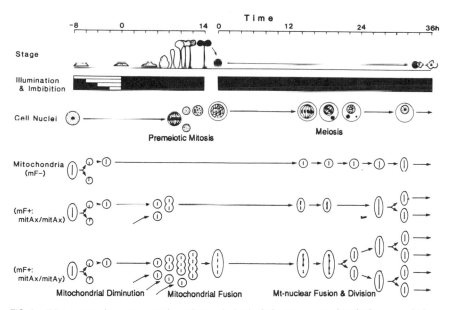

FIG. 2 Diagrammatic representation of morphological changes occurring during sporulation: behavior of cell nuclei, mitochondria and mt-nuclei during sporulation, and spore germination of *Physarum polycephalum*. Sporulation is synchronously started by a light turned off after 2–8 hr of illumination, and germination of spores almost synchronously by imbibition. Cytoplasmic cleavage for spore formation occurs around each cell nucleus after premeitotic mitosis. Cell nuclei enter into the early prophase of meiosis during head blackening of sporangia and become dormant at this stage within spores. The rest of the meiotic processes occur during germination within spores. Mitochondria divide to be spherical or oval during the starved plasmodium stage. Mitochondria fusion starts with sporangium formation in the plasmodial strain carrying the mF plasmid (mF+) while it fails to do so (<0.7%) in the plasmodial strain without the mF plasmid (mF−). The fused mitochondria enter into spores, and extensive mt-nuclear fusion and several successive mitochondrial divisions occur during spore germination. The frequency of mitochondrial fusion is controlled by a cell-nuclear locus, *mitA*. Mitochondrial fusion occurs at high frequency (59.2–80.5%) with the combination of unlike alleles (*mitAx/mitAy*) but at intermediate frequency with the combination of like alleles (*mitAx/ mitAx*).

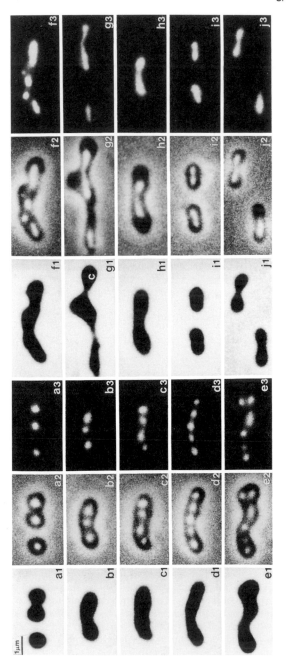

corallina (Sun *et al.,* 1988), and in plastids during the life cycle of the unicellular alga *Cyanidium caldarium* M-8 (Kuroiwa *et al.,* 1989). These results indicate that there are successive cycles of division of organelles without any concomitant duplication of DNA during the differentiation of cells and gametogenesis in some organisms.

3. Mitochondrial Fusion during Sporulation and Spore Germination

Sporulation of *P. polycephalum* occurs with a high degree of synchrony, and can be easily induced by 4–6 hr of illumination of starved plasmodia. About 8 hr after illumination, many projections form along the plasmodial strand; these increase in size and develop into sporangia over the subsequent 6 hr (Fig. 2). Nishibayashi *et al.* (1987) analyzed the behavior of cell nuclei, mitochondria, and mt-nuclei during sporulation, during which starved plasmodia differentiate into sporangia. The mitochondria also decrease in size between early and midstages of starved plasmodia to differentiate into sporangia as well as sclerotia. The mt-nuclear DNA content per mitochondrion is constant in sclerotia and during the early starved plasmodium stage, but decreases by one half at the midstarved plasmodium stage, and remains the same from the midstarved plasmodium stage to the premeiotic division stage. This process is similar to that during spherulation. At the resting stage of sporulation, dumbbell-shaped or multinucleate mitochondria which contain two or more discrete mt-nuclei frequently appear. The total fluorescence intensity of a few mt-nuclei in multinucleate mitochondria increases stepwise with the number of mt-nuclei. These results suggest that a few mt-nuclei in multinucleate mitochondria partially attach or fuse during the mature stage.

A more detailed study was carried out to determine the events leading to the appearance and subsequent disappearance of multinucleate mitochondria (Kawano *et al.,* 1991a). Mitochondrial fusion occurs at high fre-

---

FIG. 3   Representative stages in mitochondrial fusion (a–f) during sporulation and division (g–j) during spore germination in *Physarum polycephalum.* Phase-contrast (1), phase-contrast fluorescence (2), and DAPI-fluorescence (3) micrographs are shown for the same field. The mt-nuclei are visible as intense blue spots in the mitochondria by DAPI staining. Small spherical mitochondria fuse with one another to form large, knotted, multinucleate mitochondria (a–e) which subsequently undergo fusion between the mt-nuclei derived from individual mitochondria (f). Mitochondria showing various extents of mitochondrial fusion are enclosed in spores. Several successive mitochondrial divisions follow, accompanied by mt-nuclear divisions (g–j), during spore germination. Such various types of multinucleate mitochondria are also observed in zygotes. Bar = 1 μm. (From Kawano *et al.,* 1991a.)

quency during a limited period of the late sporulation stage (Fig. 3). The small, spherical mitochondria fuse with one another to form dumbbell shapes and other irregular forms (Fig. 3,a) which sometimes take on a rounded shape. Further fusions with uninucleate (Fig. 3,b and c) and multinucleate mitochondria result, by the late resting stage, in the formation of large mitochondria of irregular form (Fig. 3,d and e). Some fusions between mt-nuclei are observed at this stage, and mitochondria showing various extents of mitochondrial and mt-nuclear fusion are enclosed within the spores. Further events in the spores were investigated by suspending mature spores in a drop of distilled water on a microscope slide. Meiosis resumes and is completed 14–18 hr after wetting. Each spore then hatches to yield a single myxameba or swarm cell. When chromosome condensation occurs at the resumption of meiosis, vigorous morphological changes also begin in the multinucleate mitochondria. Irregular-shaped mitochondria undergo successive rounds of unequal division until the daughter mitochondria are similar in size to prefusion mitochondria (Fig. 3, g–j). These mitochondrial divisions are completed by the time of spore hatching.

4. Isolation of Mitochondrial Fusion and Its Defective Strains

P. polycephalum provides particularly favorable material for the analysis of mitochondrial and mt-nuclear fusions. These fusions result in the formation of multinucleate mitochondria in the spore. The mitochondria in spores of many laboratory strains of *P. polycephalum* that were derived from crosses of several different natural isolates were examined by cracking open spores in the presence of glutaraldehyde so that the mitochondria were fixed immediately and retained their normal shape (Fig. 4; Kawano *et al.,* 1993). DAPI-fluorescence microscopy revealed that the spores of strain Ng and its derivatives contained multinucleate mitochondria, as expected from our previous studies (Nishibayashi *et al.,* 1987; Kawano *et al.,* 1991a). Unexpectedly, however, the spores of all the other strains were found to contain only uninucleate mitochondria, suggesting that sporulation in most strains is not accompanied by mitochondrial fusion, which produces the multinucleate state. Spores with uninucleate mitochondria each contain approximately 22–30 small, spherical mitochondria, in which the mitochondrial nuclei are visible as intense blue spots after staining with DAPI (Fig. 4,a–c). By contrast, multinucleate mitochondria are large and irregular in shape, containing several mitochondrial nuclei, with more mitochondrial nuclei in large mitochondria. Spores containing multinucleate mitochondria were classified as being of two types. One type contained approximately 12–16 mitochondria, and somewhat more than half of these mitochondria were large, irregular in shape, and multinucleate (Fig. 4,d–f). The other type contained only a few multinucleate mitochondria (Fig. 4,g–i). Mitochon-

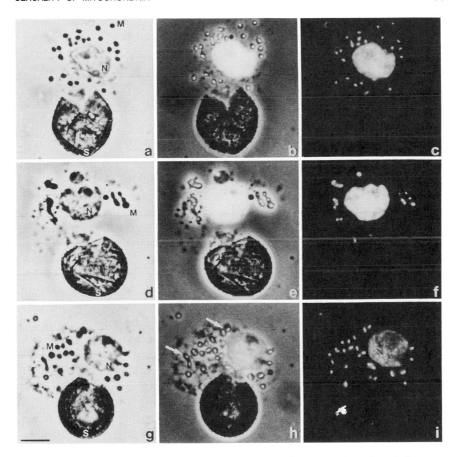

FIG. 4   Mif⁻ (mitochondrial fusion-deficient; a–c) and Mif⁺ (mitochondrial fusion; d–i) pheno-types of the spores cracked open in *Physarum polycephalum*. Phase-contrast (a, d, and g), phase-contrast fluorescence (b, e, and h), and DAPI-fluorescence (c, f, and i) micrographs are shown for the same field. The mt-nuclei are visible as intense blue spots in the mitochondria as a result of staining with DAPI. The Mif⁺ phenotype was classified as high-frequency (d–f) and low-frequency (g–i) according to the frequency of multinucleate mitochondria. M, mitochondrion; N, nucleus; S, spore wall. Bar = 5 μm. (From Kawano *et al.*, 1993.)

drial fusion occurs frequently during the late stages of sporulation. The small, spherical mitochondria fuse with one another to form dumbbells and various other irregular shapes that sometimes tend toward spherical. Further fusions of uninucleate and multinucleate mitochondria result in the formation of large mitochondria of irregular shape, and mitochondria showing evidence of mithcondrial fusion to various extents were enclosed within the spores. The frequency of multinucleate mitochondria in spores seems to reflect the frequency of mitochondrial fusions.

B. Genetic Analysis of Mitochondrial Fusion

1. Hierarchical Transmission of mtDNA during Plasmodium Formation

In *P. polycephalum,* haploid myxamebas act as isogametes, fusing in pairs of individuals of different mating types to form diploid zygotes which develop into macroscopic, diploid plasmodia by repeated mitotic cycles in the absence of cell division (Fig. 1). Thus, the random partitioning of mitochondria by cell division does not occur during formation of the plasmodium. Attempts have been made to determine whether a heteroplasmic zygote produced by mating would develop into a homoplasmic plasmodium in spite of the absence of such random partitioning. However, the inheritance of mitochondrial genes in *P. polycephalum* is difficult to study because easily scored cytoplasmic mutants have not been available. Therefore, the analysis with restriction endonucleases was performed with mtDNA from 19 plasmodial strains (Kawano *et al.,* 1987a). The extent of variation in mtDNA among these strains is high in comparison to that found in other organisms, and it provides a useful source of cytoplasmic genetic markers. Although plasmodia of *P. polycephalum* are diploid, being formed by fusion of myxamebal isogametes, each of the 19 plasmodia possesses mtDNA of only a single type. The pattern of transmission of mtDNA during formation of the plasmodium was studied by mating pairs of myxamebal strains with mtDNAs of different types (Kawano *et al.,* 1987b). Transmission is uniparental; the plasmodia that were formed carried mtDNA with the restriction pattern of only one of the two parental types. Since diploid zygotes develop into plasmodia by repeated mitotic cycles in the absence of cell division, it is clear that this uniparental transmission of mtDNA does not depend upon random partitioning either of the mitochondria or of the mtDNA molecules during cell division.

The mating-type system of *P. polycephalum* appears particularly complex and it is, therefore, of some interest to determine the way in which mitochondrial inheritance is regulated in this organism (Fig. 5). The sexual development (crossing) is under the control of a mating-type system which consists of three loci: *matA, matB,* and *matC* (Dee, 1966; Youngman *et al.,* 1979; Kirouac-Brunet *et al.,* 1981; Kawano *et al.,* 1987a; Dee, 1982, 1987). To cross efficiently, myxamebas must carry different alleles of at least *matA* and *matB*. For each of these loci, full compatibility results when any two

Genetic Control of Plasmodium Formation

matA locus controls development of zygote cells in heterozygosity
   15 alleles of *matA1, matA2, matA3, matA4, matA5, matA6, matA7,*
   matA8, matA9, matA11, matA12, matA15, matA16, matA17,
   and *matA18*

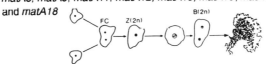

matB locus influences amoebal fusion in heterozygosity
   15 alleles of *matB1, matB2, matB3, matB4, matB5, matB6, matB7,*
      matB8, matB9, matB10,
      matB11, matB12. matB13,
      matB14, and *matB15*

matC locus accelerates amoebal fusion in heterozygosity
   3 alleles of *matC1, matC2,* and *matC3*

Genetic Control of mtDNA Transmission

Transmission pattern of mtDNA is determined hierarchically by *matA*

matA7 > matA2 > matA11 > matA12 > matA1 / matA15 > matA6

female ◄---► male

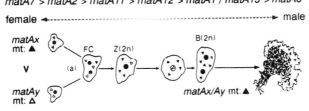

FIG. 5   Mating loci and hierarchical transmission pattern of mtDNA in *Physarum polycephalum*. Effects and alleles of *matA, matB,* and *matC* are shown. A fusion cell (FC) is heteroallelic for *matA:* plasmodium development is initiated, nuclear fusion occurs in interphase, and the zygote (Z) undergoes plasmodial mitosis without cytokinesis to a plasmodium via a binucleate cell (B). FC is homoallelic for *matA:* cells remain ameboid and may separate again. The *matB* and *matC* both regulate cell fusion, apparently by promoting the probability of cell fusion in heterozygosity. The transmission pattern of mtDNA is uniparental: the mtDNA donor in any cross would be the strain of higher status in the order of *matA* represented by a reduced hierarchy (note that *matA1* and *matB15* have not been tested against each other).

alleles are combined from among a set of at least thirteen. In addition, heteroallelism is necessary for *matC,* which has at least three alleles, when crosses are carried out under conditions of elevated pH or reduced ionic strength (Shinnick *et al.,* 1978; Kawano *et al.,* 1987a). As described above, uinparental inheritance of mtDNA is demonstrated in crosses that involved one particular pair of strains. Thus, there may be a consistent bias in favor of the mtDNA from one of these strains. If so, it is clearly necessary to investigate whether such bias could be correlated with particular mitochondrial genomes or myxamebal mating types.

Plasmodia were generated by matings between pairs of myxamebal strains that carried mtDNA molecules distinguishable by restriction endonuclease digestion (Kawano and Kuroiwa, 1989). In each mating pair, one strain consistently acted as the donor of mtDNA, but this strain does not always act as the mtDNA donor when combined in other mating pairs. The identity of the mtDNA donor in each pair is not determined by the different types of mtDNA molecule present or by the different alleles of *matB* or *matC*, two mating-type loci that regulate myxamebal fusion. The results suggest that alleles of the third mating-type locus, *matA*, which controls zygote development, might form a hierarchy such that the mtDNA donor in any cross would be the strain of higher status (Fig. 5). The deduced hierarchy is *matA2* > *matA11* > *matA12* > *matA1*. Recently, such a hierarchy has been confirmed independently by Meland *et al.* (1991). Pooling our data with those of theirs, the tentative order of the *matA* hierarchy with regard to the inheritance of mtDNA is as follows: *matA7* > *matA2* > *matA11* > *matA12* > *matA1/matA15* > *matA6* (note that *matA1* and *matA15* have not been tested against each other).

Meland *et al.* (1991) have also shown that one parental mitochondrial genome is rapidly eliminated from the plasmodium formed when two myxameba fuse and give rise to a zygote which subsequently develops into the plasmodium. Although it is not possible to rigorously exclude "outreplication," their data seem to be most consistent with the "active degradation" model. The quantitative analysis which overestimates the percentage of the "recessive" genomes present during differentiation suggests that this likely process of active degradation is initiated immediately after gamete fusion and is virtually completed by the time of the second division in the newly differentiated plasmodium.

Active degradation of organelle genomes has been proposed for plastids. An active process is indicated during sexual crosses of some isogamous green algae (Kuroiwa, 1985) and during senescence and gametogenesis in higher plants (Sun *et al.*, 1988; Sodmergen *et al.*, 1989, 1991, 1992; Kuroiwa *et al.*, 1991b). The best-studied example is the uniparental inheritance of the chloroplast genome in *Chlamydomonas reinhardtii*. One parental chloroplast genome disappears in young zygotes during the first 6 hr after mating (Kuroiwa *et al.*, 1982). Two different hypotheses have been proposed to explain this rapid elimination: the "methylation-restriction hypothesis" (Sager and Gabowy, 1985) and the "active digestion hypothesis" (Kuroiwa, 1985). The chloroplast inheritance in *C. reinhardtii* is closely related to the mating-type system. This is analogous to the situation of mtDNA in *P. polycephalum*. However, in this case, the mechanism may be much more complex since at least 13 alleles of the *matA* locus are known, compared with only two in *C. reinhardtii*. It is of interest to note that in *P. polycephalum* there is precedence for a genomic restriction system. After normal fusion of genetically compatible plasmodia (Poulter and Dee, 1968), the strain carrying dominant *kil/let* alleles apparently causes destruction of nuclei from the strain carrying recessive alleles. The nuclei of the sensitive strain are selectively destroyed, enclosed in vacuoles, and eliminated from the cytoplasm. There is no evidence that the *matA* alleles are involved in this process; nor is there any evidence for a hierarchical system of elimination. Nevertheless, there are analogies worth keeping in mind in the continuing analysis of the mtDNA inheritance system, as described by Meland *et al.* (1991).

2. A Mitochondrial Genetic Element Controlling Mitochondrial Fusion

Two distinct mitochondrial phenotypes, Mif[1] and Mif , are observed in the spores of *P. polycephalum* as described earlier (Fig. 4). The pattern of transmission of the Mif character was studied in a series of crosses between myxamebal strains (Kawano *et al.*, 1991a, 1993). The results can be explained by postulating that the occurrence of mitochondrial fusion in a plasmodium depends upon the presence of a single factor (*mif*[1]) which is transmitted to most or all of the myxamebal progeny of a Mif⁺ plasmodium. The *mif*⁺ factor is apparently present in the original natural isolate from which the plasmodial strain Ng was derived and is passed on to *mif*⁺ myxamebal strains (NG7 and OZ strains). All other isolates had the *mif*⁻ genotype (Kawano *et al.*, 1991b), for example, the myxamebal strains (e.g., OX110 and OX115) that are largely isogenic with the Colonia isolate (Kawano *et al.*, 1991b); both are classified as *mif*⁻ because plasmodia from their crosses with other *mif*⁻ strains are of the Mif⁻ phenotype and yield myxamebal progeny that all formed Mif⁻ plasmodia when mated with other *mif*⁻ strains. By contrast, NG7 was classified as *mif*⁺ because it always gave rise to Mif⁺ plasmodia, even when mated with a *mif*⁻ strain, and the progeny of these plasmodia always transmitted the Mif⁺ character. Such preferential transmission of the *mif*⁺ factor has been demonstrated over five successive sexual generations in which progeny of Mif⁺ plasmodia were mated with *mif*⁻ myxamebae.

This suggests that Mif⁺ may be of mitochondrial rather than nuclear origin. The mtDNA of appropriate myxamebal and plasmodial strains was, therefore, analyzed by agarose gel electrophoresis (Kawano *et al.*, 1991a, 1993). Figure 6 shows the patterns of *Hind*III restriction fragments of mtDNA from twenty representative strains. Twenty distinct types of mtDNA were detected. The patterns show the conservation of certain

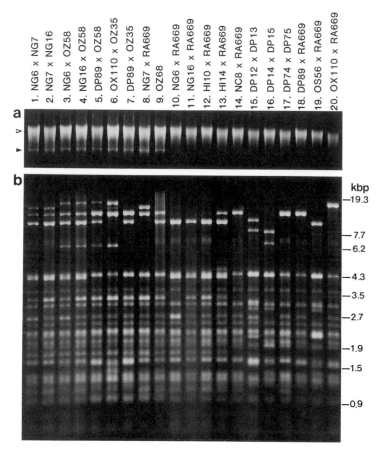

FIG. 6 Profiles after electrophoresis on an agarose gel of undigested (a) and HindIII-digested (b) mtDNAs from 20 plasmodial Mif⁺ (lanes 1–9) and Mif⁻ (lanes 10–20) strains. All Mif⁺ strains carry the mF plasmid (mF⁺) while all Mif⁻ do not carry the mF plasmid (mF⁻). Open and closed arrowheads in (a) indicate bands of mtDNA and mF plasmid, respectively. The HindIII restriction patterns indicate that high-frequency mitochondrial fusion types (lanes 1–5) are heteroplasmic, while intermediate frequency mitochondrial fusion types (lanes 6–9) and Mif⁻ strains are homoplasmic in these profiles. (From Kawano *et al.*, 1993.)

restriction fragments between the different types of mtDNA, but fragments specific for the Mif⁺ strains cannot be detected at first glance. However, from agarose gel electrophoresis of undigested mtDNA (Fig. 6a), the mtDNAs were classified into two types according to their origin in Mif⁺ or Mif⁻ strains. The mtDNA of each Mif⁻ strain formed a single band of 80–90 kbp. In addition to this main band, Mif⁺ strains always gave a second band of approxiamtely 12–16 kbp. The molar ratio of the main mitochondrial genome to the plasmid was estimated to be roughly constant from the gel profiles. These results suggest that the *mif*⁺ character might be carried on this plasmid DNA.

Electrophoretic analysis with the restriction endonuclease *Hind*III also shows that there is a direct relationship between mitochondrial fusion and reorganization of mtDNA (Fig. 7B). As described earlier, an analysis of restriction fragment length polymorphisms (RFLPs) shows that only one of the two myxamebal strains in a mating culture transmits its mtDNA to the plasmodia. These earlier studies all involved *mif*⁻ × *mif*⁻ matings (see Fig. 7A for an example of uniparental inheritance in such a mating). In contrast, results of *mif*⁻ × *mif*⁺ matings do not show simple uniparental inheritance of mtDNA; mtDNA from the plasmodia formed from such matings shows some restriction fragments characteristic of the plasmid DNA and the mtDNA, as well as two fragments of novel sizes (11.0 kbp and 6.5 kbp, Fig. 7B). This result shows that reorganization of the mtDNA is promoted by the mitochondrial and mt-nuclear fusions. Nevertheless, the pattern of restriction fragments from the plasmid DNA remained unchanged over at least five successive sexual generations derived from *mif*⁻ × *mif*⁺ matings, as shown by the restriction pattern of the plasmid DNA purified from agarose gel (Fig. 7C).

These results suggest that the *mif*⁺ character is carried on the 16-kbp plasmid DNA. Moreover, Southern hybridizations with labeled plasmid DNA as the probe demonstrate that some plasmid DNA sequences are represented in the reorganized mitochondrial genomes of the plasmodia generated by the *mif*⁻ × *mif*⁺ mating. To our surprise, not only in the case of all *mif*⁺ strains but also in the case of the *mif*⁻ strains (OX110 and NG6), the labeled plasmid probe hybridizes to bands of the mitochondrial genome (Figs. 6 and 7). The mtDNA of all myxamebal strains has been confirmed to have a duplicated region that contains certain sequences homologous to the plasmid (Section V), visible as a double-bright 15.5-kbp *Hind* III fragment. On the other hand, NG6 is unusual in that it is one of three *mif*⁻ myxamebal offspring of 50 analyzed from an Ng Mif⁺ plasmodium. More detailed information on the extent of homology between the plasmid DNA and the mitochondrial genome of NG6 is obtained by restriction endonuclease digestion of mtDNAs and further hybridization analysis (lanes 2 and 5 in Fig. 7). Only part of the sequence of the plasmid DNA was present in NG6, compared with the inbred *mif*⁺ strain NG7 (lanes 2 and 8 in Fig. 7B). This result suggests that the sequences deleted in NG6 are probably necessary for *mif*⁺ expression. On the other hand, the two fragments (11.0 kbp and 6.5 kbp) newly generated as a result of the reorganization of mtDNA hybridized strongly with labeled plasmid DNA used as the probe (lanes 10 and 11 in Fig. 7B). This result showed that the mtDNA was reorganized by insertion of certain sequences of the plasmid (see Section V).

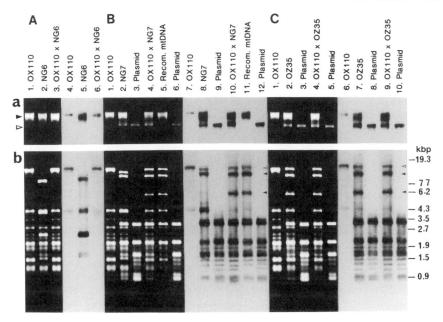

FIG. 7 Transmission of mtDNA in matings between myxamebas. (a) Undigested mtDNA.
(b) mtDNA digested with HindIII. Lanes 1–3 in panel A, lanes 1–6 in panel B, and lanes
1–5 in panel C show the results of electrophoresis on 1% agarose gels; lanes 4–6 in panel A,
lanes 7–12 in panel B, and lanes 6–10 in panel C show Southern hybridization patterns of
the same gels using labeled plasmid DNA as the probe. (A) The mtDNAs of mF⁻ myxamebal
strains (lanes 1 and 4, OX110; lanes 2 and 5, NG6) and the mating product, Mif⁻ plasmodium
(lanes 3 and 6, OX110 × NG6). (B) mF⁻ (lanes 1 and 7, OX110) and mF⁺ (lanes 2 and 8,
NG7) strains; the mating product, Mif⁺ plasmodium (lanes 4 and 10, OX110 × NG7); mF
plasmid DNAs purified from NG7 and from OX110 × NG7 (lanes 3 and 9, lanes 6 and 12)
and recombinant mtDNA purified from OX110 × NG7 (lanes 5 and 11), respectively.
(C) mF⁻ (lanes 1 and 6, OX110) and mF⁺ (lanes 2 and 7, OZ35); the mating product, Mif⁺
plasmodium (lanes 4 and 9, OX110 × OZ35); mF plasmid DNAs purified from mF⁺ myxamebal
and Mif⁺ plasmodial OX110 × OZ35 (lanes 3 and 8, lanes 5 and 10). Open and closed large
arrowheads in (a) indicate bands of mtDNA and plasmids, respectively. Open and closed
small arrowheads in panels B and C indicate the fragment (15.5 kbp) of mtDNA before
insertion of the plasmid and the newly formed fragments (11.0 and 6.5 kbp) that resulted
from insertion of the plasmid, respectively.

3. Nuclear Alleles Controlling the Frequency of Mitochondrial Fusion

The Mif⁺ strains frequently have multinucleate mitochondria in their spores
indicative of mitochondrial fusion, but these frequencies can be classified
into two types: high multinucleate (59.2%–80.5%) and intermediate multinucleate (12.8%–21.3%) (Fig. 4). The Mif⁻ strains and their progeny have

very low frequencies (<0.7%) of multinucleate mitochondria at all stages of the life cycle, and no additional mitochondrial fusion occurs during sporulation in the Mif⁻ strains. Such very low frequencies (0.1%–0.7%) of multinucleate mitochondria in spores are here designated as no mitochondrial fusion, although they all exhibit finite levels. The restriction patterns of the mtDNA of the intermediate frequency spore type indicate that they are homoplasmic: within each plasmodium, mtDNA of only a single, parental type is present. The patterns of the high-frequency type suggest that in every case they are heteroplasmic: within each plasmodium, mtDNAs of both parental types are present. The possibility that the high-frequency mitochondrial fusion type will more often be heteroplasmic is investigated by mating myxamebal strains with mtDNA of different types having quite distinct restriction patterns. The results of these matings clearly show the link between the heteroplasmic condition and the high frequency of mitochondrial fusion (Kawano et al., 1993).

The high-frequency mitochondrial fusion tends to yield the heteroplasmic condition, with two kinds of mitochondria or mtDNA present, one from each parent, in addition to the mF plasmid. However, this situation does not imply that the frequency of mitochondrial fusion is regulated by the mtDNA itself. The two phenotypes, high-frequency and intermediate-frequency mitochondrial fusion, segregated almost equally among the progeny of these crosses. These results suggest that a nuclear gene locus exists, which controls the frequency of mitochondrial fusion. Mitochondrial fusion occurs at high frequency with a combination of unlike alleles at the locus but at intermediate frequency with a combination of like alleles. These features also suggest a mating-type system that consists of three loci: *matA*, *matB*, and *matC*. To cross efficiently, myxamebas must carry different alleles of at least *matA* and *matB*. Therefore, the alleles that regulate the efficiency of the mitochondrial fusion have been tentatively designated mitochondrial mating-type alleles, *mitA1, mitA2,* and *mitA3*. To verify this idea, we isolated new tester strains carrying *mitA1, mitA2,* and *mitA3*, and classified the progeny of high-frequency mitochondrial fusion crosses into two classes, each carrying two different *mitA* alleles. However, precise frequencies of mitochondrial fusion observed in these crosses with the tester strains varied widely. This wide variation suggests the possibility that further nuclear or mitochondrial loci that control the efficiency of the mitochondrial fusion are present. To define all the details of the genetic system that controls mitochondrial fusion in *P. polycephalum,* further genetic analysis is clearly necessary. Transmission genetics is of major importance in studies of mitochondrial genetics (Birky, 1978; Birky et al., 1982). Mitochondrial fusion influences the pattern of transmission of mtDNA and the mitochondrial plasmid. Mitochondrial fusion, however, is often ignored in spite of its obvious importance. Such studies as ours on the genetic system that controls mitochondrial fusion should provide new insight in this field.

V. Mitochondrial Fusion-Promoting Plasmid

A. Structural Features of the mF Plasmid

1. Genetic Organization of the mF Plasmid

The mF plasmid is a linear molecule with telomeric-repeated structures (Fig. 8). The genetic organization of the mF plasmid is interesting because it is the longest one of the mitochondrial linear plasmids and has an obvious phenotype that causes mitochondrial fusion (Table II). To detect ORFs of

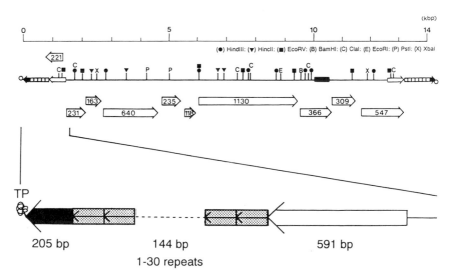

FIG. 8   Structural feature and gene organization of the mF plasmid of *Physarum polycephalum*. Complete nucleotide sequence of mF plasmid including 5 (right) and 7 (left) 144 bp repeating units totaling 14,503 bp. To facilitate the positions of ORFs, a scale and the restriction map of the mF plasmid are shown in the first and second rows, respectively. The locations and orientations of ORFs are indicated by open arrows in the third row. The numbers in open arrows show the number of amino acid residues from ORFs. The terminal structure of the mF plasmid is shown in the fourth row. There is a 205-bp TIR at the extreme ends, which is shown as a closed arrow. The 144-bp repeating units are shown by the shaded boxes. The sequences of each repeating unit were perfectly identical. Inside these repeats, a 591-bp TIR is shown as an open arrow in the lower illustration. The 5′ end of the mF plasmid is protected by putative terminal proteins (TP). (Modified from Takano *et al.*, 1994a,b.)

the mF plasmid and to determine the gene(s) associated with the mitochondrial fusion, Takano *et al.* (1994b) determined the entire DNA sequence of the mF plasmid. Since the mF plasmid has three TIRs, one of which has 144-bp repeating units, the size of the mF plasmid is estimated to vary from 13.3 to 18.2 kbp, depending on the extent of repetitions. Using certain clones, including 5 and 7 of the 144-bp repeating units for sequencing right and left ends, respectively, the size of the mF plasmid was determined to be 14,503 bp. Overall AT content of the mF plasmid is 75.0%.

The mF plasmid contains 10 ORFs which have potential methionine initiation codons. These ORFs are named according to the number of amino acid residues. All ORFs except one (ORF-221) are encoded on the same strand. The amino acid sequences derived from these ORFs, except ORF-547, do not show significant homology to any amino acid sequences in the database (SWISS-PROT compiled by EMBO). Northern hybridization and primer extension suggest that the transcription initiation site mapped near the inner end of the left TIR of 591 bp and that transcripts started at the left end and went to the lower region on the coding strand. The transcription initiation site (TTATAAG ATATA) locates near the inner end of the left TIR of 591 bp and the transcripts at 1.0, 3.4, and 4.6 knt (kilonucleotide) and longer transcripts start from this transcription initiation site. The 3.5-knt transcripts correspond to the coding region of ORF-1130 and may be derived from the longer transcripts of over 8 knt. Southern hybridization using ^{32}P-labeled mtRNA has shown that the transcripts of the upper region of the coding strand were about 500 times more abundant than those of the lower region (Takano *et al.*, 1994b).

The transcripts of the ORFs of pClK1 and *kalilo* have also been analyzed (Duvell *et al.*, 1988; Gessner-Ulrich and Tudzynski, 1992; Griffiths, 1992). In the case of pClK1, two major transcripts correspond to ORF1 (DNA polymerase) and ORF2 (RNA polymerase). These ORFs start in the TIRs of pClK1 and continue through the inner region. Both transcripts start at the end of the TIR. The concentrations of two major transcripts are almost equivalent. The same transcriptional manner is used in *kalilo*. The mode of transcription of the mF plasmid is more complex than that of other such examples and quite different from the other linear mitochondrial plasmids.

2. Terminal Structure of the mF Plasmid

The replicative completion of a linear DNA molecule has been recognized as a serious problem for a number of years. All known DNA polymerases synthesize DNA in the 5' to 3' direction and all require a primer that is usually removed; multiple rounds of DNA replication would thus result in the progressive loss of the DNA sequence from the ends. To resolve this general problem, all known linear mitochondrial plasmids have TIRs with terminal proteins covalently linked to the 5' ends of the plasmid (Meinhardt *et al.*, 1990). By contrast, the terminal structure of the mF plasmid is different from that of other linear plasmids (Figs. 8 and 9; Takano *et al.*, 1991, 1994a).

The mF plasmid has two TIRs 591 and 205 bp long, respectively, which are on either side of one or more 144-bp repeats. The number of 144-bp repeating units varies from one to more than 17 (Fig. 9). This variation results in mF plasmids of different lengths. Repeated arrays at the extreme ends of linear chromosomes are known as telomeres (Blackburn and Szostak, 1984; Zakian, 1989). In mitochondria, long repeating units, which are estimated to range in length from 31 bp to 53 bp, have been reported in *Tetrahymena* (Morin and Cech, 1986, 1988). However, since the 144-bp repeating units of the mF plasmid are not located at the extreme ends, they are quite different from telomeres. The resistance of the ends of mF plasmid to 5'-specific λ exonuclease suggests that the 5' end may be protected by terminal proteins similar to those found in other linear mitochon-

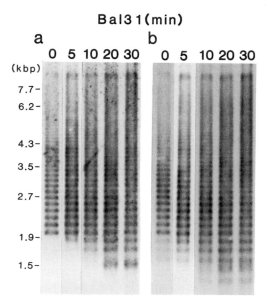

FIG. 9 The repeated structure of the mF plasmid and its sensitivity to exonuclease *Bal*31. The total DNA from mitochondria was digested with *Bal*31 for 0, 5, 10, 20, and 30 min, as indicated. The samples were then digested with *Xba*I and digests were fractionated by agarose-gel electrophoresis. The gel was blotted on a nylon membrane and hybridized with the probes, including the regions very near the left (a) and right (b) ends. The results suggest that the mF plasmid has extensive repeated sequences as a unit of 144 bp at or very near both ends, such as those degraded by exonuclease *Bal*31. The most extensive array of repeats can be estimated to consist of at least 17 repetitions of these repeating units. (From Takano *et al.*, 1991.)

drial plasmids. The structure of the extreme ends (TIR with a terminal protein) suggests that the mF plasmid utilizes a mode of replication similar to those of *Bacillus* phage φ29 and adenoviruses.

The mechanism that causes the repeating units to occur in varying numbers is still unexplained. The sequence of the last seven nucleotides of the 205-bp TIR, 5′ TATTGAA 3′, was identical to that of the 144-bp repeating unit. The heterogeneity may be the result of the "jumping" of the DNA polymerase from the 5′ TATTGAA 3′ sequence of the 144-bp repeating units to the identical sequence of the 591-bp TIR. If the replication system of the mF plasmid is similar to that of bacteriophage φ29 and adenoviruses, then replication starts at the extreme ends and progresses inward. The DNA polymerase first replicates the 205-bp TIR and then proceeds to the 144-bp repeating units. If the DNA polymerase on the 5′ TATTGAA 3′ sequence of the 144-bp repeating unit jumps to the identical sequence of the 205 bp TIR, it re-replicates the 144-bp repeating units, which means that the replicated mF plasmid will have more of the repeating units than the original plasmid. Moreover, inter- and intrarecombination between this 7-bp nucleotide sequence may also produce different numbers of repeating units.

The sequences of the three TIRs, including those of the 144-bp repeating units, have no significant sequence homology with TIRs of other linear mitochondrial plasmids. The 205-bp TIR located at the extreme ends has a higher GC content than the other region of the mF plasmid and forms three thermodynamically stable hairpin structures. It may protect against digestion from the ends by exonucleases, and/or recognition of DNA polymerase and a possible terminal protein by its replication system. However, a distinct gene that codes a terminal protein has not been disclosed in the nucleotide sequences of the mF plasmid. Chan *et al.* (1991) and Court and Bertrand (1992) have suggested that the cryptic amino-terminal domains which precede the exonuclease domains of the plasmid DNA polymerases may be parts of the terminal proteins. Within these domains, there are two SYKN sequence motifs which are composed of consistently spaced serine, tyrosine, lysine, and asparagine residues; they are also present in the terminal proteins of bacteriophages φ29 and PRD1, and adenoviruses. These sequence motifs existed in the amino acid sequence derived from ORF-309 locating just before ORF-547 (nt11110-11161 and nt11255-11318). However, the weakness in this analysis is that there is no information about the amino acid sequence of even one terminal protein from a mitochondrial plasmid, nor are there any structural or genetic data to support the notion that the SYKN motif is relevant to the function of any known terminal protein. Much more research is needed to further define the terminal proteins of the linear mitochondrial plasmids, including the mF plasmid.

3. DNA Polymerase Genes

The ORF-547 of the mF plasmid shows extensive homology with the putative DNA polymerases that are encoded on other linear mitochondrial plasmids (Fig. 10). ORFs encoding DNA polymerases have been described in other linear mitochondrial plasmids (Table I). Two conserved domains are characteristic of the proofreading and polymerization motifs of DNA polymerases in linear plasmids and linear phages. The proofreading domain, which has 3′ to 5′ exonuclease activity, is located at the N-terminal and is characterized by three amino acid sequence blocks: Exo I, Exo II, and Exo III (Bernad et al., 1989). Three strongly conserved blocks (Pol I, Pol II, and Pol III) are located in the C-terminal polymerization domain (Oeser and Tudzynski, 1989; Court and Bertrand, 1992). Figure 10 shows that ORF-547 contains these three polymerization blocks but does not contain the three conserved blocks of the proofreading domain. These results suggest that the DNA polymerase encoded by ORF-547 has DNA polymerizing activity but not the proofreading activity. The known DNA polymerases that are encoded in other linear mitochondrial plasmids all contain 3′ to 5′ exonuclease domains (Court and Bertrand, 1992; Hermanns and Osiewacz, 1992). Two hypotheses exist with regard to the proofreading activity of the DNA polymerase of the mF plasmid. One hypothesis is that the DNA polymerase encoded by ORF-547 does not have 3′ to 5′ exonuclease activity, or that it has exonuclease activity without the exonuclease domain. The

		Pol I		Pol II		Pol III
pMC3-2	526	CIKT-KSYDCNSIYPYCMLKDMPVEN	652	PTAKLLLNGLYGRFGMNP	763	KVFMTDTDCIWMNGSLS
		* . ** **.**. ** .		..* ***.* *****		.. **** * .
pAL2-1	667	YGKNLRYYDVNSTYPFVAKNTMPGHE	793	TMTKFLLNSLLGRFGMSM	913	NLYYTDTDSIVTDIDTP
		.* ****** ***.. * .**		..* .****.***** .		************. *
pClK1	697	STKSYYYDVNSLYPFASINDIPGLK	827	NIAKLILNSLIGRFGMNI	946	TLYYTDTDSIVTDLKLP
		. ********** ...*.* *		* *. .*** ****.		*******.* . **
S-1	487	YGENLYYYDVNSLYPSSMLDDMPIGK	614	FIYKITMNSLYGRFGISP	712	DCYYTDTDSVVVERELP
		... *.******** ...		.** .*.******.		* ***.. ..:*
mF	198	VAQRNYFYDVNSLYPYIMKKEKMPIG	341	DLYKKLLNTLYGRFGLVY	420	HVIYIDTDGLFLKNPIP
		.*** * * ** *		. * *.*.******.		. .****. . *
pAI2	814	EGKNIHSYDINSLYPSAMAKFDMPTG	947	FIAKLLMNSLYGRFGMDP	039	NLYAVDTDGIKVDTEID
		* **.***.***** ** * **.*		.*.*********** ..*		*.* .******** ..*
kalilo	627	FGVNIKSYDVNSLYPFAMKYFKMPSG	768	YISKLLMNSLYGRFGLNP	860	NIYYIDTDGIKVDIDLD
		** *.* ***** ** ** *		..***.*.**** *.*		* * ****:* *. **
maranhar	620	IINNIFSFDFNSLYPTAMM-MPMPVG	748	QMAKLLLNTLYGRTGMND	857	NSAYTDTDSIFVEKPLD
		..* . .* ** ** **. ** *		.*** **.***. * .		. .***. . ***.
pEM	436	LVKNGYHYDMNSQYPYAML-QSMPTG	569	YIAKLSLNSLYGKFGQKE	688	LAIASNTDSLILRKPLE

FIG. 10 Polymerizing domains of DNA polymerases coded on ORF-547. The DNA polymerases shown here are as follows: the linear mitochondrial plasmids, pMC3-2 (Rohe et al., 1991), pAL2-1 (Hermanns and Osiewacz, 1992), pClK1 (Oeser and Tudzynski, 1989), S1 (Kuzumin and Levchenko, 1987), pAI2 (Kempken et al., 1989), kalilo (Chan et al., 1991), maranhar (Court and Bertrand, 1992), and pEM (Robison et al., 1991). The sequences are aligned for maximum similarity. Identical amino acids and conserved exchanges between amino acids in neighboring sequences are indicated by asterisks and dots, respectively. The numbers of the first amino acid of each block are indicated to the left of each of the amino acid sequences. (From Takano et al., 1994b.)

other hypothesis is that extensive editing of mRNA creates the 3' to 5' exonuclease domain of the DNA polymerase. The extensive editing of the gene for the α subunit of ATP synthetases has been reported in the mitochondria of *P. polycephalum* (Mahendran *et al.*, 1991).

The left end of the mF plasmid does not contain any long ORFs (> 300 amino acids). The longest ORF is encoded in the 591-bp TIR and the 144-bp repeating unit. This ORF (ORF-211) starts in the 591-bp TIR as shown in Fig. 8, if the first ATG codon after the stop codons of the same frame can be considered the initiation codon, and if it encoded a polypeptide of 221 amino acids. Since the ORF-221 contained two conserved blocks of the polymerization domain (Pol II and Pol III) but did not include Pol I, it may not function as DNA polymerase.

B. Constitutive Recombination Mediated by the mF Plasmid

1. Constitutive Homologous Recombination between the mF Plasmid and mtDNA

To analyze the pattern of transmission of the mtDNA, the novel restriction fragments in mF$^-$ $\times$ mF$^+$ (the 11.0- and 6.5-kbp *Hind*III fragments; Fig. 7), which are not found in the restriction pattern of the mtDNA of mF$^-$, were cloned into the pBluescript vector, and compared with those of the mtDNA and the mF plasmid (Fig. 11; Takano *et al.*, 1992). These fragments consist of one part that is identical to the duplication of the mtDNA (M type) of mF$^+$ and one part that is identical to the mF plasmid. This result suggests that the novel fragments are generated by recombination between the duplication of the M type mtDNA and the mF plasmid. The recombination occurs between the region about 11.0 kbp from the left end of the duplication and the region about 3.0 kbp from the right end of the mF plasmid on the maps. The recombination seems to be caused by reciprocal crossing-over between the mtDNA of the M type and the mF plasmid.

To analyze the recombination between the mtDNA and the mF plasmid, the homologous regions of the mtDNA and the mF plasmid, and the recombination site of the recombinant mtDNA were sequenced (Fig. 12; Takano *et al.*, 1992). The mF plasmid has a 475-bp sequence that is identical to a 479-bp sequence of the mtDNA with the exception of deletions of a total of only four base pairs at three sites. The identical sequence of the mtDNA starts from nucleotides 5'TAAAAGAAA 3' and stops at 5'TTTGTTTG 3' in the duplication of the mtDNA. By contrast, the identical sequence in the mF plasmid starts at nucleotides 5' TAAAGAAA 3' with a one-base deletion and stops at 5' TTTGTTTG 3', which is the same as the sequence in the mtDNA. Other deletions in the mF plasmid correspond to positions

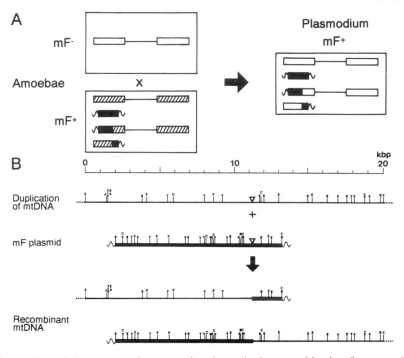

FIG. 11 Transmission patterns (upper part) and constitutive recombination (lower part) of mtDNA and the mF plasmid in the mating between mF⁻ and mF⁺ strains. The mtDNAs of two parents are shown by a thin line (single-copy region) with open or closed boxes (duplication region). In this mating, mtDNA of the mF⁻ strain is uniparentally transmitted to the resulting plasmodium since the allele of *matA* of the mF⁻ strain is chosen to be of the female type (mtDNA donor) in the hierarchical order of uniparental transmission as shown in Fig. 5. At this time, a part of the transmitted mtDNA is constitutively recombined with the mF plasmid. Restriction maps of the duplication regions clearly show that the crossing-over between mtDNA and mF plasmid occurs at a specific site (▽) in the duplication regions. The upper thin line and middle thick line show the restriction map of the total length of the duplication in the mtDNA. The middle thick line shows the restriction map of the mF plasmid. The two lower lines show the restriction map of recombinant mtDNA formed by crossing-over between mtDNA and mF plasmid. Thin and thick lines in the restriction map of recombinant mtDNA represent the duplication of mtDNA and mF plasmid, respectively. The wavy line shows the end of the mF plasmid and reflects the heterogeneity in size. The vertical lines with symbols indicate the various restriction sites. (Largely modified from Takano *et al.*, 1992.)

22 and 91–92 of the mtDNA. From these deletion sites, the recombination is estimated to occur at a region between positions 92 and 479 in the identical sequence of the mtDNA type. Such constitutive reciprocal crossing-over between the mtDNA and the mF plasmid caused structural changes in the mtDNA. The mF plasmid crosses over with the duplication of the mtDNA, so that three types of structural conversions of the mtDNA are possible:

```
                                    -50                                                          1
MT  TTAATTATAATTTTTTATAATGAAAAAGAACAAAAAGATAATTAATATTATGATAATATGGGTTTTACTTACAGATAGAGTAAAAGAAAGGAATATGCAA
     *  *  * ** *   *  *  *     * *   ** ***    *** **  *               ** *    ****-***********-******
P   gttttgcttaatgaatgttttcgtactttttatatgaattattttatataatgctatcaataaaagactttaaaaagagctaaa-gaaaggaatatgcaa

R1  TTAATTATAATTTTTTATAATGAAAAAGAACAAAAAGATAATTAATATTATGATAATATGGGTTTTACTTACAGATAGAGTAAAAGAAAGGAATATGCAA
R2  gttttgcttaatgaatgttttcgtactttttatatgaattattttatataatgctatcaataaaagactttaaaaagagcTAAA-GAAAGGAATATGCAA

                        50                                            100
MT  AATTACAGAAAATCTTCTTATTCAAAAAAACTACTTTATACAATATACATTCCAAAAATATCAGGAGTTACTCTACCCGAATATCCTTTAAAAGATGTTC
     *-********************************************************************--***************************
P   a-ttacagaaaatcttcttattcaaaaaaactactttatacaatatacattccaaaaatatcaggagttac--tacccgaatatcctttaaaagatgttc

R1  AATTACAGAAAATCTTCTTATTCAAAAAAACTACTTTATACAATATACATTCCAAAAATATCAGGAGTTACTCTACCCGAATATCCTTTAAAAGATGTTC
R2  A-TTACAGAAAATCTTCTTATTCAAAAAAACTACTTTATACAATATACATTCCAAAAATATCAGGAGTTAC--TACCCGAATATCCTTTAAAAGATGTTC

                 150                                         200
MT  AAAAAGCTAATTTTATCCAGTTCTTTAATAGTATAGTTAAAAATGACCCAATATGGGATAATTTATGTCAAACATATAATAAAGTAAATTCTTCATTTTT
     ***********************************************************************************************
P   aaaaagctaatttt atccagttctttaatagtatagttaaaaatgacccaatatgggataatttatgtcaaacatataataaagtaaattcttcatttttt

R1  AAAAAGCTAATTTTATCCAGTTCTTTAATAGTATAGTTAAAAATGACCCAATATGGGATAATTTATGTCAAACATATAATAAAGTAAATTCTTCATTTTT
R2  AAAAAGCTAATTTTATCCAGTTCTTTAATAGTATAGTTAAAAATGACCCAATATGGGATAATTTATGTCAAACATATAATAAAGTAAATTCTTCATTTTT

              250                                        300
MT  TTATCAAAAAATTAAAGATGCTTATATTAACTTAACTTTACAAACAGCTCATTATAAAGAACAAATAGCTATTTATTTATCATTTATCTTAAAATTGATT
     ***********************************************************************************************
P   ttatcaaaaaattaaagatgcttatattaacttaacttt acaaacagctcattataaagaacaaatagctatttatttatcatttatcttaaaattgatt

R1  TTATCAAAAAATTAAAGATGCTTATATTAACTTAACTTTACAAACAGCTCATTATAAAGAACAAATAGCTATTTATTTATCATTTATCTTAAAATTGATT
R2  TTATCAAAAAATTAAAGATGCTTATATTAACTTAACTTTACAAACAGCTCATTATAAAGAACAAATAGCTATTTATTTATCATTTATCTTAAAATTGATT

           350                                        400
MT  GAAATGCACCCTCAAATACCTATTAAAGATATGTGCTTAAAAGCAGTCAATTATGATGTTTATTTTGACATAAATGACCACGATAGTTTTTATGTTACAA
     *****************************************************************************************************
P   gaaatgcaccctcaaatacctattaaagatatgtgcttaaaagcagtcaattatgatgtttattttgacataaatgagcaggatagttttt atgttacaa

R1  GAAATGCACCCTCAAATACCTATTAAAGATATGTGCTTAAAAGCAGTCAATTATGATGTTTATTTTGACATAAATGAGCAGGATAGTTTTTATGTTACAA
R2  GAAATGCACCCTCAAATACCTATTAAAGATATGTGCTTAAAAGCAGTCAATTATGATGTTTATTTTGACATAAATGAGCAGGATAGTTTTTATGTTACAA

        450                    479                   500
MT  TAAAACAGAATCCTTATAGTCCTATTGAAATTAGTCTTCAACATGATAATATTTGTTTGCACACTTTCCATTTATTACTTATTGAAGAAATATTTCTACA
     **********************************************************  *   * *  **    *   ** *   * *      *  *
P   taaaacagaatccttatagtcctattgaaattagtcttcaacatgataatatttgtttgaaagatataaatcgctatattgctctattagttcaaatgga

R1  TAAAACAGAATCCTTATAGTCCTATTGAAATTAGTCTTCAACATGATAATATTTGTTTGaaagatataaatcgctatattgctctattagttcaaatgga
R2  TAAAACAGAATCCTTATAGTCCTATTGAAATTAGTCTTCAACATGATAATATTTGTTTGCACACTTTCCATTTATTACTTATTGAAGAAATATTTCTACA

     559
MT  AAACTTTCAAGATAATATTGAACCATTTTCTGATCTCTT
     ***  *   ** **    *  *  *      *   *  *
P   aaaacattttt atgaattattagaaagtatatttattat

R1  aaaacattttt atgaattattagaaagtatatttattat
R2  AAACTTTCAAGATAATATTGAACCATTTTCTGATCTCTT
```

FIG. 12 Comparison of sequences between mtDNA (MT), mF plasmid (P), and recombinant mtDNA (R1 and R2). Sequences are numbered relative to the first nucleotide of the so-called identical sequence of the mtDNA, and matches between the mtDNA (capitals) and the mF plasmid (small letters) are indicated by asterisks. Sequences are shown from −80 to 559. Recombinant molecules are classified into two types: R1 consists of the minus region of mtDNA (−0), the so-called identical region (1–479) of mtDNA type, and the plus region of mF plasmid (480−); and R2 consists of the minus region of the mF plasmid (−0), the so-called identical region (1–479) of the plasmid type, and the plus region of the mtDNA (480−). The so-called identical region in the recombinant mtDNAs (capitals) is underlined. The deletions of nucleotides in the mF plasmid and the recombinant mtDNA are shown by hyphens (-). (From Takano et al., 1992.)

crossing-over at the left duplication, at the right duplication, and at both duplications. Each crossover event creates large and small recombinant mtDNAs. As shown in Fig. 12, one-half of the duplication of the mtDNA is recombined with the mF plasmid, and the small recombinant mtDNAs

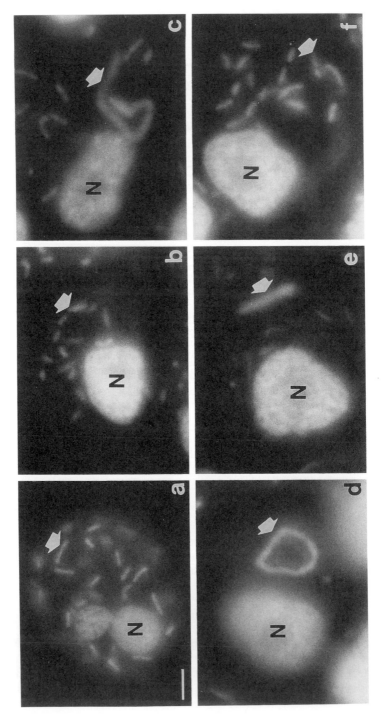

are derived from the left or right duplications. In each case, the end of the mF plasmid with extensive repeating units is located at the terminal of the recombinant mtDNA.

Integrations of linear plasmids have been also found in the mtDNAs of maize (Schardl *et al.*, 1985), *Claviceps purpurea* (Tudzynski and Esser, 1986), and *Neurospora* (Myers *et al.*, 1989; Court *et al.*, 1991). Among these plasmids, *kalilo* of *N. intermedia* has been well studied. It induces senescence in *Neurospora* by integration into the mtDNA (Bertrand *et al.*, 1985, 1986). The integration of *kalilo* occurs at seven distinct regions of the mtDNA and always generates very long inverted repeats of mtDNA flanking the two ends of the *kalilo* insertion sequence (Dasgupta *et al.*, 1988). In the case of the mF plasmid, however, the identical sequence between the mtDNA and the mF plasmid is not flanked by inverted repeats of the mtDNA, and the recombination between the mtDNA and the mF plasmid is not associated with senescence of *P. polycephalum*.

The comparison between the so-called identical sequences of the mtDNA and the mF plasmid showed a high level of similarity in nucleotide sequence (more than 99%), which suggests a common origin for this sequence. With respect to the origin of the identical sequence in mtDNA and plasmids, S1 and S2 plasmids have been well studied. The same sequences in the mtDNA of maize, which are identical to those in the S1 and S2 plasmids, are derived from R1 and R2 plasmids (Schardle *et al.*, 1984, 1985; Houchins *et al.*, 1986). The R1 and S1 plasmids exhibit approximately 70% homology and R2 and S2 exhibit complete sequence homology (Weissinger *et al.*, 1982). It is possible that the R1 and R2 plasmids presumably integrated into the mtDNA and became fixed in a sequence identical to the S1 and S2 plasmids after deletion of other sequences. This may occur with the accompanying loss of the free plasmids. From such speculation, in the mitochondria of *P. polycephalum*, the mF plasmid is thought to be recombined with the mtDNA in an ancient strain and fixed as the so-called identical sequence after deletion of the entire sequence apart from the identical sequence and the free plasmid. This hypothesis is supported by the fact that the identical

---

FIG. 13 Mt-nuclear fusion and division in the zygote formation of *Physarum polycephalum*. The zygotes which were developing to plasmodia in the mating culture of mF⁻ x mF⁺ were stained with DAPI. The rod-shaped mt-nucleus in individual mitochondria is observed in the binucleate zygote formed by myxamebal fusion (a). After cell-nuclear fusion, the mt-nuclei are queued (b) and fused to become a line (c) accompanying mitochondrial fusion. The fused mt-nuclei occasionally become circular (d) and tightly fused to become a large, rod-shaped mt-nucleus (e). Then, before the first plasmodial mitosis, the fused mt-nucleus successively divides to return to its former state (f). Arrows show mt-nuclei. N, nucleus. Bar = 1 μm.

sequence of the mF plasmid hybridized with the mtDNA of all strains
during Southern hybridization (see Fig. 6).

2. Recombination and mt-Nuclear Fusion in Zygotes

It is obvious that the constitutive homologous recombination between mF
plasmid and mtDNA occurs during plasmodium formation by a cross between mF⁻ and mF⁺myxamebal strains. This suggests that the mitochondrial
fusion accompanying mt-nuclear fusion occurs in the zygotes produced by
such a cross (Fig. 13). Thus, these results suggest a picture of mitochondrial
fusion events promoted by the plasmid, as illustrated in Fig. 14. In a zygote
formed as a result of an mF⁺ × mF⁻ mating, uninucleate mitochondria of
two parental types fuse to form multinucleate mitochondria, in which the
two main species of mtDNA and the plasmid are all present (Fig. 14A).
After fusion of mitochondria, crossing-over occurs between homologous
regions on the two main species of mtDNA or the plasmid. Recombinant
mtDNA molecules and the unchanged free plasmid are then transmitted

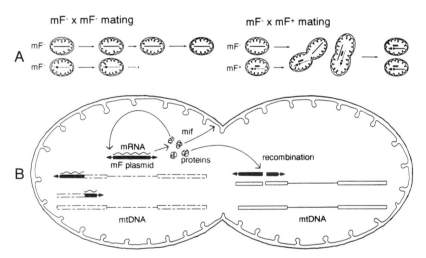

FIG. 14 Two transmission patterns of mitochondria (A) and events produced by the mF
plasmid in a fused mitochondrion (B). In mF⁻ x mF⁻, the mitochondria do not fuse but are
transmitted uniparentally. In mF⁻ x mF⁺, the mitochondria fuse and their mtDNA is recombined with the mF plasmid. The mitochondrial fusion is caused by the mF plasmid. The mF
plasmid has three TIRs and an identical sequence with mtDNA, and codes 10 ORFs. The
mF plasmid may be transcribed polycistronically from a specific site. The translated proteins
of the mF plasmid may act on the mitochondrial fusion and recombination between the
mtDNA and the mF plasmid in a fused mitochondrion.

to daughter mitochondria. A similar cycle also occurs during sporulation and the germination of spores (Fig. 1).

The transmission of the mF plasmid and the recombinant mtDNA in the mating between mF^- and mF^+ can be explained as follows (Fig. 14B). The mitochondria of two parental myxamebal strains fuse with each other in the zygote. The two parental mtDNAs of mF^+ and mF^-, and the mF plasmid can be paired in one mitochondrion, and then crossing-over occurs between the mtDNA of mF^- and the mF plasmid. In every mating between mF^- and mF^+, the recombinant mtDNA from the cross-over between the mtDNA of mF^- and the mF plasmid is transmitted to the plasmodium. It seems that the recombinant mtDNA is transmitted according to the *matA* hierarchy (*matA2* > *matA12*) with respect to the transmission of the mtDNA (Kawano and Kuroiwa, 1989; Meland *et al.*, 1991). The data in Fig. 12 suggest that the mtDNA of mF^+ (NG7) crosses over with the mF plasmid. Thus, mtDNA from mF^- or from mF^+ is transmitted to the plasmodium, and the sequence of the mF plasmid is maintained in the next generation as part of the recombinant mtDNA. Moreover, a free plasmid that does not cross over with the mtDNA in the fused mitochondrion is preferentially transmitted to the plasmodium regardless of the transmission of the mtDNA. The fusion of mitochondria is important and provides the basic mechanism for the spread of the mF plasmid and the recombinant mtDNA with the mF plasmid through the mitochondrial population. The fact that the mF plasmid is associated with mitochondrial fusion gives it an advantage in terms of the frequency with which it is transmitted.

VI. Sexuality of Mitochondria

A. Evolution of Sex

A very obvious biological consequence of sex is the generation of new combinations of genes by mixing genomes, or portions thereof, from different individuals. This "mixis" has fundamental implications for both genetics and the evolutionary process. While we may agree that mixis is a fundamental consequence of sex, there is much less agreement concerning the mechanisms responsible for the origin and maintenance of sex. Heritable variation in fitness is the fuel of adaptive evolution, and sex can, although perhaps rarely, generate new combinations of alleles that are adaptive. However, it is not at all obvious whether the new adaptive combinations of alleles produced by sex are sufficient to confer a selective advantage on sexual individuals. There seems to be so far no consensus concerning the answer to the question of the adaptive significance of sex (Maynard-Smith, 1978; Ghiselin, 1988; Hamilton *et al.*, 1990).

In contrast, Hickey and Rose (1988) have taken a different approach altogether, challenging whether sex is advantageous to the individuals practicing it, and developed a multistage scenario for the evolution of sex in which parasitic gene transfer is the dominant factor. In this schema, sex could be the result of the evolution of parasitic selfish DNA molecules that exploit the opportunities for horizontal transmission afforded by cycles of germ cell fusion and fission to spread horizontally through populations. The parasitic selfish DNA molecules could, in theory, enhance their own fitness by promoting gamete fusion (Hickey, 1982, 1984). The formal analyses in which both the origin and the maintenance of such "parasitic sex" have been considered to be of central importance (Rose, 1983; Tremblay and Rose, 1985; Krieber and Rose, 1986) suggest that such parasitic gene transfer could indeed play an important role in the evolution of eukaryotic sex, although adaptive benefits to the host cells might also play a critical role. A similar argument for the role of transmissible genes in evolution of bacterial sex has been outlined by Zinder (1985).

The important conclusion of the "parasitic origin of sex" theory is that transposable elements that have positive, neutral, or negative effects on host fitness can all spread within a sexual population in a way that would not be possible within an asexual population. Thus this advantage of sexuality for the transposable elements implicates them in the origin of sex. A coincidental evolution hypothesis for mixis is also favored for plasmid- and phage-mediated recombination (Hickey and Rose, 1988). In these cases, sex, the capacity for infectious transfer, is encoded by genes borne by the plasmid or phage, rather than by host genes, and the mixis of host genes is viewed as a coincidental by-product of evolution of the plasmid or phage. Such parasite coevolution seems to be superior to previous models of the evolution of sex by supporting the stability of sex under the following challenging conditions: very low fecundity, realistic patterns of genotype fitness and changing environment, and frequent mutation to parthenogenesis, even while sex pays the full two-fold cost (Hamilton *et al.,* 1990; Howard and Lively, 1994). Both the theoretical calculations and the experimental observations support the idea that self-replicating, intragenomic elements can gain a large advantage from the sexual mixing of their host genomes.

B. Bacterial Sex

The best clues as to the probable nature of primitive sex come from the study of living prokaryotes rather than eukaryotes (Jacob and Wollman, 1961; Margulis and Sagan, 1986; Levin, 1988). In eukaryotes, a single specialized sexual system based on cycles of syngamy and meiosis is virtually ubiquitous. In contrast, the prokaryotes still display a wide variety of sexual and asexual reproductive systems. The prokaryotic sexual systems are both conjugative and nonconjugative. The genes of prokaryotes are infected by one of three basic mechanisms—transformation, transduction, and conjugation. With transformation, free DNA is taken up by the bacterium and incorporated into its genome. In transduction, the movement of genetic material between cells is accomplished by bacteriophage vectors that pick up DNA from a donor cell and transmit it to a recipient. Conjugation is a more intimate process, with the transfer of genes requiring contact between a donor and a recipient cell.

The first genetic evidence for sex via conjugation was obtained in *Escherichia coli* (Lederberg and Tatum, 1946). The observation of conjugation between bacterial cells was taken as direct evidence for sexual mating in this species. This process is genetically controlled by a self-transmissible conjugative plasmid, the F factor (Ippen-Ihler and Minkley, 1986; Smith, 1991). More than 60 genes and sites have been located on the physical and genetic map of the F plasmid, a covalently closed, circular, double-stranded DNA molecule close to 100 kbp in length. The capacity of *E. coli* to transfer chromosomal genes by conjugation and the variety of changes in the donor cell surface associated with the transfer process are determined by genes carried by the F plasmid. The integration of the plasmid and chromosome, forming a state known as a "Hfr" (high-frequency recombination), can occur in a number of places along the circular *E. coli* chromosome. Transfer is polar, originating with a specific sequence on F, not surprisingly known as the origin of transfer (*oriT*), and immediately followed by the regions of the F plasmid associated with replication. The genes responsible for the self-transmissibility of the plasmid enter the recipient last. In Hfr, the entire bacterial chromosome (about 50 times the length of F) separates the origin of transfer from the genes required for conjugation.

The idea of a relationship between bacterial parasites and a form of bacterial sex is not a new one. These bacteriophages, which act as generalized transducing phages, are potentially important in recombination yet they are clearly bacterial parasites. Dougherty (1955) was one of the first to point out that phages, although they are parasites, could also facilitate a form of bacterial sexuality. The essence of the "parasitic origin of sex" theory is not only that intracellular parasites provide a form of sexuality among prokaryotes, but also that "conventional" eukaryotic sex itself began with a type of intracellular parasitism. As regarded with the evolution of sex as a multisite process, where each step evolved as a function of the individual selection of genes that were favored under conditions existing at a particular stage. This is not surprising if primitive eukaryotic mating, or syngamy, was the first step in a complex evolutionary process starting from such conjugation. During the course of the ensuing process, syngamy has taken on a new significance.

C. Mitochondrial Sex

As outlined in the preceding section, both mathematical and biological arguments suggest that there is no appreciable problem with hypothesizing an initially parasitic origin of sex. It is highly unlikely, however, that sex has since continued to be simply parasitic. No biological system that displays the initial stages in a parasitic origin of sex has been described. According to the endosymbiosis theory, mitochondria originated as free-living cells that were later the prey, or parasites, of the ancestors of modern eukaryotic cells. Proponents of the endosymbiotic theory would not, however, consider modern mitochondria as an illustration of the initial parasitic origin of sex; although they have obviously become an integral part of the eukaryotic cell, they are highly infectious, with mobile genetic elements such as ω-like introns and plasmids and have the ability to fuse and recombine. Mitochondria and chloroplasts have their own genomes, which are different from those of the cell nuclei. The origins of these organelles have been regarded as a line in the evolution of the aerobic bacteria or green algae that cohabited in primordial cells during the establishment of eukaryotic cells. A possibility remains that the primordial sex of the symbionts was inherited as an attribute of the organelles. If so, the recombination mediated by parasitic selfish DNA such as conjugative plasmids must take place between different organelles. This is very similar to that theory proposed by Hickey and Rose (1988) for the initial stage in the evolution of parasitic sex. The mF plasmid that promotes mitochondrial fusion in *Physarum polycephalum* provides direct support for the possibility that selfish genes are capable of manipulating their hosts and developing sexuality as pointed out by Hurst (1991). This is very similar to bacterial conjugative plasmids such as the F plasmid in *E. coli* (Fig. 15). Such similarity indicates the origin not only of themselves but also of their sex.

Recombination between mitochondrial genomes has also been found in a few other organisms, the best studied of which is the yeast *Saccharomyces cerevisiae*. This recombination can involve biased gene conversion, or a similar process at the ω locus such that in a cross between cells which are ω^+ and ω^-, nearly all of the mitochondria end up being ω^+. Unlike the case of *P. polycephalum,* ω^+ is not a plasmid but an intron. It is not known if ω^+ can induce mitochondrial fusion, but its infectious feature is well known (Dujon, 1989; Lambowitz and Belfort, 1993). Moreover, in *S. cerevisiae,* recombination of mtDNA is known to take place between mitochondria in the absence of ω^+. In contrast, the mF plasmid promotes mitochondrial fusion, but the recombination of mtDNA without the mF plasmid is not known in *P. polycephalum.* There are very serious differences in the manner of mitochondrial fusion and recombination in these two characteristic organisms. Such differences likely reflect the multistage scenario for the evolution of sex in which parasitic gene transfer is the dominant factor. The

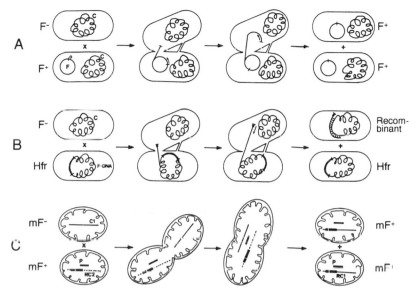

FIG. 15 Bacterial sex and mitochondrial sex: conjugation and gene transfer mediated by the F plasmid of *Escherichia coli* (A, B), and mitochondrial fusion and recombination mediated by the mF plasmid of *Physarum polycephalum* (C). Heavy lines represent plasmid genomes; light lines, the host chromosomes. The origin of conjugative DNA transfer is indicated by a vertical short line. Donor and recipient bacterial cells are shown in the process of conjugation, joined by the pilus. (A) F plasmid conjugation. Only one copy of the F plasmid is transferred. (B) Hfr-mediated host gene transfer. The F plasmid of the donor is integrated into the host chromosome. During conjugation, a copy of a portion of the donor chromosome is transferred to the recipient, where it may replace parts of the recipient chromosome by homologous recombination. (C) mF plasmid-mediated mitochondrial fusion. The thin and dotted lines represent the mtDNA in individual mitochondria of two parents. Following mitochondrial fusion and mt-nuclear fusion, crossing-over occurs between homologous regions on the mtDNA and the mF plasmid. Newly formed recombinant mtDNA and unchanged mF plasmid are then transmitted to daughter mitochondria preferentially.

mitochondria are very much infectious; numerous mobile genetic elements are known in the mitochondria of some organisms, as described in Section III. Such an infectious situation in mitochondria also suggests the existence of other types of primordial sex mediated by some mobile genetic element and provides a useful tool for analyzing the parasitic origin of sex.

VII. Concluding Remarks

In this chapter, mitochondrial fusion, recombination, and mobile genetic elements, which are essential for mitochondrial sexuality, are outlined. These phenomena seem to be ubiquitous. Nevertheless, their roles in mitochondrial biogenesis and sexuality remain somewhat obscure, as do the molecular mechanisms that control mitochondrial fusion in most organisms, with the notable exception of *Saccharomyces cerevisiae* and *Physarum polycephalum*. The existence of the mF plasmid that promotes mitochondrial fusion and recombination in *P. polycephalum* should provide direct evidence for primordial sex in mitochondria and mitochondria as biological entities in the initial stages of the evolution of parasitic sex. The processes of mt-nuclear fusion, recombination, and segregation, which are demonstrated in *P. polycephalum,* provide evidence of a kind of mitochondrial "meiotic" cycle. Cell-nuclear meiosis is a process closely related to the sexuality of eukaryotes; primitive sexuality and recombination of chromosomes are even observed in bacteria. The possibility that mitochondria possess a primitive mode of sexuality similar to that of bacteria does not conflict with the postulate that mitochondria originated as the result of bacterial endosymbiosis. Making the best use of these favorable features of *P. polycephalum* and advancing studies on the mitochondrial fusion and recombination in some other organisms, it should be possible to establish the phenomenon which we have termed "mitochondrial sex" in the field of mitochondrial biogenesis.

Acknowledgments

We wish to thank Thomas O'Keefe of the University of Wisconsin-Madison for critical reading of the manuscript, and the Ministry of Education, Science and Culture of Japan for supporting our work for this chapter with Grant No. 03640579 (S.K.), 05454020 (S.K.) and 06101002 (T.K.). This work was also supported by a grant to T.K. for Original and Creative Research Project on Biotechnology from the Research Council of the Ministry of Agriculture, Forestry and Fisheries of Japan.

References

Akins, R. A., Kelley, R. L., and Lambowitz, A. M. (1986). Mitochondrial plasmids of *Neurospora:* Integration into mitochondrial DNA and evidence for reverse transcription in mitochondria. *Cell (Cambridge, Mass.)* **47,** 505–516.

Alexopoulos, C. J. (1982). Morphology, taxonomy, and phylogeny. *In* "Cell Biology of *Physarum* and *Didymium*" (H. C. Aldrich and J. W. Daniel, eds.), vol. 1, pp. 3–23. Academic Press, New York.

Altmann, R. (1890). "Die Elementarorganismen und ihre Beziehung zu den Zelle." Veit, Leipzig.

Aufderheide, K. J., and Johnson, R. G. (1976). Cytoplasmic inheritance in *Saccharomyces cerevisiae:* Comparison of zygotic mitochondrial inheritance patterns. *Mol. Gen. Genet.* **144,** 289–299.

Bang, B. G., and Bang, F. B. (1957). Graphic reconstruction of the third dimension from serial electron microphotographs. *Ultrastruct. Res.* **1**, 138–146.

Bereiter-Hahn, J. (1983). "Mitochondrien-Bewegung, Teilung und Fusion," Publ. Wisse. Film D 1455 16 No.7.

Bereiter-Hahn, J. (1990). Behavior of mitochondria in the living cell. *Int. Rev. Cytol.* **122**, 1–63.

Bereiter-Hahn, J., and Voth, M. (1994). Dynamics of mitochondria in living cells: Shape changes, dislocations, fusion, and fission of mitochondria. *Microsc. Res. Technol.* **27**, 198–219.

Bernad, A., Blanco, L., La'zaro, J. M., Martin, G., and Salas, M. (1989). A conserved 3'-5' exonuclease active site in prokaryotic and eukaryotic DNA polymerases. *Cell* (*Cambridge, Mass.*) **59**, 219–228.

Bertrand, H., Chan, B. S.-S., and Griffiths, A. J. F. (1985). Insertion of a foreign nucleotide into mitochondrial DNA causes sequence in *Neurospora intermedia. Cell* (*Cambridge, Mass.*) **41**, 877–884.

Bertrand, H., Griffiths, A. J. F., Court, D. A., and Cheng, C. K. (1986). An extrachromosomal plasmid is the etiological precursor of kalDNA insertion sequences in the mitochondrial chromosome of senescent *Neurospora. Cell* (*Cambridge, Mass.*) **47**, 829–837.

Biermans, W., Bakker, A., and Jacob, W. (1990). Contact sites between inner and outer mitochondrial membrane—A dynamic microcompartment for creatine kinase activity. *Biochim. Biophys. Acta* **108**, 225–228.

Birky, C. W., Jr. (1978). Transmission genetics of mitochondria and chloroplasts. *Annu. Rev. Genet.* **12**, 471–512.

Birky, C. W., Jr. (1983a). The partitioning of cytoplasmic organelles at cell division. *Int. Rev. Cytol., Suppl.* **15**, 49–89.

Birky, C. W., Jr. (1983b). Relaxed cellular controls and organelle heredity. *Science* **222**, 468–475.

Birky, C. W., Jr. (1991). Evolution and population genetics of organelle genes: Mechanisms and models. *In* "Evolution at the Molecular Level" (R. K. Selander, A. G. Clark, and T. S. Whittam, eds.), pp. 112–149. Sinauer Assoc., Sunderland, MA.

Birky, C. W., Jr., Acton, A. R., Dietrich, R., and Carver, M. (1982). Mitochondrial transmission genetics: Replication, recombination, and segregation of mitochondrial DNA and its inheritance in crosses. *In* "Mitochondrial Genes" (P. Slonimski, P. Borst, and G. Attardi, eds.), pp. 333–348. Cold Spring Harbor Lab., Cold Spring Harbor, NY.

Blackburn, E. H., and Szostak, J. W. (1984). The molecular structure of centromers and telomeres. *Annu. Rev. Biochem.* **53**, 163–194.

Bolotin, M., Coen, D., Deutsch, J., Dujon, B., Netter, P., Petrochilo, E., and Slonimski, P. P. (1971). La recombinason des mitochondries chez *Saccharomyces cerevisiae. Bull. Inst. Pasteur* (*Paris*) **69**, 215–239.

Boynton, J. E., Harris, E. H., Burkhart, B. D., Lamerson, P. M., and Gillham, N. W. (1987). Transmission of mitochondrial and chloroplast genomes in crosses of *Chlamydomonas. Proc. Natl. Acad. Sci. U.S.A.* **84**, 2391–2395.

Brdiczka, D. (1991). Contact sites between mitochondrial envelope membranes—Structure and function in energy-transfer and protein transfer. *Biochim. Biophys. Acta* **1071**, 291–312.

Burland, T. G., Solnica-Krezel, L., Bailey, J., Cunningham, D. B., and Dove, W. F. (1993). Patterns of inheritance, development and the mitotic cycle in the protist *Physarum polycephalum. Adv. Microbiol. Physiol.* **35**, 1–69.

Campbell, J. L. (1986). Eukaryotic DNA replication. *Annu. Rev. Biochem.* **55**, 733–771.

Cech, T. R. (1988). Conserved sequences and structures of group I introns: Building an active site for RNA catalysis—a review. *Gene* **73**, 259–271.

Challberg, M. D., Desiderio, S. V., and Kelly, T. J. (1980). Adenovirus DNA replication *in vitro:* Characterization of a protein covalently linked to nascent DNA strands. *Proc. Natl. Acad. Sci. U.S.A.* **77**, 5105–5109.

Chan, B. B.-S., Court, D. A., Vierula, P. J., and Bertrand, B. (1991). The *kalilo* linear senescence-inducing plasmid of *Neurospora* is an invertron and encodes DNA and RNA polymerases. *Curr. Genet.* **20**, 225–237.

Chen, L. B. (1988). Mitochondrial membrane potential in living cells. *Annu. Rev. Cell Biol.* **4**, 155–181.

Chen, L. B. (1989). Fluorescent labeling of mitochondria. *Methods Cell Biol.* **29**, 103–123.

Clark-Walker, G. D. (1992). Evolution of mitochondrial genomes in fungi. *Int. Rev. Cytol.* **141**, 89–127.

Clewell, D. B., and Gawron-Burke, C. (1986). Conjugative transposons and the dissemination of antibiotic resistance in *Streptococci. Annu. Rev. Microbiol.* **40**, 635–659.

Coleman, A. W. (1978). Visualization of chloroplast DNA with two fluorochromes. *Exp. Cell Res.* **114**, 95–100.

Collins, R. A., and Saville, B. J. (1990). Independent transfer of mitochondrial chromosomes and plasmids during unstable vegetative fusion in *Neurospora. Nature (London)* **345**, 177–179.

Collins, R. A., Stohl, L. L., Cole, M. D., and Lambowitz, A. M. (1981). Characterization of a novel plasmid DNA found in mitochondria of *N. crassa. Cell (Cambridge, Mass.)* **24**, 443–452.

Court, D. A., and Bertrand, H. (1992). Genetic organization and structural features of *maranhar,* a senescence-inducing linear mitochondrial plasmid of *Neurospora crassa. Curr. Genet.* **22**, 385–397.

Court, D. A., Griffiths, A. J. F., Kraus, S. R., Russell, P. J., and Bertrand, H. (1991). A new senescence-inducing mitochondrial linear plasmid in field-isolated *Neurospora crassa* strains from India. *Curr. Genet.* **19**, 129–137.

Cowdry, E. V. (1918). The mitochondrial constituents of protoplasm. *Contrib. Embryol. Carnegie Inst.* **8**, 39–160.

Damsky, C. H. (1976). Environmentally induced changes in mitochondria and endoplasmic reticulum of *Saccharomyces carlsbergensis* yeast. *J. Cell Biol.* **71**, 123–135.

Dasgupta, J., Chan, B. S.-S., Keith, M. A., and Bertrand, H. (1988). *Kalilo* insertion sequences from the senescent strains of *Neurospora intermedia* are flanked by long inverted repeats of mitochondrial DNA. *Genome* **30S**, 318.

Dee, J. (1966). Multiple alleles and other factors affecting plasmodium formation in the true slime mould *Physarum polycephalum* Schw. *J. Protozool.* **13**, 610–616.

Dee, J. (1982). Genetics of *Physarum polycephalum. In* "Cell Biology of *Physarum* and *Didymium*" (H. C. Aldrich and J. W. Daniel, eds.), Vol. 1, pp. 211–251. Academic Press, New York.

Dee, J. (1987). Genes and development in *Physarum. Trends Genet.* **3**, 208–213.

Diboll, A. G., and Larson, D. A. (1966). An electron microscopic study of the mature megagametophyte in *Zea mays. Am. J. Bot.* **53**, 391–402.

Dougherty, E. C. (1955). Comparative evolution and the origin of sexuality. *Syst. Zool.* **4**, 145–169.

Dujon, B. (1981). Mitochondrial genetics and functions. *In* "The Molecular Biology of the Yeast *Saccharomyces:* Life Cycle and Inheritance" (J. N. Strathern, E. W. Jones, and J. Broach, eds.), pp. 505–635. Cold Spring Harbor Lab., Cold Spring Harbor, NY.

Dujon, B. (1989). Group I introns as mobile genetic elements: Facts and mechanistic speculations—a review. *Gene* **82**, 91–114.

Dujon, B., Belfort, M., Butow, R. A., Lacq, C., Lemieux, C., Perlman, P. S., and Vogt, V. M. (1989). Mobile introns: Definition of terms and recommended nomenclature. *Gene* **82**, 115–118.

Düvell, A., Hessberg-Stutzke, H., Oeser, B., Rogmann-Backwinkel, P., and Tudzynski, P. (1988). Structural and functional analysis of mitochondrial plasmids in *Claviceps purpurea. Mol. Gen. Genet.* **214**, 128–134.

Ernster, L., and Schatz, G. (1981). Mitochondria: A historical review. *J. Cell Biol.* **91**, 227s–255s.

Esser, K., Kuck, U., Lang-Hinrichs, C., Lemke, P., Osiewacz, H. D., Stahl, U., and Tudzynski, P. (1986). "Plasmids of Eukaryotes: Fundamentals and Applications." Springer-Verlag, New York.

Faure, J.-E., Mogensen, H. L., Kranz, E., Digonnet, C., and Dumas, C. (1992). Ultrastructural characterization and three-dimensional reconstruction of isolated maize (*Zea mays* L.) egg cell protoplasts. *Protoplasma* **171**, 97–103.

Gauthier, A., Turmel, M., and Lemieux, C. (1991). A group I intron in the chloroplast large subunit rRNA gene of *Chlamydomonas eugametos* encodes a double-strand endonuclease that cleaves the homing site of this intron. *Curr. Genet.* **19**, 43–47.

Gessner-Ulrich, K., and Tudzynski, P. (1992). Transcripts and translation products of a mitochondrial plasmid of *Claviceps purpurea. Curr. Genet.* **21**, 249–254.

Ghiselin, M. T. (1988). The evolution of sex: A history of competing points of view. *In* "The Evolution of Sex" (R. E. Michod and B. R. Levin, eds.), pp. 7–23. Sinauer Assoc., Sunderland, MA.

Gillham, N. W. (1978). "Organelle Heredity." Raven Press, New York.

Girbardt, M. (1970). Morphologische korrelate von transportvorgägen in sellen und organellen. *Dtsch. Akad. Wiss. Berlin,* pp. 25–39.

Griffiths, A. J. F. (1992). Fungal senescence. *Annu. Rev. Genet.* **26**, 351–372.

Grimes, G. W., Mahler, H. R., and Perlman, P. S. (1974). Nuclear gene dosage effects on mitochondrial mass and DNA. *J. Cell Biol.* **61**, 565–574.

Gutierrez, J., Garmendia, C., and Salas, M. (1988). Characterization of the origins of replication of bacteriophage φ29 DNA. *Nucleic Acids Res.* **16**, 5895–5914.

Hamilton, W. D., Axelrod, R., and Tanese, R. (1990). Sexual reproductions as an adaptation to resist parasites (a review). *Proc. Natl. Acad. Sci. U.S.A.* **87**, 3566–3573.

Hayashi, Y., and Ueda, K. (1989). The shape of mitochondria and the number of mitochondrial nucleoids during the cell cycle of *Euglena gracilis. J. Cell Sci.* **93**, 565–570.

Heinemann, J. A. (1991). Genetics of gene transfer between species. *Trends Genet.* **7**, 181–185.

Heinemann, J. A., and Sprague, G. F., Jr. (1989). Bacterial conjugative plasmids mobilize DNA transfer between bacteria and yeast. *Nature (London)* **340**, 205–209.

Hermanns, J., and Osiewacz, H. D. (1992). The linear mitochondrial plasmid pAL2-1 of a long-lived *Podospora anserina* mutant is an invertron encoding a DNA and RNA polymerase. *Curr. Genet.* **22**, 491–500.

Hickey, D. A. (1982). Selfish DNA: A sexually-transmitted nuclear parasite. *Genetics* **101**, 519–531.

Hickey, D. A. (1984). DNA can be a selfish parasite. *Nature (London)* **311**, 417–418.

Hickey, D. A., and Rose, M. R. (1988). The role of gene transfer in the evolution of eukaryotic sex. *In* "The Evolution of Sex" (R. E. Michod and B. R. Levin, eds.), pp. 161–175. Sinauer Assoc., Sunderland, MA.

Hoffman, H. P., and Avers, C. J. (1973). Mitochondrion of yeast: Ultrastructural evidence for one giant, branched organelle per cell. *Science* **181**, 749–751.

Houchins, J. P., Ginsburg, H., Rohrbaugh, M., Dale, R. M. K., Schardl, C. L., Hodge, T. P., and Lonsdale, D. M. (1986). DNA sequence analysis of a 5.27-kb direct repeat occurring adjacent to the regions of S-episome homology in maize mitochondria. *EMBO J.* **5**, 2781–2788.

Howard, R. S., and Lively, C. M. (1994). Parasitism, mutation accumulation and the maintenance of sex. *Nature (London)* **367**, 554–557.

Hurst, L. D. (1991). Sex, slime and selfish genes. *Nature (London)* **354**, 23–24.

Ippen-Ihler, K. A., and Minkley, E. G., Jr. (1986). The conjugation system of F, the fertility factor of *Escherichia coli. Annu. Rev. Genet.* **20**, 593–624.

Jacob, F., and Wollman, E. L. (1961). "Sexuality and Genetics of Bacteria." Academic Press, New York.

Jacquier, A., and Dujon, B. (1985). An intron-encoded protein is active in a gene conversion process that spreads an intron into a mitochondrial gene. *Cell (Cambridge, Mass.)* **41**, 383–394.

James, T. W., and Jope, C. (1978). Visualization by fluorescence of chloroplast DNA in higher plants by means of the DNA-specific probe 4'6-diamidino-2-phenylindole. *J. Cell Biol.* **79**, 623–630.

Johannes, H. (1941). Beiträge zur vitalfarbung von pilzmyzelien II. Die inturbanz der färbung mit rhodaminen. *Protoplasma* **36**, 181–194.

Johnson, L. V., Walsh, M. L., and Chen, L. B. (1980). Localization of mitochondria in living cells with rhodamine 123. *Proc. Natl. Acad. Sci. U.S.A.* **77**, 990–994.

Kawano, S. (1991). The life cycle of mitochondria in the true slime mould, *Physarum polycephalum. Bot. Mag. Tokyo* **104**, 97–113.

Kawano, S., and Kuroiwa, T. (1989). Transmission pattern of mitochondrial DNA during plasmodium formation in *Physarum polycephalum. J. Gen. Microbiol.* **135**, 1559–1566.

Kawano, S., Suzuki, T., and Kuroiwa, T. (1982). Structural homogeneity of mitochondrial nucleoid of *Physarum polycephalum. Biochim. Biophys. Acta* **696**, 290–298.

Kawano, S., Nishibayashi, S., Shiraishi, N., Miyahara, M., and Kuroiwa, T. (1983). Variance of ploidy in mitochondrial nucleus during spherulation in *Physarum polycephalum. Exp. Cell Res.* **149**, 359–373.

Kawano, S., Kuroiwa, T., and Anderson, R. W. (1987a). A third multiallelic mating-type locus in *Physarum polycephalum. J. Gen. Microbiol.* **133**, 2539–2546.

Kawano, S., Anderson, R. W., Nanba, T., and Kuroiwa, T. (1987b). Polymorphism and uniparental inheritance of mitochondrial DNA in *Physarum polycephalum. J. Gen. Microbiol.* **133**, 3175–3182.

Kawano, S., Takano, H., Mori, K., and Kuroiwa, T. (1991a). A mitochondrial plasmid that promotes mitochondrial fusion in *Physarum polycephalum. Protoplasma* **160**, 167–169.

Kawano, S., Takano, H., Mori, K., and Kuroiwa, T. (1991b). The oldest laboratory strain of *Physarum polycephalum. Physarum Newsl.* **22**, 70–75.

Kawano, S., Takano, H., Mori, K., and Kuroiwa, T. (1993). A genetic system controlling mitochondrial fusion in the slime mould, *Physarum polycephalum. Genetics* **133**, 213–224.

Kempken, F., Meinhardt, F., and Esser, K. (1989). *In organello* replication and viral affinity of linear, extrachromosomal DNA of the ascomycete *Ascobolus immersus, Mol. Gen. Genet.* **218**, 523–530.

Kirouac-Brunet, J., Mansson, S., and Pallotta, D. (1981). Multiple allelism at the *matB* locus in *Physarum polycephalum. Can. J. Genet. Cytol.* **23**, 9–16.

Krieber, M., and Rose, M. R. (1986). Males, parthenogenesis and the maintenance of anisogamous sex. *J. Theor. Biol.* **122**, 421–440.

Kuiper, M. T. R., and Lambowitz, A. M. (1988). A novel reverse transcriptase activity associated with mitochondrial plasmids of *Neurospora. Cell* (*Cambridge, Mass.*) **55**, 693–704.

Kuroiwa, H., and Kuroiwa, T. (1992). Giant mitochondria in the mature egg cell of *Pelargonium zonale. Protoplasma* **168**, 184–188.

Kuroiwa, T. (1982). Mitochondrial nuclei. *Int. Rev. Cytol.* **75**, 1–59.

Kuroiwa, T. (1985). Mechanisms of maternal inheritance of chloroplast DNA: An active digestion hypothesis. *Microbiol. Sci.* **2**, 267–270.

Kuroiwa, T., and Hori, T. (1986). Preferential digestion of male chloroplast nuclei and mitochondrial nuclei during gametogenesis of *Bryopsis maxima* Okamura. *Protoplasma* **133**, 85–87.

Kuroiwa, T., Kawano, S., Nishibayashi, S., and Sato, C. (1982). Epifluorescent microscopic evidence for maternal inheritance of chloroplast DNA. *Nature* (*London*) **298**, 481–483.

Kuroiwa, T., Kuroiwa, H., and Sugai, M. (1988). Behavior of chloroplasts and chloroplast nuclei during spermatogenesis in the fern, *Pteris vittata* L. *Protoplasma* **146**, 89–100.

Kuroiwa, T., Nagashima, H., and Fukuda, I. (1989). Chloroplast division without DNA synthesis during the life cycle of the unicellular alga *Cyanidium caldarium* M-8 as revealed by quantitative fluorescence microscopy. *Protoplasma* **149**, 120–129.

Kuroiwa, T., Kuroiwa, H., Mita, T., and Fujie, M. (1990). Fluorescence microscopic study of the formation of giant mitochondrial nuclei in the young ovules of *Pelargonium zonale. Protoplasma* **158**, 191–194.

Kuroiwa, T., Fujie, M., Mita, T., and Kuroiwa, H. (1991a). Application of embedding of samples in Technovit 7100 resin to observations of small amount of DNA in the cellular organelles associated with cytoplasmic inheritance. *Fluoresc. Tech.* **3,** 23–25.

Kuroiwa, T., Kawano, S., Watanabe, M., and Hori, T. (1991b). Preferntial digestion of chloroplast DNA in male gametangia during the late stage of gametogenesis in the anisogamous alga *Bryopsis maxima. Protoplasma* **163,** 102–113.

Kuroiwa, T., Ohta, T., Kuroiwa, H., and Kawano, S. (1994). Molecular and cellular mechanisms of mitochondrial nuclear division and mitochondriokinesis. *Microsc. Res. Technol.* **27,** 220–232.

Lambowitz, A. M. (1989). Infectious introns. *Cell (Cambridge, Mass.)* **56,** 323–326.

Lambowitz, A. M., and Belfort, M. (1993). Introns as mobile genetic elements. *Annu. Rev. Biochem.* **62,** 587–622.

Lederberg, J., and Tatum, E. L. (1946). Gene recombination in *Escherichia coli. Nature (London)* **158,** 558.

Levin, B. R. (1988). The evolution of sex in bacteria. *In* "The Evolution of Sex" (R. E. Michod and B. R. Levin, eds.), pp. 194–211. Sinauer Assoc., Sunderland, MA.

Lewis, M. R., and Lewis, W. H. (1914). Mitochondria (and other cytoplasmic structures) in tissue cultures. *Am. J. Anat.* **17,** 339–401.

Li, Q., and Nargang, F. E. (1993). Two *Neurospora* mitochondrial plasmids encode DNA polymerase containing motifs characteristic of family B DNA polymerases but lack the sequence Asp-Thr-Asp. *Proc. Natl. Acad. Sci. U.S.A.* **90,** 4299–4303.

Linnane, A. W., Saunders, G. W., Gingold, E. B., and Lukins, H. B. (1968). The biogenesis of mitochondria. V. Cytoplasmic inheritance of erythromycin resistance in *Saccharomyces cerevisiae. Proc. Natl. Acad. Sci. U.S.A.* **59,** 903–910.

Lonsdale, D. M. (1989). The plant mitochondrial genome. *Biochem. Plants* **15,** 229–295.

Lonsdale, D. M., Brears, T., Hodge, T. P., Melville, S. E., and Rottmann, W. H. (1988). The plant mitochondrial genome: Homologous recombination as mechanism for generating heterogeneity. *Philos. Trans. R. Soc. London, B Ser.* **319,** 149–163.

Macreadie, I. G., Scott, R. M., Zinn, A. R., and Butow, R. A. (1985). Transposition of an intron in yeast mitochondria requires a protein encoded by that intron. *Cell (Cambridge, Mass.)* **41,** 395–402.

Mahendran, R., Spottswood, M. R., and Miller, D. L. (1991). RNA editing by cytidine insertion in mitochondria of *Physarum polycephalum. Nature (London)* **249,** 434–438.

Margulis, L., and Sagan, D. (1986). "Origin of Sex: Three Billion Years of Recombination." Yale Univ. Press, New Haven, CT.

Marshall, P., and Lemieux, C. (1991). Cleavage pattern of the homing endonuclease encoded by the fifth intron in the chloroplast large subunit rRNA-encoding gene of *Chlamydomonas eugametos. Gene* **104,** 241–245.

May, G., and Taylor, J. W. (1989). Independent transfer of mitochondrial plasmids in *Neurospora crassa. Nature (London)* **339,** 320–322.

Maynard Smith, J. (1978). "The Evolution of Sex." Cambridge Univ. Press, Cambridge, UK.

Meinhardt, F., Kempken, F., Kämper, J., and Esser, K. (1990). Linear plasmids among eukaryotes: Fundamentals and application. *Curr. Genet.* **17,** 89–95.

Meland, S., Johansen, S., Johansen, T., Haugli, K., and Haugli, F. (1991). Rapid disappearance of one parental mitochondrial genotype after isogamous mating in the myxomycete *Physarum polycephalum. Curr. Genet.* **19,** 55–60.

Miyakawa, I., Aoi, H., Sando, N., and Kuroiwa, T. (1984). Fluorescence microscopic studies of mitochondrial nucleoids during meiosis and sporulation in the yeast, *Saccharomyces cerevisiae. J. Cell Sci.* **66,** 21–38.

Miyakawa, I., Higo, K., Osaki, F., and Sando, N. (1994). Double-staining of mitochondria and mitochondrial nucleoids in the living yeast cells during the life cycle. *J. Gen. Appl. Microbiol.* **40,** 1–14.

Morin, G. B., and Cech, T. R. (1986). The telomeres of the linear mitochondrial DNA of *Tetrahymena thermophila* consist of 53 bp tandem repeats. *Cell (Cambridge, Mass.)* **46,** 873–883.

Morin, G. B., and Cech, T. R. (1988). Mitochondrial telomeres: Surprising diversity of repeated telomeric DNA sequences among six species of *Tetrahymena. Cell (Cambridge, Mass.)* **52,** 367–374.

Myers, C. J., Griffiths, A. J. F., and Bertrand, H. (1989). Linear *kalilo* DNA is a *Neurospora* mitochondrial plasmid that integrates into the mitochondrial DNA. *Mol. Gen. Genet.* **220,** 113–120.

Nargang, F. E., Bell, J. B., Stohl, L. L., and Lambowitz, A. M. (1984). The DNA sequence and genetic organization of a *Neurospora* mitochondrial plasmid suggest a relationship to introns and mobile elements. *Cell (Cambridge, Mass.)* **38,** 441–453.

Natvig, D. O., May, G., and Taylor, J. W. (1984). Distribution and evolutionary significance of mitochondrial plasmids in *Neurospora* species. *J. Bacteriol.* **159,** 288–293.

Nicolay, K., Rojo, M., Wallimann, T., Demel, R., and Hovius, R. (1990). The role of contact sites between inner and outer mitochondrial membrane in energy transfer. *Biochim. Biophys. Acta* **1018,** 229–233.

Nishibayashi, S., Kawano, S., and Kuroiwa, T. (1987). Light and electron microscopic observations of mitochondrial fusion in plasmodia induced sporulation in *Physarum polycephalum. Cytologia* **52,** 599–614.

Oeser, B., and Tudzynski, P. (1989). The linear mitochondrial plasmid pClKl of the phytopathyogenic fungus *Claviceps purpurea* may code for a DNA polymerase and an RNA polymerase. *Mol. Gen. Genet.* **217,** 132–140.

Osafune, T. (1973). Three-dimensional structures of giant mitochondria, dictyosomes and "concentric lamellar bodies" formed during the cell cycle of *Euglena gracilis* (Z) in synchronous culture. *J. Electron Microsc.* **22,** 51–61.

Pande, S., Lemire, E. G., and Nargang, F. E. (1989). The mitochondrial plasmid from *Neurospora intermedia* strains Labelle 1b contains along open reading frame with blocks of amino acids characteristic of reverse transcriptase and related proteins. *Nucleic Acids Res.* **17,** 2023–2042.

Pon, L., and Schatz, G. (1991). Biogenesis of yeast mitochondria. *In* "The Molecular and Cellular Biology of the Yeast *Saccharomyces:* Genome Dynamics, Protein Synthesis, and Energetics" (E. W. Jones, J. R. Pringle, and J. R. Broach, eds.), pp. 333–406. Cold Spring Harbor Lab., Cold Spring Harbor, NY.

Poulter, R. T. M., and Dee, J. (1968). Segregation of factors controlling fusion between plasmodia of the true slime mould *Physarum polycephalum. Genet. Res.* **12,** 71–79.

Raub, T. J., and Aldrich, H. C. (1982). Sporangia, spherules, and microcysts. *In* "Cell Biology of *Physarum* and *Didymium*" (H. C. Aldrich and J. W. Daniel, eds.), Vol. 2, pp. 561–564. Academic Press, New York.

Remacle, C., Bovie, C., Michel-Wolwertz, M.-R., Loppes, R., and Matagne, R. F. (1990). Mitochondrial genome transmission in *Chlamydomonas* diploids obtained by sexual crosses and artificial fusions: Role of the mating type and of a 1 kb intron. *Mol. Gen. Genet.* **223,** 180–184.

Robison, M. M., Royer, J. C., and Horgen, P. A. (1991). Homology between mitochondrial DNA of *Agaricus bisporus* and an internal portion of a linear mitochondrial plasmid of *Agaricus bitorquis. Curr. Genet.* **19,** 495–502.

Rose, M. R. (1983). The contagion mechanism for the origin of sex. *J. Theor. Biol.* **101,** 137–146.

Sager, R. (1972). "Cytoplasmic Genes and Organelles." Academic Press, New York.

Sager, R., and Gabowy, C. (1985). Sex in *Chlamydomonas:* Sex and the single chloroplast. *MBL Lect. Biol.* **7,** 113–121.

Samac, D. A., and Leong, S. A. (1989). Mitochondrial plasmids of filamentous fungi: Characteristics and use in transformation vectors. *Mol. Plant-Microbe Interact.* **2,** 155–1559.

Sando, N., Miyakawa, I., Nishibayashi, S., and Kuroiwa, T. (1981). Arrangement of mitochondrial nucleoids during life cycle of *Saccharomyces cerevisiae*. *J. Gen. Appl. Microbiol.* **27**, 511–516.

Sasaki, N., Suzuki, T., Ohta, T., Kawano, S., and Kuroiwa, T. (1994). Behavior of mitochondria and their nuclei during amoebae cell proliferation in *Physarum polycephalum*. *Protoplasma* **182**, 115–125.

Satoh, M., and Kuroiwa, T. (1993). Organization of multiple nucleoids and DNA molecules in mitochondria of human cells. *Exp. Cell Res.* **196**, 137–140.

Scazzocchio, C. (1989). Group I introns: Do they only go home? *Trends Genet.* **5**, 168–172.

Schardl, C. L., Lonsdale, S. M., Pring, D. R., and Rose, K. R. (1984). Linearization of maize mitochondrial chromosomes by recombination with linear episomes. *Nature (London)* **310**, 292–296.

Schardl, C. L., Pring, D. R., and Lonsdale, D. M. (1985). Mitochondrial DNA rearrangements associated with fertile revertants of S-type male-sterile maize. *Cell (Cambridge, Mass.)* **43**, 361–368.

Schulte, U., and Lambowitz, A. M. (1991). The LaBelle mitochondrial plasmid of *Neurospora intermedia* encodes a novel DNA polymerase that may be derived from reverse transcriptase. *Mol. Cell Biol.* **11**, 1696–1706.

Schuster, W., and Brennicke, A. (1994). The plant mitochondrial genome: Physical structure, information content, RNA editing, and gene migration to the nucleus. *Annu. Rev. Physiol. Plant Mol. Biol.* **45**, 61–78.

Shinnick, T. M., Pallotta, D. J., Jones-Brown, Y. V. R., Youngman, P. J., and Holt, C. E. (1978). A gene, *imz*, affecting the pH sensitivity of zygote formation in *Physarum polycephalum*. *Curr. Microbiol.* **1**, 163–166.

Smith, G. R. (1991). Conjugational recombination in *E. coli*: Myths and mechanisms (review). *Cell (Cambridge, Mass.)* **64**, 19–27.

Sodmergen, Kawano, S., Tano, S., and Kuroiwa, T. (1989). Preferential digestion of chloroplast nuclei (nucleoids) during senescence of the coleoptile of *Oryza sativa*. *Protoplasma* **152**, 65–68.

Sodmergen, Kawano, S., Tano, S., and Kuroiwa, T. (1991). Degradation of chloroplast DNA in second leaves of rice (*Oryza sativa*) before leaf yellowing. *Protoplasma* **160**, 89–98.

Sodmergen, Suzuki, T., Kawano, S., Nakamura, S., Tano, S., and Kuroiwa, T. (1992). Behavior of organelle nuclei (nucleoids) in generative and vegetative cells during maturation of pollen in *Lilium longiflorum* and *Pelargonium zonale*. *Protoplasma* **168**, 73–82.

Stevens, B. J. (1977). Variation in number and volume of the mitochondria in yeast according to growth conditions. A study based on serial sectioning and computer graphics reconstruction. *Biol. Cell.* **28**, 37–56.

Stevens, B. J. (1981). Mitochondrial structure. *In* "The Molecular Biology of the Yeast *Saccharomyces*: Life Cycle and Inheritance" (J. N. Strathern, E. W. Jones, and J. R. Broach, eds.), pp. 471–504. Cold Spring Harbor Lab., Cold Spring Harbor, NY.

Stohl, L. L., Collins, R. A., Cole, M. D., and Lambowitz, A. M. (1982). Characterization of two new plasmid DNAs found in mitochondria of wild type *Neurospora intermedia* strains. *Nucleic Acids Res.* **10**, 1439–1458.

Strausberg, R. L., and Perlman, P. J. (1978). The effect of zygotic budposition on the transmission of mitochondrial genes in *Saccharomyces cerevisiae*. *Mol. Gen. Genet.* **163**, 131–144.

Sun, G.-H., Ueda, T. Q. P., and Kuroiwa, T. (1988). Destruction of organelle nuclei during spermatogenesis in *Chara corallina* examined by staining with DAPI and anti-DNA antibody. *Protoplasma* **144**, 185–188.

Suzuki, T., Kawano, S., and Kuroiwa, T. (1982). Structure of three-dimensionally rod-shaped mitochondrial nucleoids isolated from the slime mould *Physarum polycephalum*. *J. Cell Sci.* **58**, 241–261.

Takano, H., Kawano, S., Suyama, Y., and Kuroiwa, T. (1990). Restriction map of the mitochondrial DNA of the true slime mould, *Physarum polycephalum:* Linear and long tandem duplication. *Curr. Genet.* **18,** 125–131.

Takano, H., Kawano, S., and Kuroiwa, T. (1991). Telomeric structures in a linear mitochondrial plasmid from *Physarum polycephalum. Curr. Genet.* **20,** 315–317.

Takano, H., Kawano, S., and Kuroiwa, T. (1992). Constitutive homologous recombination between mitochondrial DNA and a linear mitochondrial plasmid in *Physarum polycephalum. Curr. Genet.* **22,** 221–227.

Takano, H., Kawano, S., and Kuroiwa, T. (1994a). Complex terminal structure of a linear mitochondrial plasmid form *Physarum polycephalum:* Three terminal inverted repeats and an ORF encoding DNA polymerase. *Curr. Genet.* **25,** 252–257.

Takano, H., Kawano, S., and Kuroiwa, T. (1994b). Genetic organization of a linear mitochondrial plasmid (mF) that promotes mitochondrial fusion in *Physarum polycephalum. Curr. Genet.* **26,** 506–511.

Tanaka, K., Kanbe, T., and Kuroiwa, T. (1985). Three-dimensional behaviour of mitochondria during cell division and germ tube formation in the dimorphic yeast *Candida albicans. J. Cell Sci.* **73,** 207–220.

Thomas, D. Y., and Wilkie, D. (1968). Recombination of mitochondrial drug resistance factors in *Saccharomyces cerevisiae. Biochem. Biophys. Res. Commun.* **30,** 368–372.

Tourte, Y. (1975). Etude infrastructurale de l'oogénèse chez une Pteridophyte. II. Evolution des mitochondries et des plastes. *J. Microsc. Biol. Cell.* **23,** 301–316.

Tremblay, C., and Rose, M. R. (1985). Population dynamics of gene transfer. *Theor. Popul. Biol.* **28,** 359–381.

Tudzynski, P., and Esser, K. (1986). Extrachromosomal genetics of *Claviceps purpurea.* II. Plasmid in various wild strains and integrated plasmid sequences in mitochondrial genomic DNA. *Curr. Genet.* **10,** 463–467.

Weissinger, A. K., Timothy, D. H., Levings, C. S., III, Hu, W. W. L., and Goodman, M. M. (1982). Unique plasmid-like mitochondrial DNAs from indigenous maize races of Latin America. *Proc. Natl. Acad. Sci. U.S.A.* **79,** 1–5.

Wheals, A. E. (1970). A homothallic strain of the Myxomycete *Physarum polycephalum. Genetics* **66,** 623–693.

Wilkie, D. (1983). Genetic and functional aspects of yeast mitochondria. *In* "Yeast Genetics: Fundamental and Applied Aspects" (J. F. T. Spencer, D. M. Spencer, and A. R. W. Smith, eds.), pp. 255–267. Springer-Verlag, New York.

Williamson, D. H., and Fennell, D. J. (1975). The use of fluorescent DNA-binding agent for detecting and separating yeast mitochondrial DNA. *Methods Cell Biol.* **12,** 335–351.

Williamson, D. H., and Fennell, D. J. (1979). Visualization of yeast mitochondrial DNA with the fluorescent stain "DAPI." *In* "Methods in Enzymology" (S. Fleischer and L. Packer, eds.), Vol. 56, p. 728–733. Academic Press, New York.

Yang, S., and Griffiths, A. J. F. (1993). Plasmid diversity in senescent and nonsenescent strains of *Neurospora. Mol. Gen. Genet.* **237,** 177–186.

Yotsuyanagi, Y. (1962a). Études sur le chondriome de la levure. I. Variation de l'ultrastructure du chondriome au cours du cycle de la croissance aérobie. *J. Ultrastruct. Res.* **7,** 121–140.

Yotsuyanaga, Y. (1962b). Études sur le chondriome de la levure. II. Chondriomes des mutants à déficience respiratoire. *J. Ultrastruct. Res.* **7,** 141–158.

Youngman, P. J., Pallotta, D. J., Hosler, B., Struhl, G., and Holt, C. E. (1979). A new mating-compatibility locus in *Physarum polycephalum. Genetics* **91,** 683–693.

Zakian, V. A. (1989). Structure and function of telomeres. *Annu. Rev. Genet.* **23,** 576–604.

Zinder, N. (1985). The origin of sex: An argument. *In* "The Origin and Evolution of Sex" (H. O. Halvorson and A. Monroy, eds.), pp. 7–12. Alan R. Liss, New York.

Zinn, A. R., and Butow, R. A. (1985). Nonreciprocal exchange between alleles of the yeast mitochondrial 21S rRNA gene: Kinetics and the involvement of a double-strand break. *Cell (Cambridge, Mass.)* **40,** 887–895.

Analyzing Renal Glomeruli with the New Stereology

John F. Bertram
Department of Anatomy and Cell Biology, University of Melbourne, Parkville,
Victoria 3052, Australia

The highly specialized architecture of the renal glomerulus is altered in a variety of disease states. Morphometric methods, including stereological methods, have been widely used to analyze these changes in both animal and human glomeruli. However, many of the methods available until recently were biased and provided incomplete information. The past few years have witnessed the development of a new generation of unbiased stereological methods. Another advantage of these new methods and strategies is that they are less influenced by technical artifacts than the traditional methods. This chapter describes how these new stereological methods can be used to quantify glomerular morphology. Parameters considered include glomerular number and volume; glomerular cell number and size; and the length, surface area, and number of glomerular capillaries. Methods for obtaining data for average glomeruli as well as individual glomeruli are described. Technical details are included wherever possible.
KEY WORDS: Glomerulus, Kidney, Morphometry, Stereology, Renal.

I. Introduction

Renal glomeruli have a most remarkable structure. The afferent arteriole enters at the vascular pole and branches to form a series of anastomosing capillaries, lined by fenestrated endothelial cells. The endothelial cells rest on a basement membrane, the other side of which is covered by glomerular epithelial cells (GECs, podocytes). GECs have an extraordinary architecture, with primary cytoplasmic processes extending from the cell body, and secondary or foot processes extending from the primary processes to interdigitate with the foot processes of adjacent GECs. Filtration slits, guarded by specialized diaphragms, are situated between the foot processes.

Mesangial cells and extracellular matrix (ECM) form the core of the glomerular tuft. The anastomosing capillaries reunite to form the efferent arteriole which exits the glomerulus at the vascular pole. The urinary space (Bowman's space) is located between the GECs and the parietal epithelial cells which form Bowman's capsule. The proximal tubule drains the urinary space at the urinary pole (Tisher and Madsen, 1991; Kriz and Kaissling, 1992). (*Note:* Although it is recognized that the term "renal corpuscle" is more precise anatomically than the term "glomerulus" in referring to that portion of the nephron composed of the glomerular tuft and Bowman's space, the term "glomerulus" will be used throughout this review because of common usage. The term "glomerular tuft" will be used when referring specifically to the tuft alone.)

This remarkable architecture subserves the remarkable functions of the glomerulus, but unfortunately this architecture is altered in various disease states. In many instances the pathology involves alterations in the number, size, and appearance of resident glomerular cells. In other instances, cell infiltration occurs. Glomerular size changes in numerous disease states. Alterations in the size and shape of glomerular capillaries can be observed. Accumulation of ECM, and alterations in the thickness of the glomerular basement membrane (GBM) are commonly observed. Quantitative morphological (morphometric) procedures, including stereology, have been commonly used to analyze kidney structure (Oliver, 1968; Dunnill and Halley, 1973; Knepper *et al.,* 1977; Pedersen *et al.,* 1980; Pfaller, 1982; Madsen and Tisher, 1986) as well as glomerular structure (Elias and Hennig, 1967; Shea and Morrison, 1975; Olivetti *et al.,* 1977, 1980; Larsson and Maunsbach, 1980; Steffes *et al.,* 1983; Anderson *et al.,* 1989; Denton *et al.,* 1992) in health and disease. Morphometry is defined in this chapter as the measurement of morphological features. Stereology is a subset of morphometry that concerns the quantitative analysis of three-dimensional structures. Typically in stereology, measurements made on sections are used to estimate data for three-dimensional structures. As will be described in detail later, stereology is critically concerned with sampling uniformly in three-dimensional space. Once objects are sampled uniformly, then they can be measured.

Although the application of quantitative morphological methods has often proved successful in the analysis of glomerular structure, many of the methods available until recently were slow and tedious and failed to provide full information. For example, glomerular cell "number" was often indexed to an average glomerular cross-section, and thus provided no information on the total number of cells in the glomerulus. Moreover, many previous stereological studies of renal glomeruli utilized methods that required detailed knowledge of the geometry of the particle (glomerulus, nucleus) of interest. In most instances, assumptions had to be made about this geometry; for example, that glomeruli and cell nuclei were spherical. To the extent that these assumptions deviated from the truth, the final estimates were biased.

The past 10 years have witnessed the development of a new generation of *unbiased* stereological methods (unbiased is defined as being without systematic deviation from the true value). Unlike the earlier stereological methods, these new methods do not require knowledge of or assumptions about particle geometry. The new methods can provide estimates with a high degree of statistical precision (low variance) and have proven to be remarkably cost-efficient. For example, using these techniques Braendgaard *et al.* (1990) estimated the total number of neurons in the human neocortex (13.7×10^9 in the right hemisphere) in less than 7 hr (following sampling and preparation of sections). We used these new methods to estimate the number of neurons in the lateral motor column of the developing chick (Nurcombe *et al.*, 1991). On average, following sectioning and staining, fewer than 30 min were needed to estimate the total number of neurons (approximately 19,000 at embryonic day 6). A further advantage of the new stereological methods is that the influence of technical artifacts (such as the shrinkage and swelling of tissue that occur during processing for microscopy, or section compression) that plagued many traditional stereological approaches have been minimized, and in some instances totally negated. Thus, these new methods can be claimed to have revolutionized quantitative morphological assessment of glomeruli.

This chapter describes how the new stereological methods can be used to quantify glomerular morphology. Technical details are included wherever possible. Methods for estimating the following parameters are described:

• The total number of glomeruli in a kidney
• Mean glomerular tuft (and mean renal corpuscle) volume
• The total number of cells in an average glomerulus
• The total number of each cell type (GECs, endothelial cells, mesangial cells, parietal epithelial cells) in an average glomerulus, and in the kidney
• The total length, surface area, and number of capillaries in an average glomerulus
• The mean length and surface area of glomerular capillaries
• The volume of ECM, capillary lumina, and urinary space in an average glomerulus
• The mean volumes of individual glomerular cell types
• The mean surface area of GECs
• The mean thickness of the GBM.

Many of these parameters provide average values for the entire population of glomeruli, so-called *average glomerular* values. However, physiological, biochemical, and molecular biology techniques are now available for obtaining data from single glomeruli. Moreover, many glomerular lesions are focal, involving only a proportion of glomeruli. Therefore, stereological methods for estimating many of the parameters in individual glomeruli will also be described. In this way, correlations between morphological features, or between structural and functional data in single glomeruli, may be identified. These approaches for analyzing single glomeruli may be particularly valuable when analyzing glomeruli in renal biopsies where typically only a small number are available.

II. Reasons for Quantifying Glomerular Morphology

Before one begins a quantitative study of glomerular morphology, one should ask the question: "What is to be gained from quantifying glomerular structure?" The cynical answer one often hears is that the era of qualitative morphology is over, and that if you want your grant application to be successful you had better make some measurements! While there may be some truth in this statement, there are at least five other potential advantages to be gained from quantifying glomerular structure. Firstly, slight or subtle structural differences between glomeruli may be difficult or impossible to perceive through qualitative examination of sections, whereas a quantitative examination may reveal such differences. A second potential advantage of quantitative descriptions of glomerular structure over qualitative descriptions concerns reproducibility. An objective, quantitative set of tools for assessing glomerular architecture should provide a more reproducible set of data than can be obtained through qualitative microscopy, allowing more meaningful comparison of data collected by the same individual on different occasions, and of data collected by different individuals. A third advantage to be gained from a quantitative morphological study concerns integration of morphological data with data obtained using other techniques. A variety of methods are available for obtaining physiological, biochemical, and molecular data from individual glomeruli, from purified populations of isolated glomeruli, and from whole kidneys. For example, the filtration rate in single glomeruli can be obtained using micropuncture techniques; biochemical techniques can be used to quantify amounts of specific proteins in average glomeruli, and competitive polymerase chain reactions (PCR) can be used to quantify amounts of specific mRNAs in glomeruli (Peten *et al.,* 1993). Bolender *et al.* (1993) recently described methods for linking data from immunocytochemistry and *in situ* hybridization histochemistry with quantitative morphological data. Software to link stereological data with immunogold and *in situ* hybridization data is now available and provides quantitative immunogold and *in situ* hybridization data for three-dimensional organelles, cells, tissues, and organs (Bolender, 1993). The integration of functional and molecular data with accurate and precise morphological measurements provides a powerful approach for studying glomerular cell and molecular biology.

A fourth advantage of a quantitative morphological assessment of glomeruli concerns the wealth of data that can be obtained. This chapter describes methods for estimating values for more than 20 glomerular parameters. A final advantage to be gained from a quantitative morphological assessment of glomeruli is that the full power of statistics can be used to analyze data, and thereby test hypotheses.

III. Ovorviow of Storcology

Stereology was defined by Weibel (1979) as "a body of mathematical methods relating 3-dimensional parameters defining the structure to 2-dimensional measurements obtainable on sections of the structure." As will be described below, stereology has undergone a period of unprecedented development since Weibel's definition and today is concerned as much with the practicalities of obtaining tissue for analysis, that is, sampling, as it is with the specific measurements. A stereological investigation involves much more than making measurements. Weibel's definition (1979) must also be modified because some of the latest stereological techniques involve measurements on projections from thick sections (Gokhale, 1990, 1992). An alternative definition of stereology therefore might be "the discipline concerned with the quantitative analysis of 3-dimensional structures." For reviews describing the new stereological tools see Gundersen *et al.* (1988a,b), Cruz-Orive and Weibel (1990), Bertram and Nurcombe (1992), Mayhew (1992), Bolender *et al.* (1993), Lucocq (1993), and Oorschot (1994). The earlier texts by Weibel (1979, 1980), Aherne and Dunnill (1982), Elias and Hyde (1983), and Russ (1986) also contain a wealth of information on stereological techniques and foundations and should be consulted for additional information.

It is necessary at this point to define some terms:

- Particle—A three-dimensional object, such as a glomerulus, cell, or nucleus.
- Arbitrary particle—A three-dimensional object of any size, shape, or orientation.
- Profile—The two-dimensional representation of a particle once it has been sectioned. Thus, glomerular profiles, not glomeruli, are seen in

histological sections. Similarly, nuclear profiles, not nuclei, are seen in sections.

- Estimate—Stereological methods provide estimates of true values. Rarely in stereology does one measure an entire population of features. Thus, if one wants to know the total number of glomeruli in a kidney, it is not necessary to count all of the glomeruli. Rather, an appropriate (described in detail later) sample of the kidney is analyzed, and then the total number of glomeruli in the kidney is estimated. Similarly, if we want to know mean glomerular volume (volume of an average glomerulus), it is not necessary to measure all glomeruli. Instead, measurements can be made on an appropriately collected sample of tissue and from these measurements we can estimate the mean value.
- Average glomerulus—In many instances in stereology we are not interested in, for example, the number of cells in a specific glomerulus, but rather the average number of cells in all glomeruli. The number of cells in an average glomerulus is thus determined.

Traditionally, four major stereological parameters were considered—volume density (V_V), surface density (S_V), length density (L_V), and numerical density (N_V). Each of these parameters related the measurement of the so-called feature of interest (i) to the volume of the reference (ref) compartment. These stereological densities will be discussed in turn.

A. Volume Density

Volume density (also termed volume fraction) relates the volume of the structure of interest to the volume of the reference compartment. Thus, it is a volume ratio, with units of, for example mm^3/mm^3 or cm^3/cm^3. Examples relevant to the glomerulus include:

- The volume density of glomeruli in the kidney, that is, the volume of glomeruli per unit volume of kidney $(V_{Vglom,kid}$; with this notation, the first "V" refers to the volume of the glomerulus, and the second "V" refers to the volume of the kidney),
- The volume density of glomeruli in cortex, that is the volume of glomeruli per unit volume of cortex $(V_{Vglom,cort})$,
- The volume density of ECM in glomeruli, that is, the volume of ECM per unit volume of glomerulus $(V_{Vecm,glom})$, and
- The volume density of nuclei in mesangial cells, that is, the volume of nucleus per unit volume of mesangial cell $(V_{Vnuc,mes})$, which can be used to calculate the nuclear to cytoplasmic ratio in mesangial cells.

Delesse (1847) showed that the areal proportion or areal density (A_A) of a feature of interest in the reference compartment in two-dimensional space is directly proportional to volume density. Thus:

$$A_A = V_V \tag{1}$$

or in expanded form:

$$A_{Ai,\,ref} = V_{Vi,ref} \tag{2}$$

A variety of methods are available to estimate areal densities and thereby volume density. These are described in detail in the major stereological texts (see Weibel, 1979, for example) and include paper weighing, lineal integration, random point counting, and point counting with a grid (where points are defined by the intersections between grid lines). Digitizing tablets interfaced to computers can be used to measure areas, and computerized image analysis systems are efficient if the feature of interest and the reference compartment can be easily segmented. However, most workers in most instances, including ourselves, use test grids to estimate volume densities because of their simplicity and efficiency. In stereology it is rarely necessary to precisely measure any single image (Gundersen and Osterby, 1981), making it unnecessary in most instances to use sophisticated instrumentation. Rather, measurements on numerous fields taken from numerous sections from numerous blocks are used to estimate a mean value. When using test grids, the number of points overlying the feature of interest divided by the number of points overlying the reference compartment is directly proportional to areal density and hence volume density. Thus:

$$P_{Pi,ref} = A_{Ai,ref} = V_{Vi,ref} \tag{3}$$

The use of a test grid to estimate volume density is illustrated in Fig. 1.

B. Surface Density

Surface density relates the surface area of the feature of interest to a unit volume of the reference compartment. Typical units are mm^2/mm^3 and cm^2/cm^3. Relevant examples of surface densities include:

- The surface density of capillaries in glomeruli, that is, the surface area of capillaries per unit volume of glomerulus ($S_{Vcap,glom}$), and
- The surface density of GEC plasma membrane in glomeruli, that is, the surface area of GEC plasma membrane per unit volume of glomerulus ($S_{Vgec,glom}$).

Surfaces in three dimensions are seen as boundaries in two dimensions (Fig. 2). The length of boundary (B) per unit area (A) of the reference

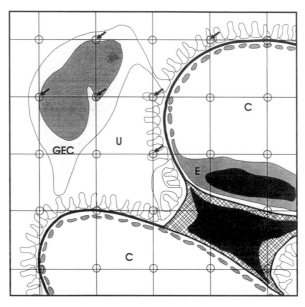

FIG. 1   The use of a stereological test grid to estimate volume density (volume per unit volume). The test grid contains 25 points (defined by the intersections between grid lines). In this example, all 25 points overlie the reference compartment, which is glomerulus (renal corpuscle)—no points lie outside the glomerulus. Six points (indicated by arrows) overlie the cytoplasm and nucleus of glomerular epithelial cells. Therefore, in this field, the volume density of GECs in the glomerulus ($V_{Vgec,glom}$) is 6/25 or 0.24 cm^3/cm^3, according to Eq. (3). Of course, in a full study, test grids are placed on numerous fields on sections from numerous glomeruli. Note that no knowledge of grid size or magnification is needed to estimate volume density. C, capillary lumen; E, endothelial cell; M, mesangial cell; U, urinary space.

compartment in a two-dimensional section is related to the surface area of that feature per unit volume of reference compartment through:

$$S_{Vi,ref} = (4/\pi) \times (B_i/A_{ref}) \tag{4}$$

A variety of methods are available for measuring boundaries in sections. Test grids again provide one of the most efficient methods for estimating boundary lengths. With test grids, the boundary is not measured per se, but rather the number of intersections (I) between the lines of the test grid and the boundary of interest are used to estimate boundary length, and the number of grid points (P) overlying the reference compartment is used to estimate the volume of the reference compartment. The equation for estimating surface density with a stereological test grid is:

$$S_{Vi,ref} = (2 \times I_i)/(P_{ref} \times k \times d) \tag{5}$$

where 2 is a constant, k is the number of test lines per grid point (2 for an orthogonal grid), and d is the length of each test line adjusted for magnifica-

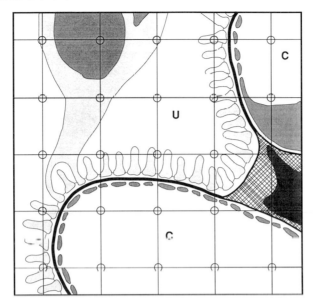

FIG. 2   The use of a stereological test grid to estimate a surface density (surface area of GEC plasma membrane per unit volume of glomerulus). The test grid contains 25 points. To estimate surface density with a stereological test grid, it is necessary to count the number of intersections between the grid and the boundary of interest as well as the number of grid points overlying the reference compartment. In this example, all 25 points overlie the reference compartment (glomerulus); there are 26 intersections between the horizontal grid lines and the GEC plasma membrane, and 30 intersections between the vertical grid lines and the GEC plasma membrane. The surface density of the GEC plasma membrane in the glomerulus ($S_{Vgec,glom}$) calculated using Eq. (5) is $S_{Vgec,glom} = (2 \times 56)/(25 \times k \times d)$. k is 2 for an orthogonal grid and let d (the length of test line between two grid points) equal 4 μm. Therefore, $S_{Vgec,glom} = 0.56$ μm^2/μm^3. If in this instance we wished to estimate the surface area of GEC plasma membrane per unit volume of GEC ($S_{Vgec,gec}$), and in this way calculate a surface area-to-volume ratio for GECs, the equation would become $S_{Vgec,gec} = (2 \times 56)/(7 \times k \times d) = 2$ μm^2/μm^3, with 7 being the number of grid points overlying GEC nuclei and cytoplasm (the redefined reference compartment). C, capillary lumen; U, urinary space.

tion. The use of a test grid to estimate surface density is illustrated in Fig. 2. The estimation of surface density critically depends on section orientation relative to the surface of interest. See Section IV,B for details.

C. Length Density

Length density (L_V) relates the length of a feature of interest to a unit volume of the reference compartment. Typical units are mm/mm^3 or cm/cm^3. An example of a length density relevant to the glomerulus is the length density of capillaries in glomeruli, that is, the length of capillaries per unit volume of glomerulus ($L_{V\text{cap,glom}}$).

Length density can be estimated in sections using:

$$L_V = 2 \times Q/A \tag{6}$$

where 2 is a constant, Q is the number of profiles of the feature of interest, and A is the area of the reference compartment analyzed. Stereological test grids can be used to estimate length density as illustrated in Fig. 3. In

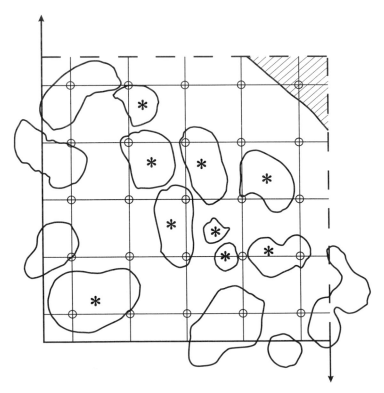

FIG. 3 The use of a stereological test grid to estimate length density (length of capillaries per unit volume of glomerulus) is illustrated. The unbiased stereological counting frame contains 25 grid points; forbidden lines are solid and inclusion lines are broken. The boundary of the renal corpuscle is shown in the top right corner; the cross-hatched tissue is located outside the renal corpuscle. Thus, 24 grid points overlie the renal corpuscle. To estimate length density, it is necessary to count the number of glomerular capillary profiles sampled by the unbiased counting frame. In this example, 9 profiles (indicated by asterisks) are sampled and therefore counted. If the area of each grid square is 50 μm^2, then A_{ref} is 1200 μm^2. Length density is estimated using: $L_{V\text{cap,glom}} = 2 \times Q/A$ [Eq. (6)]. In this instance $L_{V\text{cap,glom}} = (2 \times 9)/ 1200 \ \mu$m$^2 = 1.5 \times 10^{-2} \ \mu$m/$\mu$m^3.

this instance, one counts the number of feature profiles as well as the number of grid points overlying the reference compartment. One additional factor to consider when estimating length density concerns those profiles that cross the edge of the test grid. This problem was solved by Gundersen (1977), who introduced the unbiased counting frame (shown in Fig. 3). Profiles that touch the forbidden line (indicated by the solid line in Fig. 3) are not counted, whereas profiles that touch the inclusion line (broken line in Fig. 3) are counted provided they do not touch the forbidden line at any point. Estimates of length density are influenced by the orientation of the section relative to the features of interest. See Section IV,B for further details.

D. Numerical Density

Numerical density (N_V) relates the number of the features of interest to a unit volume of the reference compartment. Units are, for example, number per cubic millimeter. Relevant examples include

- The numerical density of glomeruli in the kidney, that is, the number of glomeruli per unit volume of kidney ($N_{Vglom,kid}$),
- The numerical density of glomeruli in the cortex, that is, the number of glomeruli per unit volume of cortex ($N_{Vglom,cort}$),
- The numerical density of cells in glomeruli, that is, the number of cells per unit volume of glomerulus ($N_{Vcells,glom}$), and
- The numerical density of endothelial cells in glomeruli, that is, the number of endothelial cells per unit volume of glomerulus ($N_{Vendo,glom}$).

It is very important when considering particle number to appreciate that the number of profiles observed in sections of a particular particle population is not only related to the number of particles in three-dimensional space (the true frequency), but also to particle size, shape, and orientation. Thus, the relationship between the number of profiles per unit area (N_A) and the number of particles per unit volume (N_V) is a complex one. Indeed, Gundersen *et al.* (1988b) point out that "the cells (particles) seen in a single section is a biased, non-representative sample of all cells." This is because the size, shape, and orientation of particles influence their chance of being contained (sampled) in a single section. Thus, the glomeruli seen in the single paraffin sections used for the diagnosis of glomerular disease constitute a biased sample of glomeruli, because larger glomeruli have a greater chance of being contained in the section than smaller glomeruli. Moreover, the glomerular cell nuclei seen in single paraffin sections of glomeruli are biased in terms of the relative frequencies of cell nuclei–GEC nuclei are larger than mesangial and endothelial cell nuclei and will therefore be overrepresented in single sections. (*Note:* nuclear counts are used to estimate cell numbers, assuming one nucleus per cell.)

Until 1984, estimating numerical density represented one of the most difficult tasks in stereology. The available stereological counting methods required detailed knowledge of the geometry (size, size distribution, shape, orientation) of the feature or particle of interest. Most stereological equations required knowledge of the mean caliper diameter $(\overline{D})$ of the particle of interest. If this was not accurately known, or if assumptions of particle size distribution, shape, or orientation were required to estimate $\overline{D}$, then the resultant estimates of numerical density were biased. The equation of Abercrombie (1946) is a well-known and, until recently, widely used traditional stereological equation for estimating numerical density:

$$N_V = N_A/(\overline{D} + T) \tag{7}$$

where N_A is the number of profiles per unit area of the sectioned reference compartment, and T is section thickness. Other traditional stereological methods for estimating numerical density include those of Wicksell (1925), Floderus (1944), Fullman (1953), DeHoff and Rhines (1961), and Weibel and Gomez (1962). The limitation common to all of these traditional stereological counting methods was that knowledge of particle geometry was required. It was no accident therefore that the cells most commonly counted prior to 1984 had nuclei that were approximately spherical, such as hepatocytes, pancreatic exocrine cells, and neurons. No methods were available for obtaining unbiased number estimates of cells with complex nuclei such as GECs, mesangial cells, endothelial cells, and parietal epithelial cells.

Researchers have been trying to count glomeruli for at least a century. Miller and Carlton (1895), for example, counted glomeruli in section pairs. The major methods used have been serial section techniques, independent section approaches, and maceration techniques. While some of the earlier estimates are remarkably similar to recent estimates obtained with the new stereological methods, many previous estimates differed widely. This was due to a variety of reasons, including inappropriate geometric assumptions, sampling procedures, tissue deformation, and edge effects. For a detailed overview of early methods used to estimate glomerular number, see Bendtsen and Nyengaard (1989).

1. The Physical Disector

The disector method of Sterio (1984) revolutionized the process of counting particles in three-dimensional space and thereby estimating numerical density. In their review of new stereological methods, Cruz-Orive and Weibel

(1990) stated that with the disector method now available to count particles, traditional biased counting methods "should no longer be used."

The disector is a three-dimensional probe that samples particles uniformly. In other words, with the disector, small particles have the same chance of being sampled as large particles; round particles have the same chance as long particles. Once particles are sampled uniformly with the disector, then they can be counted (or their volume estimated; see below). No knowledge or assumptions of particle size, size distribution, shape, or orientation are required to sample and then count particles with the disector. The only absolute requirements are that: (1) the particles of interest be unambiguously identified on every occasion—in other words, if you cannot identify the particle, then you cannot count it! and (2) the limits or boundaries of the reference compartment in three-dimensions must be identifiable. The principle of the disector method is illustrated in Fig. 4. Although larger particles have a greater chance than smaller particles of being contained within the disector sampling volume, they also have a greater chance of crossing the forbidden planes and therefore of not being counted.

Two variations of the disector counting approach are now available: the physical disector as originally described by Sterio (1984), and the optical disector (Gundersen, 1986). Two physical sections (histological sections or

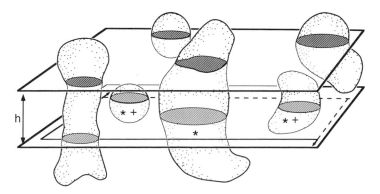

FIG. 4   The disector principle of Sterio (1984). The disector is an unbiased three-dimensional sampling probe. If we designate the bottom plane to be the sampling plane, we see an unbiased counting frame. Three particles indicated by asterisks are sampled by this sampling frame. Two (indicated with a + sign) of these three particles are not contained in the top plane (the look-up plane) and are therefore said to be sampled by the disector; somewhere between the sampling plane and the look-up plane these two particles ended. If the distance h between the two planes is known, and the area of the unbiased two-dimensional frame is also known, then the volume of the disector can be calculated. Dividing the number of counted particles (2) by disector volume gives the numerical density of the particles in the reference space.

electron micrographs) are needed to count particles with physical disectors. Corresponding fields on the two sections are compared and particles sampled by an unbiased counting frame (Gundersen, 1977) on the first "sampled" section but not present in the second "look-up" section are counted; if a particle has disappeared, a unique event has occurred (the end of the particle has been reached) and it is this unique event that is actually counted. If the distance between the two sections is known, then numerical density can be estimated through:

$$N_V = Q^-/(A \times T) \tag{8}$$

where Q^- is the number of particles actually counted using the disector principle, A is the area of sampled section analyzed, and T is the distance between the two sections (when adjacent sections are used, T is section thickness).

While publication of the disector method by Sterio (1984) unquestionably represented a landmark event in stereology, making it possible for the first time to count particles of any shape (arbitrary particles), in practical terms it is tedious to count cell nuclei, and thereby cells, with physical disectors. The major difficulty is that the same nucleus has to be searched for on adjacent sections. Nevertheless, physical disectors have been successfully used at the light microscopic level to estimate the number of neurons in human thalamus (Pakkenberg and Gundersen, 1988), the number of cells in the pars distalis of the sheep pituitary gland (Antolovich et al., 1992), and the number of neurons in the rat cortex (Bedi, 1994). Physical disectors have also been used to estimate the number of capillaries in average glomeruli. In this instance, events involving capillary lumina as defined by Nyengaard and Marcussen (1993) are counted (see Section V,G for more details). Physical disectors have also been used at the electron microscopic level to estimate the number of perforated synapses in neocortex (Calverley and Jones, 1987) and the number of individual cell types in glomeruli (Bertram et al., 1992). Full details of the method used to count glomerular cells are given in Section V,D.

Today in glomerular morphometry, physical disectors are not the method of choice for counting cells because the optical disector is available (described in detail later). However, the physical disector is the method of choice for counting glomeruli because glomeruli are too large to count with optical disectors. Another limitation of the physical disector for estimating numerical density is the need in many instances to measure section thickness. While this is not an impossible task, estimation of section thickness is invariably associated with some degree of error which will translate into errors in the final estimation of numerical density. As will be described later, the development of the fractionator principle and the use of the disector/Cavalieri combination has meant that it is now rarely necessary to measure section thickness in order to count particles.

2. The Optical Disector

The optical disector (Gundersen, 1986; Gundersen et al., 1988b) is today the method of choice for counting cell nuclei, and thereby cells. Unlike the physical disector, with the optical disector only one histological section is analyzed at a time. There is no need to align pairs of sections, and therefore this method is much more efficient than the physical disector.

With the optical disector, the limited focal depth of high numerical aperture, oil immersion lenses is used to optically section at varying depths within thick tissue sections (usually greater than 20 μm). As described earlier, with the disector principle, a unique event is counted for each sampled particle. Theoretically therefore, nuclei could be counted using optical disectors when they either appear or disappear, but in most applications it is easier in practical terms to count a nucleus when it first comes into focus. At the moment the nucleus comes into focus one checks to see if the nucleus is sampled by an unbiased counting frame (described earlier and shown in Figs. 3 and 4). If a nucleus is sampled by the unbiased counting frame when it comes into focus, then it is counted. Nuclei in focus in the top plane of the disector volume are not counted because they did not come into focus. Nuclei in focus in the bottom plane of the disector are counted because they did come into focus. If the area of the unbiased counting frame and the height or thickness of the disector is known (through use of an electronic microcator, for example), then numerical density can be determined using Eq. (8). To date, optical disectors have mostly been used to count neurons (Braendgaard et al., 1990; West and Gundersen, 1990; Nurcombe et al., 1991; West et al., 1991; Bjugn, 1993; Harding et al., 1994; Tan et al., 1995), but have also been used to count vascular smooth muscle cells (Black et al., 1992, 1995; O'Sullivan et al., 1994) and glomerular cells (Bertram et al., 1992; Marcussen, 1992). Software for counting cells with optical disectors is available (Bolender and Charleston, 1993).

E. Estimating Absolute or Total Quantities

Each of the four so-called traditional stereological parameters described earlier (volume density, surface density, length density, numerical density) relates a measurement of the feature of interest to a unit volume of the reference compartment. There are three major limitations to such stereological densities. First, we often wish to know absolute or total values, not densities in which measurements of the feature of interest are expressed relative to a unit volume. For example, we may wish to know the total volume of ECM per glomerulus, not the volume of ECM per unit volume of glomerulus. Similarly, we may wish to know a total surface area (total surface area of GBM per glomerulus), total length (total length of capillaries per glomerulus), or total number (total number of mesangial cells per glomerulus).

A second limitation with density estimates is that they are ratios and therefore difficult to interpret. As a result, stereological density estimates are frequently misinterpreted. For example, if the length density (length per unit volume) of capillaries in glomeruli increases in disease or following an experimental procedure, this does *not* mean that the length of capillaries per glomerulus has increased. Indeed, the length of capillaries per glomerulus may have increased, remained unchanged, or even decreased—it all depends on glomerular volume. If glomerular volume is unknown, there is no way of determining if the total length of capillaries per average glomerulus increased, remained unchanged, or decreased. Unfortunately, an increase in length density, for example, is often misinterpreted as signifying that the length of capillaries per glomerulus has increased.

A third difficulty associated with stereological density estimates concerns the effects of the dimensional changes that occur when tissue is fixed, processed, embedded, and sectioned for microscopy. It is well established that the volume of soft tissues such as kidney decreases by approximately 40–50% when processed and embedded in paraffin (Hanstede and Gerrits, 1983; Miller and Meyer 1990), and by about 10–15% when processed and embedded in Epon/araldite (Weibel and Knight, 1964). The shrinkage of kidney tissue embedded in glycol methacrylate is likely to be the same as that embedded in Epon/araldite (Miller and Meyer, 1990; Schmitz *et al.,* 1990). There is even evidence that different components of tissues and organs shrink and swell to varying degrees, and that different components of cells shrink and swell to varying degrees (Bloom and Friberg, 1956; Bahr *et al.,* 1957; Eisenberg and Mobley, 1975; Gerdes *et al.,* 1982; Bertram *et al.,* 1986). Many of these dimensional changes are impossible to measure and therefore impossible to correct by simply introducing correction factors into stereological formulas. The easiest and best way to overcome these dimensional artifacts is to design the stereological investigation in such a way that the effects are negated (cancel) or are at worst, minimized.

For the three reasons outlined earlier, it is best to avoid stereological density estimates. While stereological densities are often required as a means to an end, in most instances they should not represent the final data from an investigation. Braendgaard and Gundersen (1986) cleverly referred to the dangers of referencing stereological estimates to a unit volume as the "reference trap." For a comprehensive discussion of the problems associated with stereological densities, see Oorschot (1994). Fortunately, two general approaches for estimating total volumes, surface areas, lengths, and numbers, and thereby avoiding the "reference trap" are available— the Cavalieri principle and the fractionator principle.

1. The Cavalieri Principle

The Cavalieri principle provides a method for obtaining an unbiased estimate of the total volume of anything (Gundersen and Jensen, 1987; Gundersen *et al.*, 1988a). Put simply, the volume of an object can be obtained by slicing or sectioning it into slices of approximately equal thickness, estimating the sum of the areas of the slices, and multiplying the sum of the areas by mean slice or section thickness:

$$V = \Sigma A \times T \tag{9}$$

The only requirement of Eq. (9) is that the position of the first slice must be selected at random. Kidney volume and cortical and medullary volumes can easily be determined using this approach with a razor blade cutting device such as that illustrated in Fig. 7 in Gundersen *et al.* (1988b). In general terms, an estimate of volume with satisfactory precision will be obtained if a minimum of about 10 slices or sections uniformly spaced through the object are measured. Area can efficiently be measured using test grids as described earlier. The Cavalieri method is not restricted to estimating the volumes of particles with simple shapes. The object can even be represented by separated profiles in sections (see, for example, various figures in West and Gundersen, 1990). The Cavalieri method is highly efficient (Gundersen and Jensen, 1987), typically requiring only four to eight systematic sections to provide an estimate with a coefficient of error (CE) of about 5% [CE(est V) = SE(est V)/V]. Typically only about 100 grid points need to be counted on the entire object to obtain a volume estimate with this precision (Gundersen and Jensen, 1987).

Multiplication of a stereological density (volume density, surface density, length density, numerical density) by the total volume of the reference compartment (from the Cavalieri method) provides an unbiased estimate of total volume, surface area, length, and number, respectively. For example, the total number of cells per average glomerulus ($N_{cells,glom}$) can be estimated using:

$$N_{cells,glom} = N_{Vcells,glom} \times V_{glom} \tag{10}$$

where $N_{Vcells,glom}$ is the numerical density of cells in glomeruli and V_{glom} is mean glomerular volume. Thus, the stereological density $N_{Vcells,glom}$ was used to estimate a total quantity. With this approach, dimensional changes in cells and tissues that occur during processing for microscopy become irrelevant when estimating total number, provided numerical density and total volume are estimated in the same sections, because changes in the dimensions of the reference space cancel. Of course if one was estimating total lengths, surface areas, or volumes, then these would still be influenced by dimensional changes, which may require correction. Depending on experimental design, it might not even be necessary to estimate section thickness in order to estimate total quantities. Pakkenberg and Gundersen (1987) demonstrated that with a disector/Cavalieri combination, section thickness cancels and therefore need not be measured in order to estimate total number. More recently, Hinchliffe *et al.* (1991, 1992, 1993) used a physical disector/Cavalieri combination to estimate total glomerular number in human kidneys.

2. The Fractionator Principle

The fractionator principle was described by Gundersen (1986). The principle is that if the number or amount of a particular feature in a known fraction of the reference space is known, then the total number or amount of that feature in the whole reference space can be calculated using simple algebra. For example, if the number of glomeruli in a known fraction of the kidney is known, then the total number of glomeruli in the whole kidney can be easily calculated.

Typically in a stereological investigation using a fractionator design, the organ is sliced; a known fraction of the slices is cut into bars or wedges; a known fraction of the wedges or bars is embedded and exhaustively sectioned (until no tissue remains in the block); a known fraction of the sections is mounted and stained; and a known fraction of the section area is actually measured. Then the total measurement is multiplied by the inverses of the successive sampling fractions to obtain an estimate for the total value. This concept is illustrated in Fig. 5.

The physical disector/fractionator combination has been used by a number of workers to estimate the total number of glomeruli in the kidney (Nyengaard and Bendtsen, 1989, 1990, 1992; Bertram *et al.*, 1992; Berka *et al.*, 1994; Cahill, 1993; Cahill *et al.*, 1995; Kett *et al.*, 1994; McCausland *et al.*, 1994). Again, the technique involves sampling known fractions of the kidney at successive stages, and then in the final step, multiplying the number of glomeruli actually counted with physical disectors by the inverses of the various fractions. The fractionator principle has also been used to estimate numbers of neurons (Pakkenberg and Gundersen, 1988; Nairn *et al.*, 1989; Mayhew, 1991; West *et al.*, 1991; Yaegashi *et al.*, 1993) and pulmonary lymphatic valves (Ogbuihi and Cruz-Orive, 1990).

It is important to emphasize some of the advantages of the fractionator approach to stereology. First, the problem of dimensional changes in tissues (shrinkage and swelling artifacts) becomes irrelevant when utilizing the

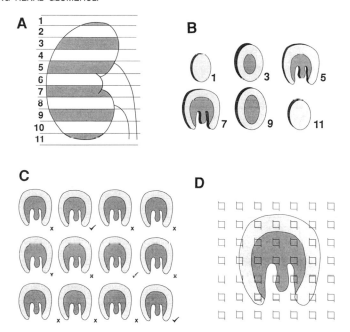

FIG. 5 Diagram illustrating the use of the fractionator principle at three successive sampling stages. (a) The kidney is macroscopically sliced and every second kidney slice (in order with a random start) is embedded (b). Fraction 1 therefore equals one half. (c) The sampled slices are exhaustively sectioned with every fifth section (in order with a random start—indicated with a tick) (fraction 2 equals one fifth) being taken for further analysis. In (d), a known fraction of the section area (fraction 3 equals 9/100) is actually used for measurement. An estimate of the total value (V, S, L, or N) of the feature of interest is obtained by multiplying the amount measured by the inverses of the three sampling fractions ($2 \times 5 \times 100/9 = 111.11$).

fractionator design to estimate number; if the fraction of kidney used for counting is known, the extent of tissue shrinkage and/or swelling is of no consequence. Second, depending on experimental design, it may not be necessary to estimate section thickness. Third, it is generally not necessary to know magnification. All that is required is correct sampling in three-dimensional space (described in greater detail later), knowledge of the sampling fractions, and the ability to identify the features of interest and the reference compartment in all sections.

F. Estimating Volume

In most glomerular stereology to date, indirect approaches have been used to estimate volumes of glomeruli and specific glomerular components (cells, extracellular matrix, capillary lumina, and so on). This typically entails dividing total volume (V) by total number (N), or dividing volume density (V_V) by numerical density (N_V). For example, mean glomerular volume (V_{glom}) can be estimated through:

$$V_{glom} = V_{Vglom,kid}/N_{Vglom,kid} \tag{11}$$

where $V_{Vglom,kid}$ is the volume density of glomeruli in the kidney, and $N_{Vglom,kid}$ is the numerical density of glomeruli in kidney. An important point to note when using equations similar to Eq. (11) is that the reference compartments for both the volume density and the numerical density must be identical. In Eq. (11) the reference compartment for both terms is kidney. A third method for indirectly estimating total volume is to multiply a volume density by an absolute volume. For example, the volume of ECM in an average glomerulus $(V_{ecm,glom})$ can be estimated through:

$$V_{ecm,glom} = V_{Vecm,glom} \times V_{glom} \tag{12}$$

where $V_{Vecm,glom}$ is the volume density of ECM in an average glomerulus.

In addition to indirect methods for estimating volume, five direct methods are available. The Cavalieri method was described earlier. The other four direct methods (point-sampled intercepts, selector, nucleator, rotator) have been developed in recent years and their application in glomerular morphometry to date is minimal. The point-sampled intercepts and selector method are described later. To the author's knowledge, the nucleator (Gundersen, 1988) and rotator (Jensen and Gundersen, 1993) have not previously been used in glomerular morphometry and are therefore not described here. For a detailed discussion of these four methods for directly estimating particle volume, see either the original papers or the reviews by Gundersen *et al.* (1988a,b), Cruz-Orive and Weibel (1990), Mayhew (1992), and Bolender *et al.* (1993).

1. Point-Sampled Intercepts

Point-sampled intercepts provide unbiased estimates of the volume-weighted mean volume $(\bar{v}_V)$ of arbitrary particles (Gundersen and Jensen, 1983, 1985). Volume-weighted mean volume is the mean particle volume obtained from the volume distribution of the particle volumes. It is different from the volume we usually think about—the number-weighted mean volume $(\bar{v}_N)$, which is derived from the numerical frequency distribution of particle volumes. The two volume estimators are related through:

$$\bar{v}_V = \bar{v}_N \times (1 + CV^2_N) \tag{13}$$

where CV_N is the coefficient of variation of the number distribution of particle volumes (Gundersen and Jensen, 1985; Cruz-Orive and Hunziker,

1986). The significant attraction of the volume-weighted mean volume is that it is the only volume estimator that can be used on independent (single) sections. Accordingly, estimates can be obtained quickly.

To estimate volume-weighted mean volume using point-sampled intercepts, a test grid is applied to sections of the tissue. Intercepts are measured through those points that hit (sample) the profiles of interest. The direction of the intercepts must be isotropic uniform random (IUR) with respect to the particles. Isotropic means that all *directions* in three-dimensional space are equally likely, while uniform implies that all *positions* in three-dimensional space are equally likely. IUR or vertical sections (as defined by Baddeley *et al.*, 1986; see Section IV,B) can be used if the particles have a preferred orientation, and systematic random sections can be used if the particles are isotropic. Clearly, larger profiles will, on average be sampled by more points and thus the final estimate of volume is volume weighted. Volume-weighted mean particle volume is estimated through:

$$\overline{v}_V = (\pi/3) \times \overline{l_{o,i}^3}$$ (14)

where $\pi/3$ is a constant and $\overline{l_{o,i}^3}$ is the average of the cubed intercept lengths.

Artacho-Perula *et al.* (1993) used point-sampled intercepts to estimate glomerular volume in streptozotocin-diabetic rats, and reported that less than 20 min was needed to obtain an estimate of volume-weighted mean glomerular volume for each rat. It is important to emphasize, however, that this technique provides volume-weighted estimates, meaning that the estimates are weighted by larger glomeruli. The influence of larger glomeruli on estimates of volume-weighted mean glomerular volume was clearly shown in the recent study by Nyengaard *et al.* (1994). This is described in greater detail in Section V,B.

2. The Selector

The selector combines point-sampled intercepts with the disector to estimate number-weighted mean volume (Cruz-Orive, 1987). IUR or vertical sections must be used if the particles have a preferred orientation, and systematic random sections can be used if the particles are isotropically arranged. In brief, stacks of serial sections are cut and physical disectors used to select particles. Thus, particles are selected according to their presence, not their size as with the point-sampled intercept approach. Then, point-sampled intercepts are measured on serial profiles of those sampled particles. The estimate of volume for a single particle (i) is obtained using:

$$v_i = (\pi/3) \times \overline{l_{o,i}^3}$$ (15)

An unbiased estimate of the mean volume of all particles (n) is obtained using:

$$\bar{v}_N = (\pi/3) \times (1/n) \times \overline{l^3_{o,i}} \qquad (16)$$

To the author's knowledge, the selector has not been used to estimate the volume of glomeruli or any glomerular components. It is unlikely to be used to estimate glomerular cell volume (because serial electron microscopic sections would be needed), but could theoretically be used to estimate nuclear volumes and glomerular volume.

IV. Sampling

Without question, sampling is one of the most important steps in every stereological study. As explained earlier, rarely in a stereological study does one measure the entire population. Rather, measurements made on a sample, typically an extremely small sample, are used to estimate the population value. Obviously, this sample must be representative of the population; if not, the stereological estimates obtained will constitute an inaccurate account of the cell or tissue morphology. While the philosophy of a representative sample is easy enough to understand, in practice it takes care to ensure that such a sample is actually obtained. For example, as already described, a single section constitutes a biased, nonrepresentative sample of cell nuclei, and for that matter glomeruli, because larger nuclei and glomeruli have a greater chance of being sampled by a single section than do smaller nuclei and glomeruli (Gundersen *et al.*, 1988b).

Three issues must be considered when sampling for stereology—sample distribution, sample orientation, and sample size. These topics are now considered.

A. Sample Distribution

The underlying philosophy of sampling in stereology, if not in all biological studies, is that every part of the cell, tissue, or organ being analyzed must have the same chance of being selected for measurement—the sampling must be uniform. The most efficient approach for obtaining a uniform sample is to use systematic random sampling. Systematic random sampling is more efficient, and in many instances easier in practical terms than purely random sampling (Gundersen and Jensen, 1987). Systematic random sampling means that at each sampling stage—for example, selecting kidney slices, selecting sections, selecting fields for measurement, selecting glomeruli for measurement—the tissue must be sampled in a systematic fashion with a random component. For example, if every third kidney slice is to be taken for analysis, the first slice (slice 1, 2, or 3) should be selected at random (using a random number table, for example) and then every third slice taken in order (systematically). Thus, if the first slice chosen is slice number 2, the slices taken for further analysis would be slices 2, 5, 8, and so on. If every tenth section is to be analyzed, the first section should be selected at random between sections 1 and 10 (inclusive). After that, every tenth section is taken. To select fields on sections, a random position on the microscope stage can be chosen as a starting point, and then a motorized stage used to select fields in a systematic pattern. Thus, at every sampling stage, a random start is used and subsequent sampling is systematic.

B. Sample Orientation

Most stereological procedures require some kind of isotropy. In general, only number and indirect volume estimators do not require isotropy. Thus, length, surface, and direct volume estimators (point-sampled intercepts, selector) all require isotropy. Isotropic sections can be obtained in three ways: either the specimen is isotropic, or IUR or vertical sections must be used. IUR or vertical sections are used today in most stereological studies. (IUR sections were defined in Section III,F.) A vertical section is a plane section perpendicular to a given horizontal plane. The horizontal plane is only a plane of reference, and can be given by the tissue itself or generated artificially by the investigator. Full details on obtaining IUR and vertical sections are given in Baddeley *et al.* (1986), Gundersen *et al.* (1988a), Cruz-Orive and Weibel (1990), and Mayhew (1992).

Fortunately, glomeruli can reasonably be considered to be isotropic, that is, to have no preferred orientation in three dimensions. Therefore, any section through a glomerulus can be considered to be isotropic. To be doubly sure of glomerular isotropy, one can employ the isector technique of Nyengaard and Gundersen (1992). With the isector, specimens are embedded in small spherical moulds. Once embedded, any handling of the sphere independent of the specimen will induce isotropy of the final sections. An alternative approach for generating isotropic sections is the orientator (Mattfeldt *et al.,* 1990).

C. Sample Size

The rule of thumb when considering sample size is to sample lightly (Gundersen and Osterby, 1981). For example, to use disectors to estimate the number of particles of interest in a reference space (for example, the number of glomeruli in the kidney), it is unnecessary to count more than about 150 glomeruli per kidney. It has been shown repeatedly that to obtain an estimate with a coefficient of error of less than 15%, fewer than 200 particles per object (in this case glomeruli per kidney) need to be counted (Gundersen *et al.*, 1988a,b). Similarly, to estimate the volume of anything using the Cavalieri principle, only about 10–15 slices or sections are required from the object to obtain an estimate with a coefficient of error of about 5%. Unfortunately, the field of stereology is haunted by stories of investigators counting tens of thousands, or even hundreds of thousands of grid points. In most cases this is totally unnecessary. The new methods are not only unbiased, but are almost without exception extremely efficient. It is best to sample and measure lightly at each stage, and if necessary put more effort into increasing sample size by analyzing, for example, more animals.

The coefficient of variation of a stereological estimate represents the sum of the biological variance (variance between animals) and technical variance (variance resulting from the stereological technique). Procedures are available to estimate the contributions of these two sources of variance to the total variance (Kroustrup and Gundersen, 1983). Knowledge of the relative contributions of biological and technical variance can be used to optimize the design of stereological investigations. For guidelines on optimizing sampling strategies for estimating total glomerular number, see for example, Nyengaard and Bendtsen (1992). For optimizing sampling for estimating glomerular capillary number, see Nyengaard and Marcussen (1993). For additional details on optimizing sampling schemes, see West and Gundersen (1990) and West *et al.* (1991).

V. Glomerular Stereology

In this section, unbiased stereological methods for estimating a range of glomerular structural parameters are described. Although many of these methods have only recently been developed, applications to the glomerulus are appearing, and technical details and results from these studies are included wherever possible.

A. Glomerular Number

Two unbiased stereological methods, the disector/fractionator combination and the disector/Cavalieri combination, have been used to estimate the total number of glomeruli in the kidney. The disector/fractionator combination has been most commonly used and involves the use of physical disectors to estimate the number of glomeruli in a known fraction of the kidney.

Once the number of glomeruli in a known fraction of the kidney is known, simple algebra can be used to estimate the number of glomeruli in the whole kidney. With the disector/Cavalieri combination, physical disectors are used to estimate glomerular numerical density, and this is multiplied by the volume of the kidney or cortex (depending on which reference compartment is being utilized), which is estimated using the Cavalieri principle. It is important to emphasize that with both of these methods, all glomeruli, regardless of size, have the same chance of being sampled, and subsequently counted. The physical disector/fractionator combination has been used to estimate glomerular number in a range of species, including humans (Nyengaard and Bendtsen, 1989, 1990, 1992; Nyengaard et al., 1994; Bertram et al., 1992; Cahill, 1993; Berka et al., 1995; Cahill et al., 1995; Kett et al., 1994; McCausland et al., 1994; Soosaipillai, 1994; Soosaipillai et al., 1994), while the physical disector/Cavalieri combination has mostly been used in human studies (Hinchliffe et al., 1991, 1992, 1993). As noted earlier, glomeruli are too large to be counted using optical disectors.

We used a physical disector/fractionator combination to count glomeruli in the kidneys of seven normal adult Sprague-Dawley rats (Bertram et al., 1992). This method is described in detail here. When dealing with smaller or larger kidneys, modifications to the method detailed are required (Nyengaard and Bendtsen, 1990, 1992).

Following perfusion-fixation, the rat kidneys were cut into 1-mm-thick slices using a razor blade cutting device. The position of the first slice must be random as described earlier. Approximately 15–20 slices were obtained from each kidney, and every second slice (in order) was taken for glomerular counting. These slices were embedded in glycol methacrylate. The blocks were then mounted in a microtome chuck and the distance from the base of the chuck to the block face recorded with a digital micrometer with a precision of 1 μm. The slices were then exhaustively sectioned (from one end to the other) at a nominal thickness of 20 μm, and the total number of sections cut was recorded. We use a Reichert Supercut microtome fitted with glass knives for this purpose. The distance from the base of the chuck to the block face was remeasured, and average section thickness (t) calculated for each block. It is important to note that knowledge of section thickness is not required to estimate glomerular number using a physical disector/fractionator combination. However, section thickness is required to estimate some other glomerular parameters and is easily determined during the sectioning stage. During serial sectioning, every tenth ("sample section," see below) and eleventh ("look-up section") sections were mounted, with the first chosen at random using a random number table. Sections were stained with hematoxylin and eosin. Further details on section staining are given in the following paragraphs.

When the kidney was cut into slices before embedding, the razor blades produced artificial surfaces. More artificial surfaces were produced by the glass knives during sectioning. These artificial surfaces can interfere with glomerular counting. For example, parts of glomeruli may be absent, or some glomeruli may be difficult to recognize owing to mechanical damage. To overcome the problems associated with such artificial surfaces, we use only those sections between the first and last sections of a block containing the complete circumference of the kidney. It is therefore necessary to estimate the fraction of the sampled section area used to count glomeruli. We use a Fuji Minicopy Reader and project the sections at a final magnification of 19.2× onto an orthogonal grid. The number of grid points overlying all kidney sections (P_s) as well as the number of points overlying complete sections (P_f) are counted. The fraction of the sectioned tissue used for glomerular counting is given by P_f/P_s.

To count glomeruli, two light microscopes equipped for projection are required. A sampled section is placed in the first microscope and the corresponding "look-up section" in the second microscope. The fields of vision are projected side by side onto a table in a semidarkened room (Fig. 6).

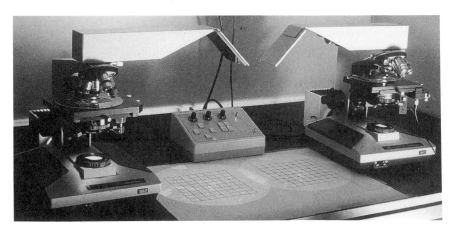

FIG. 6 Apparatus used in our laboratory to count glomeruli with physical disectors. The sampled section is placed on the right microscope and the look-up section on the left microscope. The images are projected side by side at a final magnification of approximately 150×. An unbiased counting frame in the field of the sampled section is used to sample glomeruli. Those sampled glomeruli not present in the look-up section are counted (Q^-). Then, those glomeruli sampled by an unbiased sampling frame in the look-up section are searched for in the sampled section, and counted if they have disappeared. This approach doubles the efficiency of glomerular counting. The control box for operating the motorized stage (right microscope) is seen between the microscopes. The left microscope is fitted with a rotating stage attachment to facilitate section alignment.

We use a final magnification of approximately 150×. An unbiased counting frame (Gundersen, 1977) is applied to the field of vision of the first microscope. To count glomeruli, corresponding regions of the section pairs are found and those glomeruli sampled by the unbiased counting frame of the first microscope that are not present in the look-up section are counted (Q^-; Fig. 7). To double the efficiency, those glomeruli in the look-up section that are sampled by a counting frame, but which are not present in the first section, are also counted.

As described above, it is critically important that fields used for glomerular counting be selected using a systematic uniform random scheme. After all, only 100 to 200 glomeruli will actually be counted in each kidney. Therefore, each glomerulus must have an equal chance of being included in the sample. To select fields on sections, we use a motorized stage fitted to one of the microscopes (see Fig. 7). An orthogonal grid [a(p) of approximately 0.02 mm^2] within the unbiased counting frame was used to count points overlying kidney sections and this allowed calculation of the fraction of the section area (f_a) used to count glomeruli. The formula for estimation of the total number of glomeruli in a kidney ($N_{glom,kid}$) was:

$$N_{glom,kid} = 2 \times 10 \times (P_s/P_f) \times [1/(2f_a)] \times Q^- \qquad (17)$$

where 2 was the inverse of the slice sampling fraction (every second 1-mm slice was taken for glomerular counting), 10 was the inverse of the section sampling fraction (every tenth section, and the adjacent section was mounted), P_s/P_f and $1/(2f_a)$ gave the fraction of the total section area used to count glomeruli, and Q^- was the actual number of glomeruli counted.

In our initial studies of the normal rat kidney (Bertram *et al.*, 1992; Soosaipillai *et al.*, 1994; Soosaipillai, 1994) we typically counted approximately 125 glomeruli per kidney. Glomeruli were counted in approximately 100 fields from approximately 10–15 section pairs per kidney. The time required to estimate the total number of glomeruli in a kidney, using the physical disector/fractionator experimental design described earlier, was approximately 2.5 hr. This does not take into account the time needed to process, section, and stain tissue. Data from our study on glomerular number in normal adult rat kidney are shown in Table I. The mean number of glomeruli per kidney is 31,764 with a standard deviation of 3,667. The coefficient of variation (standard deviation as a percentage of the mean) which takes into account both true biological variation among animals as well as technical sources of variation, is 11.5%, illustrating the precision of the approach. Estimates for the seven rats studied ranged from 25,960 to 36,109 glomeruli.

In more recent studies requiring estimation of glomerular number in the rat kidney, we have doubled the number of fields used to count glomeruli, and now count approximately 200–250 glomeruli in each kidney on 10–15

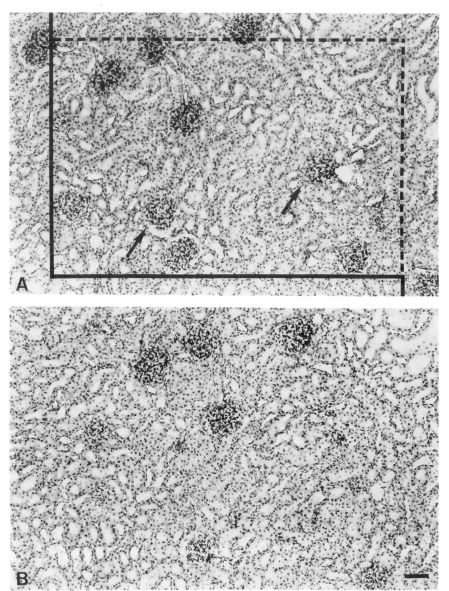

FIG. 7 Photomicrographs showing a visual field from a sampled kidney section (A) and the corresponding field on the look-up section (B). An unbiased counting frame is superimposed on the sampled section. The forbidden line is solid. Glomerular profiles are counted if they are sampled in A and are not present in B. In this case, the two glomeruli indicated by arrows are counted; $Q^- = 2. \times 150$. Bar $= 100\ \mu m$. (Reproduced with permission from Bertram *et al.*, 1992.)

TABLE I

Body Weights and Stereological Estimates for Seven Normal Sprague-Dawley Rats[a]

Animal	Body weight g	Total volume of kidney V_{kid} (mm^3)	Average renal corpuscle volume V_{corp} (μm$^3 \times 10^5$)	Average glomerular volume V_{glom} (μm$^3 \times 10^5$)	Total number of glomeruli per kidney $N_{glom,kid}$
1	218	649.8	8.448	7.140	31912 (134)
2	234	819.5	8.672	6.341	32440 (127)
3	201	942.0	9.053	5.249	35920 (132)
4	189	822.2	6.928	5.542	31445 (87)
5	210	819.0	10.534	7.882	25960 (98)
6	228	1032.6	8.137	6.604	36109 (126)
7	227	854.7	8.343	7.636	28563 (157)
Mean	215	848.6	8.588	6.628	31764
S.D.	16	118.7	1.084	1.001	3667
C.V. (%)[b]	7.4	14.0	12.6	15.0	11.5

[a] Values of Q^- (the actual number of glomeruli counted) are given in parentheses. Data reproduced from Bertram et al. (1992) and Soosaipillai et al. (1994) with permission.

[b] CV (coefficient of variation) is the standard deviation expressed as a percentage of the mean. CV is the sum of true biological variation (between rats) and variation introduced by the scientific techniques.

section pairs. This takes approximately 5 hr per kidney. Our purpose in increasing the sample size was to decrease even further the variance of the estimates of glomerular number since these estimates are used in many other stereological equations (for example, to estimate glomerular volume, glomerular cell number and size, glomerular capillary length, surface area, and number), and variances accumulate at each stage of a multistaged stereological experiment. The coefficients of variation for our most recent estimates of glomerular number in the rat kidney have been as low as 8% (Kett et al., 1994; McCausland et al., 1994).

The physical disector/fractionator combination has been used extensively by Nyengaard and his co-workers to estimate glomerular number in rat kidneys. Their estimates for the adult rat typically range between 25,000 and 30,000 glomeruli per kidney. The kidneys of 5-day Wistar rats contain 21,000 glomeruli (Nyengaard, 1993a). For additional information on techniques and results, see Nyengaard (1993a,b), Nyengaard and Rasch (1993), and Nyengaard et al. (1994).

Nyengaard and Bendtsen (1992) used the physical disector/fractionator combination to estimate glomerular number in 37 normal, mature human kidneys. The mean estimate was 617,000, with individual values ranging widely, from 331,000 in an 80-year-old man to 1,424,000 in a 50-year-old woman. Total glomerular number was negatively correlated with age and positively correlated with kidney weight, but was not correlated with body surface area. Bendtsen and Nyengaard (1992) used the physical disector/ fractionator combination to estimate glomerular number in 25 Type 1 (insulin-dependent) and 39 Type 2 (noninsulin-dependent) diabetic patients. The number of glomeruli in kidneys from Type 1 and Type 2 diabetic patients with mild glomerulopathy or none did not differ significantly from nondiabetic control subjects. However, a subgroup of the Type 1 diabetic patients who died with severe nephropathy had fewer glomeruli than Type 1 diabetic patients with mild glomerulopathy or none. With the recent interest in the relationship between glomerular number, hypertension, and the progression of renal disease (Brenner *et al.,* 1988, 1992; Lafferty and Brenner, 1990; Fogo and Ichikawa, 1992), and the concept of "glomerular endowment," the emergence of the physical disector/fractionator combination as an unbiased, robust, and efficient method for estimating total glomerular number is timely.

Hinchliffe *et al.* (1991, 1992, 1993) also obtained unbiased estimates of glomerular number in human kidneys, but instead of a disector/fractionator combination, used a disector/Cavalieri combination. With this approach, physical disectors are used to estimate glomerular numerical density in cortex, and the Cavalieri principle is used to estimate the volume of the cortex. The product of these two estimates provides an unbiased estimate of total glomerular number provided both estimates are made in the same, identical volume, in this case paraffin-embedded cortex, and provided, of course, that the glomeruli can be unambiguously identified. The main difference between the unbiased method used by Hinchliffe *et al.* (1991, 1992, 1993) and the disector/fractionator combination used by other workers is that the disector/Cavalieri combination requires estimation of section thickness. Both methods can be performed in paraffin or glycol methacrylate-embedded tissue, although if it is intended to estimate volumes of glomeruli or glomerular compartments, paraffin should be avoided because of the considerable changes in volume that occur during processing and embedding (see Section V,B).

Hinchliffe *et al.* (1991) estimated glomerular number in humans between 15 and 40 weeks of gestation. Mean glomerular number increased from 15,000 at 15 weeks of gestation to 740,000 at 40 weeks. Hinchliffe *et al.* (1992) showed that Type II (asymmetrical) intrauterine growth retardation can have a marked effect on renal development in humans, particularly on total glomerular number. Most recently, Hinchliffe *et al.* (1993) estimated glomerular number in the kidneys of 24 victims of the sudden infant death syndrome (SIDS). They found reduced glomerular numbers in SIDS subjects with both low and normal birth weights. For example, the kidneys of

SIDS subjects ($n = 15$) with normal birth weights contained approximately 690,000 glomeruli whereas control kidneys from subjects with the same age range ($n = 16$) contained 903,000 glomeruli (range 740,000 to 1,060,000).

B. Kidney and Glomerular Volume

If total glomerular number is estimated as in our study on the normal rat kidney (Bertram *et al.*, 1992), then the same glycol methacrylate sections can be used to provide unbiased estimates of mean glomerular volume (V_{glom}). Only one extra measurement is required, and this can be made while glomeruli are being counted. Quite simply, grid points overlying profiles of glomeruli and renal corpuscles and total kidney in sampled fields (described earlier) are counted. Division of the total number of points overlying glomerular profiles by the number of points overlying kidney sections (P_s) gives the volume density of glomeruli in kidney ($V_{Vglom,kid}$). This is not glomerular volume, but is needed to estimate glomerular volume with Eq. (11). Glomerular volume may be overestimated if 20-μm sections are used owing to the problem of overprojection or the so-called Holmes effect (Weibel, 1979). To overcome this effect $V_{Vglom,kid}$ could be estimated in 2-μm sections. When exhaustively sectioning the kidney slices at 20 μm, a series of ten 2-μm sections could be cut and used to estimate $V_{Vglom,kid}$. The effects of depth of focus and overprojection on estimation of $V_{Vglom,kid}$ were considered by Marcussen (1992).

The volume of the embedded kidney (V_{kid}) can be easily estimated using the Cavalieri principle. We make the necessary measurements when counting glomeruli. The formula used in Bertram *et al.* (1992) was:

$$V_{kid} = 2 \times 10 \times t \times a(p) \times P_S \qquad (18)$$

The terms in Eq. (18) were described earlier.

We estimate mean glomerular volume (V_{glom}) using Eq. (11), with the numerical density of glomeruli in kidney being obtained by dividing $N_{glom,kid}$ by V_{kid}. Our estimate of mean glomerular tuft volume in normal adult Sprague-Dawley rats was $6.63 \pm 1.00 \times 10^5$ μm^3 (mean $\pm$ standard deviation; see Table I for individual values) (Bertram *et al.*, 1992). If points on the entire renal corpuscle are counted, then renal corpuscle volume can be estimated; our estimate for the normal adult rat was $8.59 \pm 1.08 \times 10^5$ μm^3 (Soosaipillai *et al.*, 1994). Mean glomerular tuft volume in Wistar Kyoto rats and spontaneously hypertensive rats at 10 weeks of age is $7.21 \pm 0.66 \times 10^5$ μm^3 and $7.49 \pm 0.72 \times 10^5$ μm^3, respectively (Kett *et al.*, 1994). Other estimates for glomerular volume in normal rats obtained using the new stereological methods include those given by Nyengaard (1993a), Nyengaard and Rasch (1993), and Nyengaard *et al.* (1994). Nyengaard

(1993a) estimated glomerular tuft volume in rats aged from 5 days to 540 days. The estimates for tissue embedded in glycol methacrylate were (2 rats per time point; mean ± standard deviation): $0.75 \pm 0.07 \times 10^5 \ \mu m^3$ (5 days), $1.20 \pm 0.14 \times 10^5 \ \mu m^3$ (10 days), $4.35 \pm 1.06 \times 10^5 \ \mu m^3$ (30 days), $8.35 \pm 0.21 \times 10^5 \ \mu m^3$ (60 days), $9.00 \pm 0.42 \times 10^5 \ \mu m^3$ (135 days), $11.30 \pm 2.26 \times 10^5 \ \mu m^3$ (270 days) and $12.25 \pm 0.07 \times 10^5 \ \mu m^3$ (540 days).

Nyengaard and Bendtsen (1992) estimated mean glomerular tuft volume in human kidneys from subjects ranging in age from 16 years to 87 years. Tissue was embedded in glycol methacrylate. The mean estimate was $59.8 \times 10^5 \ \mu m^3$, with individual estimates ranging from 38.5 to $89.7 \times 10^5 \ \mu m^3$. Glomerular volume, like glomerular number, was negatively correlated with age and positively correlated with kidney weight.

The approach described here for estimating glomerular volume provides an unbiased estimate of the number-weighted mean glomerular volume. In other words, all glomeruli, regardless of size, make the same contribution to the estimate. As described earlier, Artacho-Perula et al. (1993) used point-sampled intercepts to estimate the volume-weighted mean volumes of glomeruli in control rats and streptozotocin-diabetic rats. With this approach, larger glomeruli influence the final estimate more than smaller glomeruli—the estimate is volume weighted. The estimates of glomerular volume for superficial, midcortical, and juxtamedullary glomeruli in control rats were approximately $2.65 \pm 0.32 \times 10^5 \ \mu m^3$ (mean ± standard deviation), $2.76 \pm 0.35 \times 10^5 \ \mu m^3$, and $2.96 \pm 0.45 \times 10^5 \ \mu m^3$, respectively. While these data illustrate that juxtamedullary glomeruli are the largest glomeruli in the kidney, the estimates are considerably smaller than those cited earlier. This is most likely because Artacho-Perula et al. (1993) embedded the kidney samples in paraffin. As discussed in Section III,E, soft tissues such as kidney shrink by approximately 40 to 50% when processed and embedded in paraffin. Far less shrinkage occurs when kidney tissue is embedded in glycol methacrylate or Epon-araldite. Schmitz et al. (1990), for example, found that the volume of glomeruli embedded in glycol methacrylate was approximately double that of the volume of glomeruli embedded in paraffin. As mentioned previously, volume-weighted estimates of volume are more easily obtained than number-weighted volume estimates. Artacho-Perula et al. (1993) took just 20 min to obtain estimates for each rat once sections were obtained.

Recently, Nyengaard et al. (1994) estimated the number-weighted mean volume as well as the volume-weighted mean volume of glomeruli in normal and lithium-treated rats. The number-weighted mean glomerular volume was reduced by more than 40% in lithium-treated rats, whereas the volume-weighted mean glomerular volume was similar in the two groups. Nyengaard et al. (1994) attributed this discrepancy in volume estimates to the fact that the kidneys of the lithium-treated rats contained many small glomeruli and a few very large glomeruli. Clearly, these larger glomeruli in lithium-treated rats were preferentially sampled when estimating volume-weighted mean volume with point-sampled intercepts. This result nicely illustrates how measurements of size on a biased sample of particles (those contained in a single section for example) can lead to erroneous conclusions.

With the current level of interest in the role of glomerular hypertrophy in progressive glomerular disease (Mauer *et al.*, 1984; Fries *et al.*, 1989; Yoshida *et al.*, 1989; Fogo *et al.*, 1990; Fogo and Ichikawa, 1992; Nagata *et al.*, 1992; Schmidt *et al.*, 1992) and in the role of growth factors and glomerular growth in renal function and disease, it would be highly desirable if a uniform set of quantitative tools could be used to estimate glomerular volume. Differences between glomerular volume estimates from different laboratories are notorious. Such disparity does little to advance our understanding of glomerular biology.

Schmidt *et al.* (1992) recently stated that the "reported values for glomer ular size vary widely" and suggested three main reasons for this variation, these being: (1) sampling of glomeruli, (2) tissue fixation and processing, and (3) the measurements and formulas employed. Adoption of a uniform method for estimating glomerular volume should minimize the current disparity in results. The methods described in this chapter constitute an unbiased and cost-efficient set of tools for estimating glomerular volume, with the effects of technical artifacts minimized, and in some cases totally negated. However, it should also be borne in mind that the methods described here for estimating glomerular volume may not be ideal when dealing with biopsy specimens containing a small number of glomeruli. In this instance the Cavalieri method may be better suited (Lane *et al.*, 1992). Moreover, the Cavalieri method applied to individual glomeruli would provide information on the distribution of glomerular size, not simply a mean value for a population of glomeruli. Methods for estimating morphometric parameters for individual glomeruli and glomeruli in biopsy samples are considered in greater detail in Section V,D.

C. Total Number of Cells in an Average Glomerulus

Two strategies, both employing optical disectors, have been used to estimate the total number of cells in an average glomerulus. These methods are described in Bertram *et al.* (1992) and Marcussen (1992). First I will describe the method used in our earlier study (Bertram *et al.* 1992), and then the approach used by Marcussen (1992).

In our study of the normal rat kidney, the first step in counting glomerular cells was to obtain an unbiased sample of glomerular volume. Accordingly, the glycol methacrylate sections previously used to count glomeruli were placed on a 1-mm × 1-mm grid in order to select a set of fields in which to choose glomeruli. Every ninth field was selected, with the first chosen at random using a random number table. This gave about two fields per section. The fields were outlined with a pen on the underside of the slides, and were then viewed on a microscope at low magnification. One glomerulus in each field was selected for cell counting, this being the glomerulus whose center was closest to the center of the field. Alternatively, the single glomerulus could be selected using random numbers. If more than one glomerulus is required per section, then the modulus sampling approach described by Nyengaard and Marcussen (1993) can be used. A third scheme for sampling glomeruli which is particularly useful for sampling in small specimens such as renal biopsies has been provided by Bangstad *et al.* (1993). In brief, with the specimen embedded in an epoxy resin, 1-μm sections are cut and stained. The first section in the block is not used for sampling, but any new glomerulus appearing in subsequent sections (cut, say, at 10-μm intervals) would be selected (sampled) for analysis. In this way, glomeruli are sampled according to their incidence, not their size or shape. Then sections are collected at regular intervals through the sampled glomeruli, say, every 50 μm.

In our 1992 study (Bertram *et al.,* 1992), to count cells in sampled glomeruli with optical disectors, sections were viewed on an Olympus BH-2 microscope equipped with a 63× oil immersion lens (numerical aperture = 1.4) and an electronic microcator which gave the position of the z-axis with a precision better than 0.5 μm (Fig. 8). Observation of fields was via a color video camera and the image was displayed on a monitor at a final magnification of 1,280×. The projection lens to the video camera was equipped with a square graticule. Optical disectors were sampled in sampled glomeruli by choosing every 14th grid square with a random start between 1 and 14. About 50 grid squares, and thereby disectors, were sampled in approximately 25 glomeruli per rat.

An optical disector for measurement was defined by focusing on the top of the section, zeroing the microcator, and then moving 3 μm into the section. At this time, the number of grid points overlying the renal corpuscle was counted. The microcator was rezeroed, the microscope was focused through the section for 10 μm, and cells were counted when the nuclear edge and associated chromatin came into focus. A series of optical sections through a glomerulus is shown in Fig. 9.

The numerical density of cells in an average glomerulus was calculated using:

$$N_{V\text{cells,glom}} = Q^-/[10 \times P_{\text{glom}} \times a(p)] \tag{19}$$

where Q^- was the total number of glomerular cells counted per kidney, 10 μm was the thickness of the optical disector, P_{glom} was the number of

FIG. 8   Photograph showing modifications needed to a light microscope in order to count cell nuclei with optical disectors. The electronic microcator (arrow) measures movement in the z-axis. The motorized stage is used to obtain a uniform sample of fields on sections and within glomeruli as depicted in Fig. 5(d). To count nuclei with optical disectors, it is necessary to use a high numerical aperture (preferably 1.4) oil immersion lens and a matching condensor lens.

grid points lying on the sampled disectors, and $a(p)$ was the area of each grid square (approximately 160 μm^2).

The mean number of cells (nuclei) in an average glomerulus ($N_{\text{cells,glom}}$) was obtained using:

$$N_{\text{cells,glom}} = N_{V\text{cells,glom}} \times V_{\text{glom}} \tag{20}$$

Approximately 170 ± 30 cells were counted in each kidney in order to estimate $N_{\text{cells,glom}}$. Approximately 2 hr were needed to count this number of cells per kidney using optical disectors. Our estimates for the mean number of cells in glomeruli of adult rats are given in Table II. The mean number of cells in an average glomerulus was 674 cells, with values ranging from 536 to 913 cells.

When counting cell nuclei in thick sections with optical disectors, it is important to ensure that nuclei throughout the section thickness are stained and therefore visible (and countable). In pilot studies it is advisable to stain trial sections for varying periods of time and check stain penetration. We stain our glycol methacrylate sections with Mayer's hematoxylin for 30 min and with 0.1% eosin for 15 min. Of course, optimum staining times vary according to the fixative and processing schedule employed. Unless tissue

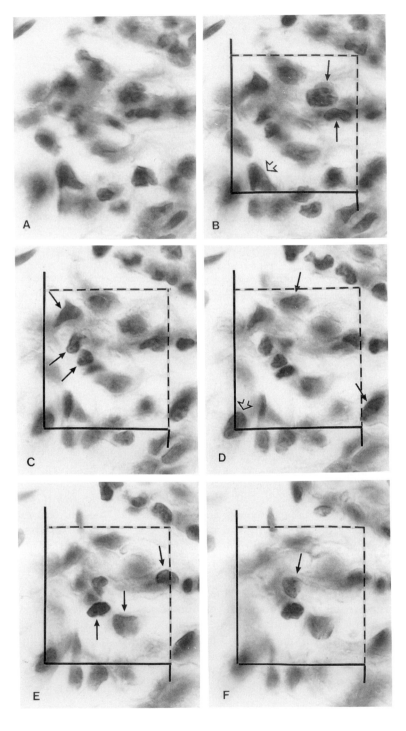

TABLE II

Estimates of the Total Number of Cells in an Average Glomerulus and the Number of Individual Cell Types for Seven Normal Sprague-Dawley Rats[a]

Animal	Total number of cells per glomerulus	Glomerular epithelial cells (podocytes)	Endothelial cells	Mesangial cells	Parietal epithelial cells
1	646 (145)	158 (50)	238 (75)	250 (79)	85 (30)
2	609 (145)	175 (54)	249 (77)	185 (57)	118 (34)
3	655 (202)	133 (44)	288 (95)	234 (77)	157 (39)
4	536 (173)	139 (55)	200 (79)	197 (78)	95 (31)
5	913 (218)	291 (72)	315 (78)	307 (76)	150 (34)
6	583 (140)	173 (66)	165 (63)	245 (93)	107 (40)
7	775 (172)	197 (57)	284 (82)	294 (85)	104 (32)
Mean	674	181	248	245	117
S.D.	129	53	53	45	27
C.V. (%)	19.1	29.3	21.4	18.3	23.1

[a] Values of Q^- (the actual number of cells counted) are given in parentheses. Data reproduced from Bertram *et al.* (1992) with permission.

FIG. 9   A series of six optical sections through a 20-μm glycol methacrylate section of rat glomerulus is shown. Each optical section is separated by 2 μm. Unbiased counting frames are superimposed on the optical sections (the forbidden line is solid). Nuclei that are sampled by the disector are counted when they come into focus. It is important to note that division of the total disector height of 10 μm into 2 μm is arbitrary and only for the purpose of illustration. In practice, one focuses down through the section and counts nuclei as they come into focus. Optical section A defines the top surface of the disector; nuclei in focus here are not counted because they did not come into focus. Therefore, an unbiased counting frame is not depicted in A. In B, three nuclei come into focus. Two of the nuclei (small arrows) do not touch the forbidden line and are therefore counted. The third nucleus (large arrow) touches the forbidden line and therefore is not counted. Three more nuclei come into focus in C. All three are counted. In D, three more nuclei come into focus. Two do not touch the forbidden line and are therefore counted (small arrows), whereas the third nucleus (large arrow) touches the bottom left corner of the counting frame and is therefore not counted. Note that one of the nuclei in D is mostly outside the sampling frame, but it is touched by the inclusion line and is therefore counted. Three more nuclei come into focus in E, and another nucleus comes into focus in F. All four of these nuclei are counted. Thus between sections A and F, a total of 11 nuclei have been counted. If the area of the sampling frame is, say, 2000 μm^2, then the numerical density of nuclei in this optical disector is obtained using: $N_{Vnuc,glom} = 11/10 \ \mu m \times 2000 \ \mu m^2$, which equals 0.55 nuclei/1000 μm^3.

is required for electron microscopy, it is best to avoid glutaraldehyde, and use standard histological fixatives such as neutral buffered formalin or Bouin's fixative. These give optimum nuclear staining, which is helpful when counting with optical disectors.

Marcussen (1992) also employed optical disectors to count glomerular cells. His so-called "double disector" approach does not require estimation of section thickness and is not affected by tissue shrinkage provided the same sections are used for all measurements. The equation used by Marcussen (1992) to estimate the number of cells in an average glomerulus was:

$$N_{Ncells,glom} = [Q^-_{cells} \times P_{Pglom,ref} \times a_{frame\ 1}]/[Q^-_{glom} \times a_{frame\ 2}] \qquad (21)$$

where Q^-_{cells} was the number of cell nuclei counted in glomeruli using optical disectors, $P_{pglom,ref}$ was the proportion of grid points overlying glomeruli; $a_{frame\ 1}$ was the area of the test frame used to count glomeruli; Q^-_{glom} was the number of glomeruli counted using physical disectors, and $a_{frame\ 2}$ was the area of the test frame used to count glomerular cell nuclei. Although it is not necessary to measure section thickness in order to estimate glomerular cell number with Eq. (21), if an estimate of average glomerular volume is required, then section thickness must be measured. Marcussen (1992) estimated glomerular cell number in three human kidneys and in one male Wistar rat. Mean values were 2,853 cells and 576 cells, respectively.

D. Number of Individual Cell Types in an Average Glomerulus

An unbiased stereological method for estimating the number of individual cell types in average renal glomeruli was described by Bertram *et al.* (1992). The method relies on transmission electron microscopy for accurate identification of glomerular cells because many glomerular cells are difficult if not impossible to identify in standard light microscopic preparations. The method employs physical disectors of unknown thickness at the electron microscopic level to estimate the relative frequencies of cells. We have used this method to estimate the numbers of GECs, endothelial cells, mesangial cells, and parietal epithelial cells in average glomeruli in the normal rat kidney (Bertram *et al.,* 1992) as well as in the kidneys of rats with acute puromycin aminonucleoside nephrosis (Soosaipillai, 1994). In these studies, mesangial cells were defined as cells present in the glomerular mesangium. No attempt was made to differentiate between different populations of mesangial cells, or infiltrating cells. It is worth emphasizing at this point that although the disector originally presented by Sterio (1984) is often thought of and described as a counting method, it is more correctly defined as a sampling method, specifically, a three-dimensional probe that samples particles with uniform probability, irrespective of their size, shape, or orientation. When estimating cell or nuclear numerical density (number per volume) with the disector (physical or optical), the height or thickness of the disector must be known. However, when estimating relative frequencies of cells, knowledge of section thickness is not required. This fact was employed in our study of glomerular cell number, as we wished to avoid the well-known difficulties associated with estimating the thickness of electron microscopic thin sections. For this reason, we described our counting technique as a "physical disector of unknown thickness."

The first step in counting individual glomerular cell types with electron microscopy is to obtain a sample of kidney blocks. Two approaches are available. First, the renal cortex can be diced into small blocks suitable for electron microscopy, and a random sample of blocks selected for analysis. A drawback with this approach is that it is theoretically possible that blocks might come from adjacent sites in the cortex. An alternative approach yields a systematic uniform random sample of blocks for electron microscopy. A plastic sheet containing small holes in a uniform pattern is randomly placed over kidney slices. A biopsy needle is used to collect samples of cortex from those sites corresponding to the holes in the plastic sheet.

Once blocks have been sampled and embedded, a semithin section is cut from each block and stained with methylene blue. Glomerular profiles are sampled as described in Section V,C. Two thin sections approximately 1.5 μm apart were cut from these sampled glomeruli, mounted on a single slot grid coated with a film of 0.6% Butvar, and stained with uranyl acetate and lead citrate. Sections were viewed in a Philips EM 400, and the entire cross-section of the renal corpuscle photographed. Montages, each consisting of about 10 micrographs, were constructed at a final magnification of approximately 3,900×. Then the physical disector principle of Sterio (1984) was used to count cell nuclei. Accordingly, those nuclei present in the sample section of the disector pair but not in the look-up section were counted, and vice versa (to double sample size). Corresponding fields from a disector pair are shown in Fig. 10. With this approach, unbiased counting frames (Gundersen, 1977) are not required when counting nuclei because the entire cross-section of the renal corpuscle is analyzed—it is not subsampled. In our study on the normal rat kidney, we counted approximately 250 nuclei (from a total of 6 glomeruli) with electron microscopy from each kidney (Bertram *et al.*, 1992). The counting took about 1.5 hr per kidney; this excluded the time taken to cut thin sections, take and print electron micrographs, and construct montages.

The following equation was used to estimate, for example, the total number of GECs in an average glomerulus:

$$N_{gec,glom} = (Q^-_{gec}/Q^-_{all\ glom\ cells}) \times N_{cells,glom} \qquad (22)$$

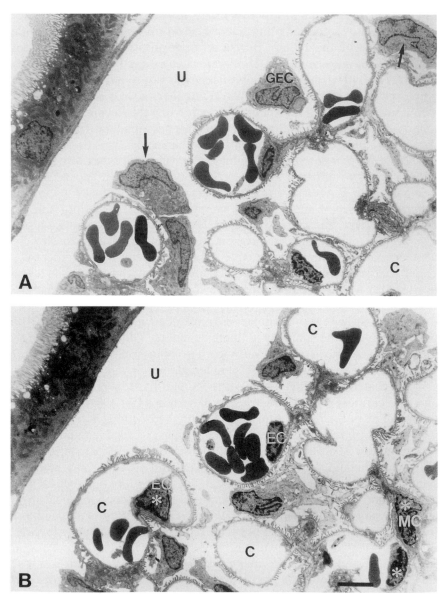

FIG. 10 A pair of transmission electron micrographs used to estimate the number of individual cell types in the rat glomerulus. The micrographs come from two thin sections approximately 1.5 μm apart. Nuclei are counted using the physical disector principle of Sterio (1984); they are counted when they are present in one section but not the other. The two GECs indicated by arrows in A are counted because they are not present in B. Those nuclei present in B but not present in A are also counted. These nuclei are indicated by asterisks. Counting frames

where Q^-_{gec} was the number of GECs counted, $Q^-_{all\ glom\ cells}$ was the total number of glomerular cells counted with electron microscopic disectors, and $N_{cells,glom}$ was the total number of cells in an average glomerulus, previously obtained using light microscopic optical disectors. Our estimates for the total number of GECs, mesangial cells, endothelial cells, and parietal epithelial cells in an average glomerulus of the normal rat kidney are given in Table II.

E. Total Number of Each Glomerular Cell Type in a Kidney

Unbiased estimates of the total number of each glomerular cell type in a kidney can be obtained by multiplying $N_{glom,kid}$ by the total number of each cell type in an average glomerulus. For example, the total number of GECs in a kidney can be estimated using:

$$N_{gec,kid} = N_{gec,glom} \times N_{glom,kid} \qquad (23)$$

The normal adult rat kidney contains $5.614 \pm 1.060 \times 10^6$ GECs, $7.793 \pm 1.444 \times 10^6$ glomerular endothelial cells, $7.685 \pm 1.125 \times 10^6$ mesangial cells, and $3.699 \pm 0.988 \times 10^6$ parietal epithelial cells (Bertram *et al.*, 1992).

F. Average Volumes and Surface Areas of Individual Glomerular Cell Types

The electron microscopic sections used to count individual glomerular cell types can be used to estimate the absolute volumes and surface areas of those cell types (Bertram *et al.*, 1992; Soosaipillai, 1994; Soosaipillai *et al.*, 1994). Again, the strategy is to estimate stereological densities (volume densities and surface densities) as a means to an end in order to determine final estimates of absolute volumes and surface areas.

In our studies we use thin sections from six glomeruli from each rat for this purpose. These glomeruli can be sampled as described in Section V,C. Approximately 20 electron micrographs are obtained from a single section

---

are not used by us when counting nuclei at the electron microscopic level because we construct montages of complete glomerular profiles. If fields within glomeruli are to be analyzed, then it would be necessary to use unbiased two-dimensional counting frames. U, urinary space; C, capillary lumen; GEC, glomerular epithelial cell; MC, mesangial cell; EC, endothelial cell. Bar = 5 μm. (Reproduced with permission from Bertram *et al.*, 1992.)

from each renal corpuscle using a systematic uniform random sampling scheme. The micrographs are printed at a final magnification of approximately 23,500×. A micrograph of a carbon replica grating (21,600 lines/ cm) is obtained during each microscope session and used to calibrate micrographs. Point counting methods are used to estimate volume densities in renal corpuscles of the nuclei and cytoplasm of GECs, endothelial cells, mesangial cells, and parietal epithelial cells. An orthogonal test grid drawn on transparent plastic and comprising 84 points separated by 2 cm is used. We use PCS System II software (Bolender and Pentcheff, 1985; Pentcheff and Bolender, 1985; Pentcheff, 1987) to collect and store data.

The volume density of, for example, mesangial cells in an average renal corpuscle is obtained using:

$$V_{Vmes,glom} = P_{mes}/P_{glom} \qquad (24)$$

where P_{mes} is the number of grid points overlying the nuclei and cytoplasm of mesangial cells, and P_{glom} is the total number of grid points overlying the glomerulus. Approximately 9,500 ± 300 points were counted in each kidney (approximately 1600 points per glomerulus) in order to estimate the 11 volume densities (nuclear and cytoplasmic volume densities of the four cell types, as well as volume densities of ECM, urinary space, and capillary lumina) (Bertram et $al.$, 1992; Soosaipillai, 1994; Soosaipillai et $al.$, 1994).

The total volume of, for example, mesangial cells in an average glomerulus was obtained using:

$$V_{total\ mes} = V_{Vmes,glom} \times V_{glom} \qquad (25)$$

The mean absolute volume of an average mesangial cell was obtained through:

$$V_{mes} = V_{total\ mes}/N_{mes,glom} \qquad (26)$$

where $N_{mes,glom}$ was the total number of mesangial cells in an average glomerulus. Estimates for the volume densities and absolute volumes of the four cell types in the kidneys of seven normal adult rats are presented in Table III. The largest cell in the normal rat glomerulus is the GEC with a mean volume of 584.2 μm^3. GECs are approximately four times larger than both endothelial cells and mesangial cells (151.1 μm^3 and 149.0 μm^3, respectively), and three times larger than parietal epithelial cells (182.0 μm^3).

As described earlier, intersections between grid test lines and membranes as observed in electron micrographs can be used to estimate the surface density (surface area per unit volume) of those membranes. Theoretically, such methods could be used to estimate the surface densities of average GECs, mesangial cells, endothelial cells, and parietal epithelial cells. However, estimating the surface area of glomerular endothelial cells would be

TABLE III

Volume Densities (in Renal Corpuscles) and Absolute Volumes of Glomerular Epithelial Cells, Endothelial Cells, Mesangial Cells, and Parietal Epithelial Cells in Seven Normal Sprague-Dawley Rat Kidneys[a]

Animal	$V_{V(gec,corp)}$	V_{gec} (μm^3)	$V_{V(endo,corp)}$	V_{endo} (μm^3)	$V_{V(mes,corp)}$	V_{mes} (μm^3)	$V_{V(pec,corp)}$	V_{pec} (μm^3)
1	0.126	733.5	0.056	217.4	0.041	153.6	0.025	273.5
2	0.149	687.8	0.049	157.5	0.043	189.2	0.031	224.1
3	0.105	535.5	0.049	116.9	0.048	140.3	0.028	161.3
4	0.127	523.3	0.042	119.9	0.055	159.8	0.022	162.9
5	0.129	428.7	0.046	140.4	0.045	144.1	0.020	140.7
6	0.136	625.2	0.039	186.8	0.052	167.6	0.022	167.3
7	0.124	555.3	0.038	118.7	0.029	88.5	0.018	144.4
Mean	0.128	584.2	0.045	151.1	0.045	149.0	0.024	182.0
S.D.	0.013	104.7	0.007	38.8	0.008	31.3	0.004	48.8
C.V. (%)	10.5	17.9	14.3	25.7	18.6	20.9	18.6	26.8

[a] Data reproduced from Soosaipillai et al. (1994).

difficult because correction factors would be needed to correct for the small diameter of endothelial cell fenestrae compared with section thickness. For further consideration of corrections for surface density estimates, see Weibel (1979, 1980) and Mayhew and Reith (1988).

We used intersection counting methods to estimate the surface density of, and thereby the absolute surface area of average GECs (Soosaipillai, 1994; Soosaipillai et al., 1994). We used the electron micrographs and test grid previously used to estimate cytoplasmic and nuclear volume densities. The surface density of GEC plasma membrane in glomeruli ($S_{Vgec,glom}$) was estimated using:

$$S_{Vgec,glom} = 2I/(P_{glom} \times 2 \times d) \tag{27}$$

where I was the total number of intersections counted, P_{glom} was the total number of points counted on the glomerulus, and d was the length of each test line corrected for magnification.

The mean surface area of GECs was then estimated using:

$$S_{gec} = (S_{Vgec,glom} \times V_{glom})/N_{gec,glom} \tag{28}$$

The surface area of the plasma membrane in GECs in the normal adult rat kidney is $5211 \pm 967 \ \mu m^2$ (Soosaipillai et al., 1994). Estimates for seven normal rats are given in Table IV. In conditions such as human minimal change disease, which is characterized by effacement of GEC foot processes, S_{gec} would be expected to decrease. In an animal model of human minimal change disease (puromycin aminonucleoside nephrosis in rats), we have

TABLE IV

Surface Density (in Renal Corpuscle) and Absolute Surface Area of Glomerular Epithelial Cell Plasma
Membrane in Seven Normal Rat Kidneys[a]

Animal	1	2	3	4	5	6	7	Mean	S.D.	C.V. (%)
$S_{V(gec,corp)}$ $\mu m^2/\mu m^3$	0.971	1.212	0.928	0.980	0.971	1.160	0.946	1.024	0.113	11.0
S_{gec} μm^2	5,657	5,591	4,746	4,041	6,865	5,333	4,243	5,211	967	18.6

[a] Data reproduced from Soosaipillai *et al.* (1994).

shown that S_{gec} decreases by approximately 25% 10 days after a single
intravenous injection of puromycin aminonucleoside (Soosaipillai, 1994).

G. Total Length, Surface Area, and Number of Capillaries in an Average Glomerulus

1. Length of Capillaries in an Average Glomerulus

An estimate of the total length of capillaries in an average glomerulus can
be obtained by multiplying the length density of capillaries in glomeruli
(length per unit volume) by mean glomerular volume. Capillary length
density can easily be estimated at the light microscopic level using semithin
Epon sections from perfusion-fixed kidneys. Glomeruli should be sampled
as described in Section V,C. To estimate capillary length density, capillary
profiles can be counted in semithin plastic sections at a final magnification
of approximately 1500×. Nagata *et al.* (1992) used photomicrographs for
this purpose whereas Nyengaard (1993a,b) projected the images. An unbiased two-dimensional counting frame (Gundersen, 1977) can be used to
count capillary profiles in subsamples of sectioned glomeruli, or alternatively, the total number of capillary profiles in complete glomerular profiles
can be counted, thereby negating the use of an unbiased counting frame.
The length of capillaries per unit volume of glomerulus ($L_{Vcap,glom}$) is estimated using:

$$L_{Vcap,glom} = (2 \times Q_{cap})/a_{glom} \qquad (29)$$

where Q_{cap} is the number of capillary profiles in a known area of sectioned
glomerulus (see Osterby and Gundersen, 1988 for further details).

The total length of capillaries in an average glomerulus is obtained using:

$$L_{cap,glom} = L_{Vcap,glom} \times V_{glom} \qquad (30)$$

Nagata *et al.* (1992), Nyengaard (1993a), and Nyengaard and Rasch (1993) recently estimated the length of capillaries in normal rat glomeruli and in rats subjected to various treatments. Nagata *et al.* (1992) reported that in 4-week, 12-week, and 24-week male Sprague-Dawley rats, capillary length in average glomeruli was 6.0 mm, 12.2 mm, and 16.3 mm, respectively. In female Wistar rats weighing 250 g, capillary length per average glomerulus was 12.5 mm (Nyengaard and Rasch, 1993). In 540-day female Wistar rats, the average length of capillaries in glomeruli is 12.1 mm (Nyengaard, 1993a). Nyengaard (1993b) also used this approach to estimate glomerular capillary length in growing rats.

2. Surface Area of Capillaries in an Average Glomerulus

To estimate the surface area of capillaries per unit volume of glomerulus ($S_{Vcap,glom}$) it is necessary to estimate the boundary of capillary walls per unit area of sectioned glomerulus. This can be done using semithin Epon sections stained with toluidine blue or methylene blue and viewed at a magnification of around 1500×. The boundary of capillary wall can be obtained using a digitizing tablet or a stereological test grid. If an orthogonal test grid is used, then $S_{Vcap,glom}$ is obtained using:

$$S_{Vcap,glom} = (2 \times I_{cap})/(P_{glom} \times k \times d) \tag{31}$$

which is identical to Eq. (5). I_{cap} is the number of intersections between the orthogonal test grid and the capillary wall, and P_{glom} is the number of grid points overlying glomerular profiles. The denominator in Eq. (31) gives the total length of test line overlying the reference compartment (glomerulus).

The total surface area of capillaries in an average glomerulus is obtained using:

$$S_{cap,glom} = S_{Vcap,glom} \times V_{glom} \tag{32}$$

When estimating the surface density of glomerular capillaries, as for the estimation of glomerular capillary length, special steps must be taken to obtain an unbiased sample of glomeruli and an unbiased sample of fields on sections from glomeruli. See Section V,C for details.

In many previous studies, the surface area of the glomerular capillary wall has been estimated at the electron microscopic level. This enables subdivision of the capillary wall into the peripheral GBM and the mesangial-capillary interface. In this instance Eqs. (31) and (32) are used, with intersections between the stereological grid and the two subdivisions of the capillary wall counted separately. It is well known that estimates of surface density and thereby estimates of absolute surface areas are influenced by magnification (Weibel, 1979). With the so-called "coast-of-England" effect, small

protruberances and invaginations in surfaces are seen at higher magnification and therefore measured, but are not seen or measured at lower magnification. One should therefore exercise caution when comparing surface
density and absolute surface area estimates obtained at widely differing
magnifications.

3. Number of Capillaries in an Average Glomerulus

Nyengaard and Marcussen (1993) recently described an unbiased method
for estimating the total number of capillaries in an average glomerulus.
The method is based on estimation of the Euler-Poincaré characteristic or
Euler number (χ) with the disector. Roughly speaking, the Euler number
is equivalent to connectivity. The generation of a new capillary creates a
new loop in the capillary network and this results in a change of one unit
in the Euler number of the network.

In practice, the Nyengaard and Marcussen (1993) method involves the
use of physical disectors at the light microscopic level to estimate the mean
Euler number per volume, which is then multiplied by mean glomerular
volume to obtain the total Euler number and thereby the total number
of capillaries per glomerulus. Three kinds of topological events involving
capillary lumina are counted using physical disectors in pairs of adjacent
semithin sections at a magnification of approximately 1500×. These topological events are (1) the appearance of a capillary lumen (implying a
luminal fragment); (2) the appearance of an isolated island of capillary wall
inside the lumen of a capillary profile (implying a luminal lagoon—this is
rarely observed); and (3) the division of a sample capillary lumen into two
or more capillary lumina (defined as a luminal connection). These three
luminal events are schematically illustrated in Fig. 11. The Euler number
is estimated using:

$$\Sigma_\chi \text{ (cap)} = 1/2 \times (\Sigma \text{luminal fragments} + \Sigma \text{luminal lagoons} - \Sigma \text{luminal connections}) \qquad (33)$$

The constant 1/2 appears because each topologically defined capillary has
two luminal connections. The capillary numerical density in glomeruli
($W_{V\text{cap,glom}}$) is obtained using:

$$W_{V\text{cap,glom}} = -\Sigma_\chi \text{ (cap)}/(2 \times t \times a_{\text{glom}}) \qquad (34)$$

where t is section thickness, 2 refers to the fact that for efficiency reasons
the disector is used in both directions (that is, from the top section to
the bottom section, and vice versa), and a_{glom} is the area of the reference
compartment (glomerular tuft) in the sampled section. The average total
number of capillaries per glomerulus ($w_{\text{cap,glom}}$) is obtained using:

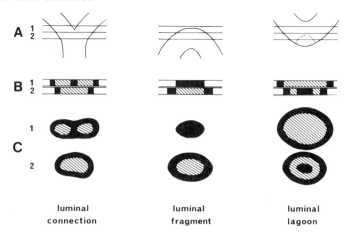

luminal   luminal   luminal
connection   fragment   lagoon

FIG. 11   Schematic diagram showing the three different capillary events that are counted to obtain an unbiased estimate of the total number of capillaries in a glomerulus. The different capillary events are shown in three different ways. The capillary lumina are hatched in order to emphasize that it is the topological events involving the capillary lumina which are used to estimate the Euler number. In A, the three section planes, indicated as parallel lines, are placed with respect to the capillaries. In B, two thick, transparent sections are viewed from the side, and in C the sections are viewed from the top. A luminal connection is shown at the left, a luminal fragment in the center, and a luminal lagoon at the right. It is mainly the luminal connections and fragments that are seen in sections, with the luminal lagoons rarely observed. In practice, physical disectors are used to estimate the numerical density of these luminal events using pairs of semithin Epon sections from perfusion-fixed glomeruli. Multiplication of numerical density by glomerular volume gives an estimate of the total number of capillaries in an average glomerulus. (Figure reproduced with permission from Nyengaard and Marcussen, 1993.)

$$w_{\text{cap,glom}} = (V_{\text{glom}} \times W_{V\text{cap,glom}}) + 2 \qquad (35)$$

Nyengaard (1993a) used this method to estimate glomerular capillary number in female Wistar rats ranging in age from 5 days to 18 months. Average glomeruli in two 5-day rats contained 13 and 14 capillaries, whereas the number in two 18-month rats was 175 and 234 capillaries (Nyengaard, 1993a). Nyengaard (1993b) used this method to estimate the number of capillaries in average rat glomeruli following unilateral nephrectomy. The average total number of capillaries per glomerulus in 18-month control, neonatal nephrectomized, and adult nephrectomized rats was 213, 357, and 260, respectively, indicating true angiogenesis (formation of new capillaries) following nephrectomy. Nyengaard and Rasch (1993, 1994) used this method to estimate glomerular capillary numbers in streptozotocin diabetic rats. The average number of capillaries per glomerulus increased from 215 in control rats to 260 and 316 in the 10-day diabetic and 50-day diabetic rats, respectively.

4. Mean Length and Surface Area of Glomerular Capillaries

The mean length of glomerular capillaries can be estimated using:

$$L_{cap} = L_{cap,glom}/w_{cap,glom} \qquad (36)$$

The mean surface area of glomerular capillaries can be estimated using:

$$S_{cap} = S_{cap,glom}/w_{cap,glom} \qquad (37)$$

H. Volumes of Capillaries, Matrix, and Urinary Space in an Average Glomerulus

1. Capillaries and Urinary Space

The volume density of capillary lumina and urinary space in renal corpuscles can be estimated at both the light and electron microscopic levels. For light microscopic estimation, the glomeruli must be perfusion fixed to ensure that no collapsed capillaries are present. Sampled glomeruli can be examined at a magnification of around 1500×. Section thickness should be less than approximately 2 μm, to guard against the Holmes section thickness effect. Grid points overlying capillary lumina and urinary space are counted. Of course, some urinary space located within the glomerular tuft (or within the minimal convex polygon that encloses the glomerular tuft) may not be visualized by light microscopy, and so the estimate of urinary space volume density and absolute volume will tend to be underestimated by light microscopy.

A systematic uniform random sample of electron micrographs such as those used by Bertram *et al.* (1992) to estimate the volume and surface densities of glomerular cells can also be used to estimate the volume densities of capillary lumina and urinary space. Usually, the volume densities of these compartments are estimated at a magnification between 2500× and 10,000×, although we used micrographs with a magnification of 23,500× (Soosaipillai, 1994; Soosaipillai *et al.*, 1994). We used an orthogonal test grid, and the volume density of, for example, capillary lumina in renal corpuscle ($V_{Vcap,corp}$) was obtained using:

$$V_{Vcap,corp} = P_{cap}/P_{corp} \qquad (38)$$

The absolute volume of capillary lumina in an average renal corpuscle ($V_{cap,corp}$) was obtained using:

$$V_{cap,corp} = V_{Vcap,corp} \times V_{corp} \qquad (39)$$

Estimates for the volume densities and absolute volumes of capillaries and urinary space in seven normal rats are presented in Table V. These estimates were obtained using point counts on approximately 20 electron micrographs from each of six glomeruli from each of seven rats.

The author is unaware of any studies in which the volume densities of urinary space and capillary lumina in glomeruli have been estimated using both light and electron microscopy, enabling comparison of the two sets of estimates. However, as mentioned earlier, light microscopic estimates for urinary space volume density may be low because of limited resolution and the complex architecture of the glomerular tuft.

2. Matrix

Glomerular ECM has been measured in numerous studies. After glomerular volume, glomerular ECM is perhaps the most widely reported glomerular parameter. However, the ECM parameters estimated, and the methods used to measure these parameters, vary widely among studies. Morphometric parameters concerned with glomerular ECM include matrix volume density (volume fraction; referenced to either a unit volume of the glomerular tuft, a unit volume of the minimal convex polygon drawn around the glomerular tuft, or a unit volume of the renal corpuscle), GBM thickness, mesangial volume density (which takes into account both the cellular and noncellular components of the mesangium), and matrix star volume. Methods for estimating matrix volume density and GBM thickness are described here. Methods for estimating matrix star volume, and derived structural quantities

TABLE V

Estimates of Volume Densities (in Renal Corpuscle) and Absolute Volumes of Glomerular Matrix, Capillary Lumina, and Urinary Space in Seven Normal Rat Kidneys[a]

Animal	$V_{V(ecm,corp)}$	$V_{V(cap,corp)}$	$V_{V(us,corp)}$	V_{ecm} ($\mu m^3 \times 10^4$)	V_{cap} ($\mu m^3 \times 10^5$)	V_{us} ($\mu m^3 \times 10^5$)
1	0.043	0.332	0.378	3.616	2.803	3.190
2	0.050	0.275	0.403	4.371	2.387	3.500
3	0.048	0.288	0.434	4.372	2.610	3.926
4	0.045	0.313	0.397	3.090	2.165	2.753
5	0.038	0.320	0.403	3.950	3.370	4.245
6	0.055	0.317	0.380	4.443	2.580	3.091
7	0.037	0.327	0.427	3.045	2.731	3.562
Mean	0.045	0.310	0.403	3.841	2.664	3.467
S.D.	0.007	0.021	0.021	0.604	0.378	0.509
C.V. (%)	14.9	6.7	5.3	15.7	14.2	14.7

[a] Data reproduced from Soosaipillai et al. (1994).

such as matrix "thickness" and mesangial surfaces are given in Osterby (1992), Walker *et al.* (1992), and Bangstad *et al.* (1993).

a. Matrix Volume Density   Estimation of matrix volume density, like all glomerular stereology, should begin with obtaining an unbiased sample as described in Section V,C. Once glomeruli have been sampled, thin sections for electron microscopy are cut and stained. Electron micrographs are then obtained from each glomerular profile, either as a systematic uniform random sample of micrographs across the profile, or as a photomontage. As for estimation of the volume densities of capillary lumina and urinary space, the magnification of electron micrographs used to estimate matrix volume density varies widely among laboratories. For example, we used a final magnification of about 23,500, whereas Mauer *et al.* (1984) used 18,000×, Remuzzi *et al.* (1992) used 3400×, and Bangstad *et al.* (1993) used 2350×. To estimate matrix volume density, a stereological test system is placed over the micrographs and grid points overlying the matrix are counted, together with grid points overlying the glomerular profile (however defined). For further details on grid dimensions used to estimate matrix volume density, see the original papers. Caution must be exercised when defining the reference compartment if matrix volume density is to be multiplied by glomerular volume to obtain an estimate of absolute mesangial matrix in the glomerulus—the definition of glomerulus must be identical in both parts of the investigation. Matrix volume density is estimated using a modification of Eq. (38), and absolute matrix volume in glomeruli is obtained using Eq. (39).

One final point should be made with regard to estimating matrix volume density, and this concerns variation in matrix volume density within individual glomeruli. Several groups have found that the intragroup coefficient of variation for estimates of matrix volume density are generally higher than the coefficients of variation for the estimates of GBM thickness (Osterby, 1990; Steffes *et al.,* 1983; Osterby *et al.,* 1990). Similarly, a high degree of intraindividual variation has been observed (Osterby *et al.,* 1988). Therefore, measurements of matrix volume density based on one plane from a few glomeruli may lead to misleading estimates. To overcome this intraglomerular variation, Walker *et al.* (1992) proposed that thin sections from several levels from each glomerulus should be analyzed. For example, thin sections could be obtained at 60μm intervals. Provided the first level is chosen at random, which is readily done when disectors are used to select glomeruli, sections obtained constitute an unbiased sample of the glomerulus. For further discussion of this technique, see Osterby (1992) and Bangstad *et al.* (1993).

Oftentimes in glomerular morphometry, mesangial volume density (in glomeruli) is measured in addition to matrix volume density. To estimate mesangial volume density with a stereological test grid, points overlying all components of the glomerular mesangium (cellular and noncellular components) are counted, and divided by the number of points overlying the glomerulus. Basgen *et al.* (1988) compared the accuracy and efficiency of three methods for estimating mesangial volume density. These methods were a digitizing tablet, a 16-point grid, and an 81-point grid. Basgen *et al.* (1988) found the three techniques to give comparable results, but the 81-point grid and the 16-point grid were, respectively, 1.9 and 3.2-fold faster than the digitizing tablet.

b. Thickness of the GBM   The apparent thickness of the GBM as seen in electron micrographs depends on the true width of the GBM as well as on the angle of sectioning. Therefore any method for estimating GBM thickness must take into account the angle of sectioning. In practice, the only distribution that can easily be obtained is that of a random orientation distribution. Again, such a distribution can be obtained by either assuming that the distribution of the GBM in glomeruli is isotropic, or alternatively, using approaches such as the isector to guarantee isotropy (Nyengaard and Gundersen, 1992).

The method most commonly used to estimate the thickness of the GBM is the orthogonal intercept method described by Jensen *et al.* (1979). Again, sampling of glomeruli is the first and most critical step. Once an unbiased sample of glomeruli has been obtained, a representative sample of electron micrographs must be obtained. These are typically printed at a magnification between 10,000 and 30,000. A stereological test grid is placed over the micrographs and GBM thickness is measured where grid lines intersect the endothelial aspect of the peripheral GBM. The harmonic mean GBM thickness (T_h) is calculated using:

$$T_h = (8/3\pi) \times l_h$$

(40)

where $8/3\pi$ corrects for variation in the angle of section through the GBM, and l_h is the harmonic mean apparent thickness calculated as:

$$l_h = n/\Sigma(1/L) \qquad (41)$$

where n is the number of measurements made in tissue and L is the measured GBM thickness in absolute units. The harmonic mean thickness described above is a more stable estimate of GBM thickness than is arithmetic mean thickness, and therefore is the more efficient estimator. Only approximately 150 measurements of GBM thickness are needed per glomerulus in order to obtain an estimate of harmonic mean thickness with a standard error of around 2% (Jensen *et al.*, 1979). As with all morphometry conducted at the electron microscopic level, it is highly desirable to record a micrograph of a calibration grating during each microscope session in order to control for lens hysteresis. For further descriptions of test grids and strategies for estimating the thickness of the GBM, see Hirose *et al.* (1982), Mauer *et al.* (1984), and Miller and Meyer (1990).

VI. Stereology of Individual Glomeruli

Most of the methods described here are designed to provide stereological estimates for average glomeruli. In some instances, however, investigators may wish to estimate values for individual glomeruli, enabling correlation of morphological features within single glomeruli (for example, cell number to glomerular volume, capillary number to endothelial cell number, ECM volume to mesangial cell number). Alternatively, the investigator may wish to correlate morphological and functional data from the same glomerulus.

Another reason for considering glomeruli individually, rather than as a homogeneous population is the well-known fact that not all glomeruli are the same. For example, juxtamedullary glomeruli are considerably larger than cortical glomeruli. Moreover, many glomerulopathies are focal, and therefore stereological estimation of values for average glomeruli may fail to identify significant changes in a small subpopulation of glomeruli. In this section methods that might be used to obtain unbiased estimates for single, perhaps designated glomeruli are presented. These methods may also be particularly useful when dealing with renal biopsies in which the number of glomeruli available for analysis is low.

If a designated glomerulus, or piece of cortex containing a designated glomerulus was embedded whole in glycol methacrylate, and then exhaustively sectioned at, say, 12 μm (measured using a micrometer as described in Section V,A), it would be a relatively straightforward matter to estimate renal corpuscle and glomerular tuft volume, and the total number of cells in the glomerulus. An estimate of volume with acceptable precision could be obtained using the Cavalieri principle with about eight systematic random sections from the glomerulus. High numerical aperture objective lenses could be used to focus at a predefined depth in each section, enabling estimation of glomerular profile area, and estimation of glomerular volume with Eq. (9). The total number of cells in the glomerulus could be estimated using an optical disector/Cavalieri combination. Optical disectors could be used to estimate nuclear numerical density using Eq. (8), and total glomerular cell number estimated with Eq. (10). With all measurements made on the same sections, the effects of shrinkage and swelling artifacts and section compression on cell number estimation would be negated, but of course glomerular volume would be affected.

An alternative scheme would see the glomerulus or cortex sample embedded in glycol methacrylate or Epon/araldite and exhaustively sectioned at 2 μm. A systematic random sample of, say, ten to fifteen section pairs would be collected. Glomerular volume could be estimated using the Cavalieri principle and glomerular cell number estimated using a physical disector/Cavalieri combination. Of course, estimating cell numbers with physical disectors is much slower than using optical disectors, but for single glomeruli, nuclear alignment should not be too difficult. Section pairs could be projected side by side at a final magnification of about 1500×. These sections could also be used to estimate capillary length density, capillary surface density, and capillary numerical density. Multiplication of these densities by total glomerular volume would provide estimates of absolute capillary length, surface area, and number in the glomerulus. Thus, with this design it would be possible to estimate values for five absolute glomerular parameters as well as three stereological densities. Again, with all measurements made on the same sections, for many parameters the effects of shrinkage and swelling artifacts and section compression would be negated, and depending on the parameter of interest, it may not be necessary to estimate section thickness.

It might also be possible in the light microscopic schemes described to estimate the number of parietal epithelial cells, and perhaps GECs, in single renal corpuscles. As described earlier, we previously used physical disectors at the electron microscopic level to identify and count individual glomerular cell types (Bertram *et al.*, 1992). However, we have also tried to count GECs and parietal epithelial cells with optical disectors in glycol methacrylate sections. Our light microscopic estimates of GEC number agreed quite well with the estimates obtained with electron microscopy, on average being only 10% less (J. F. Bertram, M. C. Soosaipillai, and G. B. Ryan, unpublished data). We consider that with practice even better agreement between light and electron microscopic estimates for GEC number will be obtained. Of course, a great advantage of the optical disector method compared with traditional cell counting methods is that one gets to see many sections (optical) through each cell. This facilitates cell identification in many instances. Our optical disector estimate for the number of parietal epithelial cells was 29% less than the estimate obtained with electron microscopy, but this most likely reflects the low number of parietal epithelial cells actually counted rather than difficulty in cell identification. An alternative light microscopic method for counting GECs and parietal epithelial cells would involve the use of physical disectors on pairs of semithin Epon/araldite or glycol methacrylate sections. Electron microscopy will probably still be required to identify, and thereby count, glomerular endothelial and mesangial cells, unless immunohistochemical methods are employed (see

Section, VII). With regard to estimating the volumes of individual glomerular cell types, electron microscopy will always be required.

Remuzzi *et al.* (1990) recently illustrated the importance of examining serial sections from glomeruli when studying segmental lesions. The systematic random sample of semithin sections described earlier would be ideally suited to studies of segmental lesions, and to correlating glomerular size, cellularity, and capillarity with the degree of sclerosis.

VII. Future Directions

The past 10 years since the publication of the disector method by Sterio (1984) have witnessed a revolution in stereology. As described in this chapter, many of the unbiased methods developed since 1984 have been used to analyze glomerular structure. Although it is difficult to predict future directions in the quantitative analysis of glomerular morphology, it seems likely that three areas will see developments in the near future: (1) the application of confocal microscopy to glomerular stereology, (2) the combination of immunohistochemical methods with stereology, and (3) the development of noninvasive methods for estimating glomerular number.

Confocal microscopy provides an ideal environment for stereology through the provision of thin, perfectly registered (aligned) optical sections. The relevant question then is why cut physical sections with a microtome if the confocal microscope can provide optical sections? Stereology has been used with confocal microscopy to estimate the volume, surface area, and dendritic length of designated neurons (Howard *et al.,* 1992, 1993). It appears to be ideally suited to estimation of glomerular volume, cellularity, and capillarity (length, surface area, and number). Presumably, thick resin, frozen, or vibratome sections could be used, or even unembedded slices of fixed kidney.

The identification and therefore counting of individual glomerular cell populations at the light microscopic level is difficult in standard histological sections. Although thick glycol methacrylate sections can be used to count GECs and parietal epithelial cells with optical disectors as described earlier, identification of endothelial cells and mesangial cells remains difficult. One solution to this problem would involve using immunohistochemical methods to identify endothelial cells or mesangial cells in 3-μm or 4-μm paraffin sections. If adjacent sections were so prepared, then the nuclei of cells with specific cytoplasmic immunostaining could be counted using physical disectors. Ideally however, optical disectors or confocal microscopy could be used to count these cells, thereby overcoming the tedious task of section alignment. Vibratome or frozen sections may be useful here. Immunohisto-chemistry performed on free-floating sections may also be advantageous. One problem associated with quantitation using frozen sections has been section collapse. However, this problem has recently been solved (Tolcos, 1994), allowing estimation of the total number of neurons in the hypoglossal nucleus of the guinea pig by using an optical disector/fractionator combination.

All of the methods described in this chapter require fixation and sectioning of kidney tissue. Most of the methods require uniform sampling from an entire kidney. While the entire kidney is usually available in studies on laboratory animals, this is rarely the situation when dealing with patients. In this regard, the recent study by Basgen *et al.* (1994) is of interest. They compared estimates of total glomerular number in 10 dog kidneys obtained using a combined *in situ* magnetic resonance imaging/biopsy method with estimates obtained using a disector/fractionator combination. Overall, the agreement between estimates obtained with the two methods was very good. Such a method, once fully validated, clearly will be of great use in renal medicine.

Acknowledgments

The author thanks Assoc. Prof. Daine Alcorn, Meroë Cahill, Michelle Kett, Prof. Graeme Ryan and Mary Soosaipillai for their comments on the manuscript. Michelle Gough, Karen Spencer, and Stuart Thyer provided expert assistance with illustrations. I am indebted to Dr. Agnes Fogo, Department of Pathology, Vanderbilt University, Nashville, Tennessee, for discussions of stereological assessment of individual glomeruli.

References

Abercrombie, M. (1946). Estimation of nuclear populations from microtomic sections. *Anat. Rec.* **94,** 239–247.

Aherne, W. A., and Dunnill, M. S. (1982). "Morphometry." Edward Arnold, London.

Anderson, W. P., Alcorn, D., Gilchrist, A. I., Whiting, J. M., and Ryan, G. B. (1989). Glomerular actions of ANG II during reduction of renal artery pressure: A morphometric analysis. *Am. J. Physiol.* **256,** F1021–F1026.

Antolovich, G. C., Perry, R. A., and Bertram, J. F. (1992). Estimates of cell number in the developing sheep pars intermedia obtained using the physical disector/Cavalieri combination. *Acta Stereol.* **11,** Suppl. 1, 735–740.

Artacho-Perula, E., Roldan-Villalobos, R., Salcedo-Leal, I., and Vaamonde-Lemos, R. (1993). Stereological estimates of volume-weighted mean glomerular volume in streptozotocin diabetic rats. *Lab. Invest.* **68,** 56–61.

Baddeley, A. J., Gundersen, H. J. G., and Cruz-Orive, L.-M. (1986). Estimation of surface area from vertical sections. *J. Microsc.* (*Oxford*) **142,** 259–276.

Bahr, G. F., Bloom, G., and Friberg, U. (1957). Volume changes of tissues in physiological fluids during fixation in osmium tetroxide or formaldehyde and during subsequent treatment. *Exp. Cell Res.* **12**, 342–355.

Bangstad, H.-J., Osterby, R., Dahl-Jorgensen, K., Berg, K. J., Hartmann, A., Nyberg, G., Bjorn, S. F., and Hanssen, K. F. (1993). Early glomerulopathy is present in young, Type 1 (insulin-dependent) diabetic patients with microalbuminuria. *Diabetologia* **36**, 523–529.

Basgen, J. M., Rich, S. S., Mauer, S. M., and Steffes, M. W. (1988). Measuring the volume density of the glomerular mesangium. *Nephron* **50**, 182–186.

Basgen, J. M., Steffes, M. W., Stillman, A. E., and Mauer, S. M. (1994). Estimating glomerular number in situ using magnetic resonance imaging and biopsy. *Kidney Int.* **45**, 1668–1672.

Bedi, K. S. (1994). Undernutrition of rats during early life does not affect the total number of cortical neurons. *J. Comp. Neurol.* **342**, 596–602.

Bendtsen, T. F., and Nyengaard, J. R. (1989). Unbiased estimation of particle number using sections—an historical perspective with special reference to the stereology of glomeruli. *J. Microsc. (Oxford)* **153**, 93–102.

Bendtsen, T. F., and Nyengaard, J. R. (1992). The number of glomeruli in Type 1 (insulin-dependent) and Type 2 (noninsulin-dependent) diabetic patients. *Diabetologia* **35**, 844–850.

Berka, J. L., Alcorn, D., Bertram, J. F., Ryan, G. B., and Skinner, S. L. (1994). Effects of angiotensin converting enzyme inhibition on glomerular number, juxtaglomerular cell activity and renin content in experimental unilateral hydronephrosis. *J. Hypertens.* **12**, 735–743.

Bertram, J. F., and Nurcombe, V. (1992). Counting cells with the new stereology. *Trends Cell Biol.* **2**, 177–180.

Bertram, J. F., Sampson, P., and Bolender, R. P. (1986). Influence of tissue composition on the final volume of rat liver blocks prepared for electron microscopy. *J. Electron Microsc. Tech.* **4**, 303–314.

Bertram, J. F., Soosaipillai, M. C., Ricardo, S. D., and Ryan, G. B. (1992). Total numbers of glomeruli and individual glomerular cell types in the normal rat kidney. *Cell Tissue Res.* **270**, 37–45.

Bjugn, R. (1993). The use of the optical disector to estimate the number of neurons, glial and endothelial cells in the spinal cord of the mouse—with a comparative note on the rat spinal cord. *Brain Res.* **627**, 25–33.

Black, M. J., Bertram, J. F., and Campbell, J. H. (1992). Effect of angiotensin II on the number of smooth muscle cells in the SHR aorta. *Genet. Hypertens.* **218**, 277–279.

Black, M. J., Bertram, J. F., Campbell, J. H., and Campbell, G. R. (1995). Angiotensin II induces cardiovascular hypertrophy in perindopril-treated rats. *J. Hypertens.* (accepted for publication).

Bloom, G., and Friberg, U. (1956). Shrinkage during fixation and embedding of histological specimens. *Acta Morphol. Neerl.-Scand.* **1**, 12–20.

Bolender, R. P. (1993). Software for quantitative immunogold and in situ hybridization. *Microsc. Res. Tech.* **25**, 304–313.

Bolender, R. P., and Charleston, J. S. (1993). Software for counting cells and estimating structural volumes with the optical disector and fractionator. *Microsc. Res. Tech.* **25**, 314–324.

Bolender, R. P., and Pentcheff, N. D. (1985). "Computer Programs for Biological Stereology: PCS System 1." Washington Research Foundation, Seattle.

Bolender, R. P., Hyde, D. M., and DeHoff, R. T. (1993). Lung morphometry: A new generation of tools and experiments for organ, tissue, cell, and molecular biology. *Am. J. Physiol.* **265**, L521–L548.

Braendgaard, H., and Gundersen, H. J. G. (1986). The impact of recent stereological advances on quantitative studies of the nervous system. *J. Neurosci. Methods* **18**, 39–78.

Braendgaard, H., Evans, S. M., Howard, C. V., and Gundersen, H. J. G. (1990). The total number of neurons in the human neocortex unbiasedly estimated using optical disectors. *J. Microsc. (Oxford)* **157**, 285–304.

Brenner, B. M., Garcia, D. L., and Anderson, S. (1988). Glomeruli and blood pressure. Less of one, more the other? *Am. J. Hypertens.* **1**, 335–347.

Brenner, B. M., Cohen, R. A., and Milford, E. L. (1992). In renal transplantation, one size may not fit all. *J. Am. Soc. Nephrol.* **3**, 162–169.

Cahill, M. M. (1993). Glomerular hypertrophy and fibroblast growth factors in a model of focal and segmental glomerulosclerosis and hyalinosis. B. Sc. Hons. Thesis, University of Melbourne.

Cahill, M. M., Bertram, J. F., and Ryan, G. B. (1995). Glomerular hypertrophy in a puromycin aminonucleoside model of focal and segmental glomerulosclerosis. *Proc. 31st Annu. Sci. Meet. Aust. N.Z. Soc. Nephrol.*, Canberra Abstract 36.

Calverley, R. K. S., and Jones, D. G. (1987). Determination of the numerical density of perforated synapses in rat neocortex. *Cell Tissue Res.* **248**, 399–407.

Cruz-Orive, L.-M. (1987). Particle number can be estimated using a disector of unknown thickness: the selector. *J. Microsc. (Oxford)* **145**, 121–142.

Cruz-Orive, L.-M., and Hunziker, E. B. (1986). Stereology for anisotropic cells: Application to growth cartilage. *J. Microsc. (Oxford)* **143**, 47–80.

Cruz-Orive, L.-M., and Weibel, E. R. (1990). Recent stereological methods for cell biology· A brief survey. *Am. J. Physiol.* **258**, L148–L156.

DeHoff, R. T., and Rhines, F. N. (1961). Determination of number of particles per unit volume from measurements made on random plane sections: the general cylinder and the ellipsoid. *Trans. Metall. Soc. AIME* **221**, 975–982.

Delesse, M. A. (1847). Procédé mécanique pour déterminer la composition des roches. *C. R. Hebd. Seances Acad. Sci.* **25**, 544.

Denton, K. M., Fennessy, P. A., Alcorn, D., and Anderson, W. P. (1992). Morphometric analysis of the actions of angiotensin II on renal arterioles and glomeruli. *Am. J. Physiol.* **262**, F367–372.

Dunnill, M. S., and Halley, W. (1973). Some observations on the quantitative anatomy of the kidney. *J. Pathol.* **110**, 113–121.

Eisenberg, B. R., and Mobley, B. A. (1975). Size changes in single muscle fibers during fixation and embedding. *Tissue Cell* **7**, 383–387.

Elias, H., and Hennig, A. (1967). Stereology of the human renal glomerulus. *In* "Quantitative Methods in Morphology" E. R. Weibel and H. Elias, (eds.), pp. 130–166. Springer-Verlag, New York.

Elias, H., and Hyde, D. M. (1983). "A Guide to Practical Stereology." Karger, Basel.

Floderus, S. (1944). Untersuchungen uber den Bau der menschlichen Hypophyse mit besonderer Berucksichtigung der quantitativen mikromorphologischen Verhaltnisse. *Acta Pathol. Microbiol. Scand., Sect. B: Microbiol.* **53B**, Suppl. 1, 1–26.

Fogo, A., and Ichikawa, I. (1992). Glomerular growth promoter—The common channel to glomerulosclerosis. *Contemp. Issues Nephrol.* **26**, 23–54.

Fogo, A., Hawkins, E. P., Berry, P. L., Glick, A. D., and Ichikawa, I. (1990). Glomerular hypertrophy in minimal change disease predicts subsequent progression to focal glomerular sclerosis. *Kidney Int.* **38**, 115–123.

Fries, J. W. U., Sandström, D. J., Meyer, T. W., and Rennke, H. G. (1989). Glomerular hypertrophy and epithelial cell injury modulate progressive glomerulosclerosis in the rat. *Lab. Invest.* **60**, 205–218.

Fullman, R. L. (1953). Measurement of particle sizes in opaque bodies. *Trans. Am. Inst. Min. Metall. Engrs.* **197**, 447–452.

Gerdes, A. M., Kriseman, J., and Bishop, S. P. (1982). Morphometric study of cardiac muscle. The problem of tissue shrinkage. *Lab. Invest.* **46**, 271–274.

Gokhale, A. M. (1990). Unbiased estimation of curve length in 3-D using vertical slices. *J. Microsc. (Oxford)* **159**, 133–141.

Gokhale, A. M. (1992). Estimation of length density L_V from vertical slices of unknown thickness. *J. Microsc. (Oxford)* **167**, 1–8.

Gundersen, H. J. G. (1977). Notes on the estimation of the numerical density of arbitrary profiles: the edge effect. *J. Microsc. (Oxford)* **111**, 219–223.

Gundersen, H. J. G. (1986). Stereology of arbitrary particles: A review of unbiased number and size estimators and the presentation of some new ones, in memory of William R. Thompson. *J. Microsc. (Oxford)* **143**, 3–45.

Gundersen, H. J. G. (1988). The nucleator. *J. Microsc. (Oxford)* **151**, 3–21.

Gundersen, H. J. G., and Jensen, E. B. (1983). Particle sizes and their distributions estimated from line- and point-sampled intercepts. Including graphical unfolding. *J. Microsc. (Oxford)* **131**, 291–310.

Gundersen, H. J. G., and Jensen, E. B. (1985). Stereological estimation of the volume-weighted mean volume of arbitrary particles observed on random sections. *J. Microsc. (Oxford)* **138**, 127–142.

Gundersen, H. J. G., and Jensen, E. B. (1987). The efficiency of systematic sampling in stereology and its prediction. *J. Microsc. (Oxford)* **147**, 229–263.

Gundersen, H. J. G., and Osterby, R. (1981). Optimizing sampling efficiency of stereological studies in biology: or "Do more less well!" *J. Microsc. (Oxford)* **121**, 65–73.

Gundersen, H. J. G., Bendsten, T. F., Korbo, L., Marcussen, N., Moller, A., Nielsen, K., Nyengaard, J. R., Pakkenberg, B., Sorensen, F. B., Vesterby, A., and West, M. J. (1988a). Some new, simple and efficient stereological methods and their use in pathological research and diagnosis. *Acta Pathol. Microbiol. Immunol. Scand.* **96**, 379–394.

Gundersen, H. J. G., Bagger, P., Bendtsen, T. F., Evans, S. M., Korbo, L., Marcussen, N., Moller, A., Nielsen, K., Nyengaard, J. R., Pakkenberg, B., Sorensen, F. B., Vesterby, A., and West, M. J. (1988b). The new stereological tools: disector, fractionator, nucleator and point sampled intercepts and their use in pathological research and diagnosis. *Acta Pathol. Microbiol. Immunol. Scand.* **96**, 857–881.

Hanstede, J. G., and Gerrits, P. O. (1983). The effects of embedding in water-soluble plastics on the final dimensions of liver sections. *J. Microsc. (Oxford)* **131**, 79–86.

Harding, A. J., Halliday, G. M., and Cullen, K. (1994). Practical considerations for the use of the optical disector in estimating neuronal number. *J. Neurosci. Methods* **51**, 83–89.

Hinchliffe, S. A., Sargent, P. H., Howard, C. V., Chan, Y. F., and van Velzen, D. (1991). Human intrauterine renal growth expressed in absolute number of glomeruli assessed by the disector method and Cavalieri principle. *Lab. Invest.* **64**, 777–784.

Hinchliffe, S. A., Lynch, M. R. J., Sargent, P. H., Howard, C. V., and van Velzen, D. (1992). The effect of intrauterine growth retardation on the development of renal nephrons. *Br. J. Obstet. Gynaecol.* **99**, 296–301.

Hinchliffe, S. A., Howard, C. V., Lynch, M. R. J., Sargent, P. H., Judd, B. A., and van Velzen, D. (1993). Renal developmental arrest in sudden infant death syndrome. *Pediat. Pathol.* **13**, 333–343.

Hirose, K., Osterby, R., Nozawa, M., and Gundersen, H. J. G. (1982). Development of glomerular lesions in experimental long-term diabetes in the rat. *Kidney Int.* **21**, 689–695.

Howard, C. V., Cruz-Orive, L.-M., and Yaegashi, H. (1992). Estimating neuron dendritic length in 3D from total vertical projections and from vertical slices. *Acta Neurol. Scand., Suppl.* **137**, 14–19.

Howard, C. V., Jolleys, G., Stacey, D., Fowler, A., Wallen, P., and Browne, M. A. (1993). Measurement of total neuronal volume, surface area, and dendritic length following intracellular physiological recording. *Neuroprotocols* **2**, 113–120.

Jensen, E. B., and Gundersen, H. J. G. (1993). The rotator. *J. Microsc. (Oxford)* **170**, 35–44.

Jensen, E. B., Gundersen, H. J. G., and Osterby, R. (1979). Determination of membrane thickness distribution from orthogonal intercepts. *J. Microsc. (Oxford)* **115**, 19–33.

Kett, M. M., Alcorn, D., Bertram, J. F., and Anderson, W. P. (1994). Glomerular numbers and dimensions in SHR: effects of chronic angiotensin II receptor antagonism. *Proc. Sci. Meet. High Blood Pressure Res. Counc. (Australia), 16th* Brisbane, p. 99.

Knepper, M. A., Danielson, R. A., Saidel, G. M., and Post, R. S. (1977). Quantitative analysis of renal medullary anatomy in rats and rabbits. *Kidney Int.* **12**, 313–323.

Kriz, W., and Kaissling, B. (1992). Structural organization of the mammalian kidney. *In* "The Kidney: Physiology and Pathophysiology" D. W. Seldin and G. Giebisch, (eds.), 2nd ed., pp. 707–778. Raven Press, New York.

Kroustrup, J. P., and Gundersen, H. J. G. (1983). Sampling problems in an heterogenous organ: Quantitation of relative and total volume of pancreatic islets by light microscopy. *J. Microsc. (Oxford)* **132**, 43–55.

Lafferty, H. M., and Brenner, B. M. (1990). Are glomerular hypertension and "hypertrophy" independent risk factors for progression of renal disease? *Semin. Nephrol.* **10**, 294–304.

Lane, P. H., Steffes, M. W., and Mauer, S. M. (1992). Estimation of glomerular volume: A comparison of four methods. *Kidney Int.* **41**, 1085–1089.

Larsson, L., and Maunsbach, A. B. (1980). The ultrastructural development of the glomerular filtration barrier in the rat kidney: A morphometric analysis. *J. Ultrastruct. Res.* **72**, 392–406.

Lucocq, J. (1993). Unbiased 3-D quantitation of ultrastructure in cell biology. *Trends Cell Biol.* **3**, 354–358.

Madsen, K. M., and Tisher, C. C. (1986). Structural-functional relationships along the distal nephron. *Am. J. Physiol.* **250**, F1–F15.

Marcussen, N. (1992). The double disector: Unbiased stereological estimation of the number of particles inside other particles. *J. Microsc. (Oxford)* **165**, 417–426.

Mattfeldt, T., Mall, G., Gharehbaghi, H., and Moller, P. (1990). Estimation of surface area and length with the orientator. *J. Microsc. (Oxford)* **159**, 301–317.

Mauer, S. M., Steffes, M. W., Ellis, E. N., Sutherland, D. E. R., Brown, D. M., and Goetz, F. C. (1984). Structural-functional relationships in diabetic nephropathy. *J. Clin. Invest.* **74**, 1143–1155.

Mayhew, T. M. (1991). The accurate prediction of Purkinje cell number from cerebellar weight can be achieved with the fractionator. *J. Comp. Neurol.* **308**, 162–168.

Mayhew, T. M. (1992). A review of recent advances in stereology for quantifying neural structure. *J. Neurocytol.* **21**, 313–328.

Mayhew, T. M., and Reith, A. (1988). Practical ways to correct cytomembrane surface densities for the loss of membrane images that results from oblique sectioning. *In* "Stereology and Morphometry in Electron Microscopy: Problems and Solutions" A. Reith and T. M. Mayhew, (eds.), pp. 99–110. Hemisphere, New York.

McCausland, J. E., Bertram, J. F., Mendelsohn, F. A. O., Ryan, G. B., Zhuo, J., and Alcorn, D. (1994). Angiotensin receptors and glomerular number in rats treated postnatally with enalapril. *Proc. Sci. Meet. High Blood Pressure Res. Counc. (Australia), 16th,* Brisbane, p. 102.

Miller, P. L., and Meyer, T. W. (1990). Effects of tissue preparation on glomerular volume and capillary structure in the rat. *Lab. Invest.* **63**, 862–866.

Miller, W. S., and Carlton, E. P. (1895). The relation of the cortex of the cats kidney to the volume of the kidney, and an estimation of the number of glomeruli. *Trans. Wis. Acad. Sci.* **10**, 525–538.

Nagata, M., Scharer, K., and Kriz, W. (1992). Glomerular damage after uninephrectomy in young rats. 1. Hypertrophy and distortion of capillary architecture. *Kidney Int.* **42**, 136–147.

Nairn, J. G., Bedi, K. S., Mayhew, T. M., and Campbell, L. F. (1989). On the number of Purkinje cells in the human cerebellum: Unbiased estimates obtained by using the "fractionator." *J. Comp. Neurol.* **290**, 527–532.

Nurcombe, V., Wreford, N. G., and Bertram, J. F. (1991). The use of the optical disector to estimate the total number of neurons in the developing chick lateral motor column: Effects of purified growth factors. *Anat. Rec.* **231**, 416–424.

Nyengaard, J. R. (1993a). The quantitative development of glomerular capillaries in rats with special reference to unbiased stereological estimates of their number and sizes. *Microvasc. Res.* **45,** 243–261.

Nyengaard, J. R. (1993b). Number and dimensions of rat glomerular capillaries in normal development and after nephrectomy. *Kidney Int.* **43,** 1049–1057.

Nyengaard, J. R., and Bendtsen, T. F. (1989). The practical use of the fractionator in glomerular counting. *Acta Stereol.* **8,** 275–279.

Nyengaard, J. R., and Bendtsen, T. F. (1990). A practical method to count the number of glomeruli in the kidney as exemplified in various animal species. *Acta Stereol.* **9,** 243–258.

Nyengaard, J. R., and Bendtsen, T. F. (1992). Glomerular number and size in relation to age, kidney weight, and body surface in normal man. *Anat. Rec.* **232,** 194–201.

Nyengaard, J. R., and Gundersen, H. J. G. (1992). The isector: A simple and direct method for generating isotropic, uniform random sections from small specimens. *J. Microsc. (Oxford)* **165,** 427–431.

Nyengaard, J. R., and Marcussen, N. (1993). The number of glomerular capillaries estimated by an unbiased and efficient stereological method. *J. Microsc. (Oxford)* **171,** 27–37.

Nyengaard, J. R., and Rasch, R. (1993). The impact of experimental diabetes mellitus in rats on glomerular capillary number and sizes. *Diabetologia* **36,** 189–194.

Nyengaard, J. R., and Rasch, R. (1994). The practical use of stereology in characterising capillary networks. *Acta Stereol.* **13,** 39–42.

Nyengaard, J. R., Bendtsen, T. F., Christensen, S., and Ottosen, P. D. (1994). The number and size of glomeruli in long-term lithium-induced nephropathy in rats. *Acta Pathol. Microbiol. Immunol. Scand.* **102,** 59–66.

Ogbuihi, S., and Cruz-Orive, L.-M. (1990). Estimating the total number of lymphatic valves in infant lungs with the fractionator. *J. Microsc. (Oxford)* **158,** 19–30.

Oliver, J. (1968). "Nephrons and Kidneys." Harper and Row, Hoeber Medical, New York.

Olivetti, G., Anversa, P., Rigamonti, W., Vitali-Mazza, L., and Loud, A. V. (1977). Morphometry of the renal corpuscle during normal postnatal growth and compensatory hypertrophy. A light microscope study. *J. Cell Biol.* **75,** 573–585.

Olivetti, G., Anversa, P., Melissari, M., and Loud, A. V. (1980). Morphometry of the renal corpuscle during postnatal growth and compensatory hypertrophy. *Kidney Int.* **17,** 438–454.

Oorschot, D. E. (1994). Are you using neuronal densities, synaptic densities or neurochemical densities as your definitive data? There is a better way to go. *Prog. Neurobiol.* **44,** 233–247.

Osterby, R. (1990). Basement membrane morphology in diabetes mellitus. *In* "Diabetes Mellitus: Theory and Practice" H. Rifkin and D. Porte, (eds.), 4th ed., pp. 220–233. Elsevier, New York.

Osterby, R. (1992). Glomerular structural changes in Type 1 (insulin-dependent) diabetes mellitus: Causes, consequences, and prevention. *Diabetologia* **35,** 803–812.

Osterby, R., and Gundersen, H. J. G. (1988). Stereological estimation of capillary length exemplified by changes in renal glomeruli in experimental diabetes. *In* "Stereology and Morphometry in Electron Microscopy: Problems and Solutions" (A. Reith and T. M. Mayhew, eds.), pp. 113–122. Hemisphere, New York.

Osterby, R., Nyberg, G., Andersson, G., and Frisk, B. (1988). Glomerular structural quantities in baseline biopsies from cadaveric donor kidney pairs. *Acta Pathol. Microbiol. Immunol. Scand. (Suppl.)* **4,** 130–136.

Osterby, R., Bangstad, H.-J., Hanssen, K. F., Mauer, S. M., Steffes, M. W., Viberti, G. C., and Walker, J. (1990). Stereological studies of early phases of diabetic glomerulopathy. *Diabetologia* **33,** Suppl., A147.

O'Sullivan, J. B., Black, M. J., Bertram, J. F., and Bobik, A. (1994). Cardiovascular hypertrophy in one-kidney, one clip renal hypertensive rats: A role for angiotensin II? *J. Hypertens.* **12,** 1163–1170.

Pakkenberg, B., and Gundersen, H. J. G. (1987). Disector-Cavalieri combination providing unbiased and efficient estimators of total number of particles unaffected by tissue shrinkage. *Acta Stereol.* **6,** 49–52.

Pakkenberg, B., and Gundersen, H. J. G. (1988). Total number of neurons and glial cells in human brain nuclei estimated by the disector and the fractionator. *J. Microsc. (Oxford)* **150,** 1–20.

Pedersen, J. C., Persson, A. E., and Maunsbach, A. B. (1980). Ultrastructure and quantitative characterization of the cortical interstitium in the rat kidney. *In* "Functional Ultrastructure of the Kidney" A. B. Maunsbach, T. S. Olsen, and E. I. Christensen, (eds.), pp. 443–456. Academic Press, London.

Pentcheff, N. D. (1987). Guidelines for developing data collection and analysis systems for stereology: A case study and proposed standards. *Acta Stereol.* **6,** 257–269.

Pentcheff, N. D., and Bolender, R. P. (1985). PCS System 1: Point counting stereology programs for cell biology. *Comput. Methods Programs Biomed.* **20,** 173–187.

Peten, E. P., Striker, L. J., García-Perez, A., and Striker, G. E. (1993). Studies by competitive PCR of glomerulosclerosis in growth hormone transgenic mice. *Kidney Int.* **43,** S55–S58.

Pfaller, W. (1982). "Structure Function Correlation on Rat Kidney. Quantitative Correlation of Structure and Function in the Normal and Injured Rat Kidney." Springer Verlag, Berlin.

Remuzzi, A., Pergolizzi, R., Mauer, S. M., and Bertani, T. (1990). Three-dimensional morphometric analysis of segmental glomerulosclerosis in the rat. *Kidney Int.* 38, 851 856.

Remuzzi, A., Puntirieri, S., Alfano, M., Macconi, D., Abbate, M., Bertani, T., and Remuzzi, G. (1992). Pathophysiologic implications of proteinuria in a rat model of progressive glomerular injury. *Lab. Invest.* **67,** 572–579.

Russ, J. C. (1986). "Practical Stereology." Plenum, New York.

Schmidt, K., Pesce, C., Liu, Q., Nelson, R. G., Bennett, P. H., Karnitschnig, H., Striker, L. J., and Striker, G. E. (1992). Large glomerular size in Pima indians: Lack of change with diabetic nephropathy. *J. Am. Soc. Nephrol.* **3,** 229–235.

Schmitz, A., Nyengaard, J. R., and Bendtsen, T. F. (1990). Glomerular volume in type 2 (Non-Insulin-Dependent-Diabetes) diabetes estimated by a direct and unbiased stereological method. *Lab. Invest.* **62,** 108–113.

Shea, S. M., and Morrison, A. B. (1975). A stereological study of the glomerular filter in the rat. Morphometry of the slit diaphragm and basement membrane. *J. Cell Biol.* **67,** 436–443.

Soosaipillai, M. C. (1994). Quantitative morphology of glomeruli in normal rats and in rats with acute puromycin aminonucleoside nephrosis. M.Sc. Thesis, University of Melbourne.

Soosaipillai, M. C., Bertram, J. F., and Ryan, G. B. (1994). Absolute volumes of glomerular cells and glomerular compartments in the normal rat kidney. *Acta Stereol.* **13,** 63–68.

Steffes, M. W., Barbosa, J., Basgen, J. M., Sutherland, D. E. R., Najarian, J. S., and Mauer, S. M. (1983). Quantitative glomerular morphology of the normal human kidney. *Lab. Invest.* **49,** 82–86.

Sterio, D. C. (1984). The unbiased estimation of number and sizes of arbitrary particles using the disector. *J. Microsc. (Oxford)* **134,** 127–136.

Tau, S-S., Faulkner-Jones, B., Breen, S. J., Walsh, M., Bertram, J. F., and Reese, B. E. (1995). Cell dispersion patterns in different cortical regions studied with an X-inactivated transgenic marker. *Development* **121,** 1029–1039.

Tisher, C. C., and Madsen, K. M. (1991). Anatomy of the kidney. *In* "The Kidney" B. M. Brenner and F. C. Rector, (eds.), 4th ed.; pp. 3–75. Saunders, Philadelphia.

Tolcos, M. (1994). The effect of chronic placental insufficiency on the structural and neurochemical development of the brainstem in the fetal guinea pig. B.Sc. Honours Thesis, University of Melbourne.

Walker, J. D., Close, C. F., Jones, S. L., Rafftery, M., Keen, H., Viberti, G. C., and Osterby, R. (1992). Glomerular structure in Type-1 (insulin-dependent) diabetic patients with normo- and microalbuminuria. *Kidney Int.* **41,** 741–748.

Weibel, E. R. (1979). "Stereological Methods," Vol. 1. Academic Press, London.

Weibel, E. R. (1980). "Stereological Methods," Vol. 2. Academic Press, London.

Weibel, E. R., and Gomez, D. M. (1962). A principle for counting tissue structures on random sections. *J. Appl. Physiol.* **17,** 343–348.

Weibel, E. R., and Knight, B. W. (1964). A morphometric study on the thickness of the pulmonary air-blood barrier. *J. Cell Biol.* **21,** 367–384.

West, M. J., and Gundersen, H. J. G. (1990). Unbiased stereological estimation of the number of neurons in the human hippocampus. *J. Comp. Neurol.* **296,** 1–22.

West, M. J., Slomianka, L., and Gundersen, H. J. G. (1991). Unbiased stereological estimation of the total number of neurons in the subdivisions of the rat hippocampus using the optical fractionator. *Anat. Rec.* **231,** 482–497.

Wicksell, S. D. (1925). The corpuscle problem. A mathematical study of a biometric problem. *Biometrika* **17,** 84–99.

Yaegashi, H., Howard, C. V., McKerr, G., and Burns, A. (1993). Stereological estimation of the total number of neurons in the asexually dividing tetrathyridium of *Mesocestoides corti.* *Parasitology* **106,** 177–183.

Yoshida, Y., Fogo, A., and Ichikawa, I. (1989). Glomerular hemodynamic changes vs. hypertrophy in experimental glomerular sclerosis. *Kidney Int.* **35,** 654–660.

Membrane Mechanisms and Intracellular Signalling in Cell Volume Regulation

Else K. Hoffmann* and Philip B. Dunham†

* Biochemical Department, August Krogh Institute, University of Copenhagen, DK-2100 Copenhagen 0, Denmark and † Biological Research Laboratories, Syracuse University, Syracuse, New York 13244

Recent work on selected aspects of the cellular and molecular physiology of cell volume regulation is reviewed. First, the physiological significance of the regulation of cell volume is discussed. Membrane transporters involved in cell volume regulation are reviewed, including volume-sensitive K^+ and Cl^- channels, K^+, Cl^- and Na^+, K^+, $2Cl^-$ cotransporters, and the Na^+, H^+, Cl^-, HCO_3^-, and K^+, H^+ exchangers. The role of amino acids, particularly taurine, as cellular osmolytes is discussed. Possible mechanisms by which cells sense their volumes, along with the sensors of these signals, are discussed. The signals are mechanical changes in the membrane and changes in macromolecular crowding. Sensors of these signals include stretch-activated channels, the cytoskeleton, and specific membrane or cytoplasmic enzymes. Mechanisms for transduction of the signal from sensors to transporters are reviewed. These include the Ca^{2+}–calmodulin system, phospholipases, polyphosphoinositide metabolism, eicosanoid metabolism, and protein kinases and phosphatases. A detailed model is presented for the swelling-initiated signal transduction pathway in Ehrlich ascites tumor cells. Finally, the coordinated control of volume-regulatory transport processes and changes in the expression of organic osmolyte transporters with long-term adaptation to osmotic stress are reviewed briefly.

KEY WORDS: Volume regulation, Signal transduction, Calcium–calmodulin, Stretch-activated channels, Eicosanoids, Macromolecular crowding, Cytoskeleton, Protein phosphorylation and dephosphorylation.

I. Introduction

Two fundamental cellular homeostatic mechanisms are the regulation of cell volume and cellular pH. The major theme in this chapter is cell volume control, but since there clearly is an interplay between the two cellular homeostatic mechanisms, pH regulation is discussed where it is important for our central theme. Even under physiological steady-state conditions, the cell volume must be maintained constant by a "pump-leak mechanism," by which the osmotic pressure arising from the cytoplasmic impermeable solutes is counteracted by the Na^+, K^+ pump with the constant expenditure of metabolic energy (Hoffmann and Ussing, 1992).

In addition to maintaining a constant cell volume at physiological steady state, most cells investigated have mechanisms that become activated after volume perturbations, resulting in volume recovery processes. Swelling causes cells to activate transport systems for K^+, Cl^-, and certain organic molecules, resulting in a loss of osmolytes and the concomitant osmotically obliged loss of cell water (regulatory volume decrease, or RVD). On the other hand, osmotic shrinkage can activate uptake systems for ions or trigger the expression of transporters for organic osmolytes. Shrunken cells can thereby increase their volume toward the initial level by net uptake of Na^+, Cl^-, and often K^+ as well, and concomitant uptake of water (regulatory volume increase, or RVI).

In cells lacking rigid walls, water is in thermodynamic equilibrium across the surface membrane and the cell volume is determined by the relationship between the concentrations of cellular osmotically active solutes and solutes in the extracellular fluid. Thus changes in cell volume can result from net gain or loss in the cellular content of osmolytes or from changes in the osmolality of the extracellular fluid.

Because of complex homeostatic mechanisms for maintaining the tonicity of body fluids, most of the cells of the body do not experience changes in osmolality of their surroundings. There are exceptions: (1) The normally high osmolality of the mammalian renal medulla increases during antidiuresis, when the osmolality in the medulla can reach levels above 1200 mosmol (Lise and de Rouffinac, 1985). Red blood cells experience a hyperosmotic shock each time they pass through the renal medulla owing to the high salt and urea concentrations. Shrinkage of the cells is minimized by rapid equilibration of the cells with urea through urea channels (Macey, 1984). In the same manner, swelling of the red cells as they leave the medulla is minimized by the rapid loss of urea. (2) Intestinal epithelial cells are exposed to hypertonicity after intake of, for example, inappropriately diluted juices or hypotonicity after excessive water intake.

Pathophysiologic disturbances of body fluid homeostasis are relatively common, and both hypertonicity and hypotonicity of the plasma are seen clinically; plasma hypotonicity, usually hyponatremia, is among the conditions most frequently observed (McManus and Churchwell, 1994). As stated by these authors, "Anisotonic volume disturbances occur under conditions such as the syndrome of inappropriate antidiuretic hormone secretion (SI-

ADH), water intoxication, congestive heart failure, inappropriate diuretic use, dehydration, diabetes mellitus, and diabetes insipidus." Such anisotonic volume disturbances will have serious pathological consequences, including seizures and severe cerebral edema.

Changes in cellular solute content are a common physiological cause for cell volume changes. Net accumulation of osmotically active substances occurs in certain epithelia during absorption of sugars and amino acids (Lau et al., 1984; Schultz et al., 1985). This results in cell swelling (MacLeod, 1994). In contrast, depletion of osmotically active substances occurs during the initiation of secretion in salivary glands and pancreatic acinar cells (Petersen and Gallacher, 1988; Petersen, 1988; Foskett and Melvin, 1989) and shark rectal glands (Greger et al., 1988). Simultaneous opening of basolateral K^+ channels and apical Cl^- channels results in K^+ and Cl^- loss (Petersen and Gallacher, 1988) and a resultant cell shrinkage (Foskett and Melvin, 1989; Foskett, 1990; Nakahari et al., 1990; Wong and Foskett, 1991). This cell shrinkage, or events associated with it, for example the Cl^- loss (Robertson and Foskett, 1994), leads to activation of Na^+, H^+ exchange and the Na^+, K^+, $2Cl^-$ cotransport system. Even though there is a rapid uptake of Na^+, K^+, and Cl^-, most cells remain shrunken as long as the gland is stimulated. In other cell types, volume is partly recovered (Foskett et al., 1994).

In glandular cells, the same processes are involved in secretion and regulation of cell volume, and the volume at which the cells are maintained is correlated with their secretory rate. Hoffmann and Ussing (1992) offer a detailed discussion of the role of the cell volume regulatory processes in absorbing and secreting epithelia, with special emphasis on isotonic secretion. It should be mentioned, however, that the mechanisms involved in volume regulation after alterations of the intracellular solute content under isosmotic conditions ("isosmotic volume regulation") may be different from the mechanisms seen during "anisosmotic volume regulation" (MacLeod, 1994; Foskett, 1994).

Changes in cell water content are an integrated part of the function of certain hormones (Häussinger and Lang, 1991; Häussinger et al., 1993). Thus insulin causes liver cells to swell by activating Na^+/H^+ exchange and Na^+, K^+, $2Cl^-$ cotransport. This in turn inhibits proteolysis and glycolysis and stimulates protein and glycogen synthesis (Häussinger et al., 1994; Agius et al., 1994). On the other hand, glucagon activates K^+ and Cl^- channels, and the cells shrink. Liver cell volume can change up to 20% in the intact liver in response to hormones (Häussinger et al., 1994; Agius et al., 1994). The metabolic events in exercising skeletal muscle can also lead to an increase in the content of cellular organic solutes (predominantly lactate), concomitant cell swelling, and activation of volume regulatory responses (Saltin et al., 1987). Thus, cell volume and the activity of the synthetic and the catabolic enzymes are intimately interrelated.

Intracellular pH also plays a role in determining cell volume. At low intracellular pH (pH_i, cellular acidosis), the negative charge on the impermeant anions, particularly proteins, is decreased. This causes the cell chloride content and therefore cell volume to increase provided the cellular K^+ and Na^+ content remains essentially constant. Furthermore, acidification in most cells will activate the Na^+/H^+ exchanger, leading to additional cell swelling (Hoffmann and Simonsen, 1989). Intracellular alkalosis promotes the opposite effects and cells shrink. For a detailed discussion of these interrelations, see Hladky and Rink (1977) and Hoffmann and Ussing (1992).

Following ischemic stroke and hypoxic/anoxic insult in the mammalian brain, there is marked cell swelling (Friedman and Haddad, 1993). This is mainly a consequence of uptake of Na^+ and Cl^-. There follows a compensatory volume regulatory response during which K^+ is lost from the neurons, and $[K^+]_o$ rises to high levels. This extracellular K^+ increase is partially compensated by uptake into glial cells (Kimelberg, 1991; Walsh, 1989). Glutamate may play an important role in cytotoxic cell swelling after brain cell injury. Increase in glutamate will stimulate the N-methyl-D-aspartate (NMDA) receptor and open nonselective cation channels. This leads to depolarization, NaCl uptake, and swelling followed by a toxic Ca^{2+} influx (Choi, 1988; Choi and Rothman, 1990; Baker et al., 1991). Serious cerebral cell swelling occurs during treatment of diabetic ketoacidosis, possibly due to activation of the Na^+/H^+ exchanger by intracellular acidosis following correction of the extracellular pH by therapy (Vander Meulen et al., 1987). Also, anoxia leads to a decrease in intracellular pH (Bickler, 1992), and this intracellular acidosis can lead to cell swelling (Kimelberg, 1991). Some examples of cell volume perturbation are listed in Table 1 of Hoffmann et al. (1993).

For reviews on cell volume regulation in a variety of cells and tissues, see Chamberlin and Strange (1989), Hoffmann and Simonsen (1989), Okada and Hazama (1989), Schultz (1989a,b), Lewis and Donaldson (1990), Geck (1990), and Lang et al. (1990); a series of reviews edited by Gilles et al. (1991), Hoffmann and Ussing (1992), Spring and Hoffmann (1992), Parker (1993b), Hoffmann et al. (1993), and Al-Habori (1994); and a recent collection of articles edited by Strange (1994a).

Volume regulatory processes are thus an integral part of a variety of cellular functions in absorbing and secreting epithelia, especially salt and water transport. They control cell metabolism, cell growth, and proliferation. The study of cell volume regulatory mechanisms can thus provide useful knowledge with important implications.

II. Volume-Regulatory Ion Transport Systems Involved in Cell Volume Regulation

The following types of transport pathways contribute to regulation of cell volume: the Na^+, K^+ pump in many cell types (and the Ca^{2+} pump in a few), channels for Ca^{2+}, K^+, Cl^- and organic osmolytes and the electroneutral Na^+, K^+, $2Cl^-$ cotransporter in many cell types (and the K^+, Cl^- and Na^+, Cl^- cotransporters in a few), the electroneutral $Na^+$$H^+$, $Cl^-$$HCO_3^-$ exchangers in many cell types (and the K^+/H^+ and Na^+/Ca^{2+} exchangers in a few). In addition there are Na^+-coupled cotransporters for organic solutes and for HCO_3^-. This chapter will emphasize the roles of the channels and the Na^+, K^+, $2Cl^-$ cotransporters. Attention will also be devoted to the various pathways of transport of taurine, an amino acid with importance as an organic osmolyte, and to K^+, Cl^- cotransport. Less attention will be given to the ATP-driven pumps, the exchange pathways, and other Na^+-coupled cotransporters.

The fluxes mediated by the Na^+, K^+ pump contribute to maintenance of constant cell volume by offsetting the tendency of cells to swell because of impermeant cellular solutes. However, the pump is generally considered not to be responsive to changes in cell volume (except as a consequence of changes in cell Na^+ concentration as a substrate of the pump). [A recent report showed significant changes in the $K_{1/2}$ of the Na^+, K^+ pump for cellular Na^+ in cardiac myocytes in response to cell volume changes. However, the authors concluded that resultant changes in fluxes through the pump were unlikely to contribute to short-term volume regulation (Whalley et al., 1993)]. Therefore regulation of cell volume requires, in addition to the pump, membrane transporters which are coupled through signal transduction pathways to the mechanisms (discussed later) that detect changes in cell volume and respond to them with high gain. There are a number of such transporters, both channels and coupled transport pathways. None is fueled by ATP, but some of the coupled transporters are capable of active transport, the energy coming from the flow of Na^+, Cl^-, or K^+ down their chemical gradients.

Large perturbations in cell volume are not necessarily required to activate volume regulatory transporters. Minor cell swelling, resulting from Na^+-dependent nutrient uptake after the cells have been exposed to elevated concentrations of L-alanine or D-glucose can activate both K^+ and Cl^- conductive pathways (MacLeod, 1994). Rabbit kidney tubule cells exposed to gradual changes in extracellular osmolality are capable of maintaining nearly constant cell volume over a wide range of osmolalities, a process termed "isovolumetric regulation" by Lohr and Grantham (1986). The volume-sensitive transporters responsible for the regulation are responding to changes in volume which are both small and gradual.

As stated above, animal cells suspended in hypotonic media initially swell by osmotic water uptake, but subsequently undergo a compensatory shrinkage through a regulatory volume decrease, or RVD (for early reviews describing RVD in different cells, see Kregenow, 1981; Cala, 1983a; Hoffmann, 1977, 1983, 1985; Grinstein et al., 1984; Lauf, 1985a,b; Siebens 1985; Larson and Spring, 1987; Eveloff and Warnock, 1987). Partial restoration of cell volume usually results from a loss of KCl and certain organic osmolytes.

After shrinking in a hypertonic environment, a regulatory volume increase (hypertonic RVI) can be observed in certain cell types, e.g., *Amphiuma* red blood cells (Cala, 1980) and *Necturus* gall bladder epithelial cells (Larson and Spring, 1987), whereas other cells, including Ehrlich ascites tumor cells (Hoffmann and Simonsen, 1989), lymphocytes (Grinstein et al., 1984), frog skin epithelial cells (Ussing, 1982a), and frog urinary bladder epithelial cells (Davis and Finn, 1985), show no indication of hypertonic RVI. However, nearly all cells show a volume recovery when cells that have undergone a RVD in hypotonic medium are transferred back to an isotonic medium, which is now hypertonic to them, called the "RVI-after-RVD protocol" or "Post-RVD RVI." After the initial osmotic shrinkage, the cells recover their volume with an associated KCl uptake—actually an uptake of Na^+, K^+, and Cl^-—followed by extrusion of Na^+ by the Na^+, K^+ pump.

It is not clear why some cells show an RVI response only following the "RVI-after-RVD protocol." When shrunk in an isotonic rather than a hypertonic medium, as in the "RVI-after-RVD" protocol, more water will be taken up with a given amount of salt, and there will be a greater increase in cell volume (Cala and Maldonado, 1994). However, this does not explain the complete lack of volume recovery in certain cells after hypertonic shrinkage. It has been suggested that the reduced intracellular Cl^- concentration after RVD plays a permissive role in the activation of Na^+, K^+, $2Cl^-$ cotransport in frog skin (Ussing, 1982b), in Ehrlich cells (Hoffmann et al., 1983), and in lymphocytes (Grinstein et al., 1983). In squid giant axons, cell shrinkage at high external osmolarity has recently been reported to stimulate influx via the Na^+, K^+, $2Cl^-$ cotransporter by shifting the relationship between internal Cl^- concentration and influx via the cotransport system, and hence reduce inhibition of the cotransporter by internal Cl^- (Breitwieser et al., 1990).

There have been several other suggestions for a role of cellular Cl^- in regulating Na^+, K^+, $2Cl^-$ cotransport (Levinson, 1990; Lytle and Forbush, 1992a). Robertson and Foskett (1994) have recently demonstrated that activation by the agonist carbachol of the Na^+, K^+, $2Cl^-$ cotransporter as well as the Na^+/H^+ exchanger in salivary gland epithelial cells indeed requires an agonist-induced decline in intracellular Cl^-. Also, in these cells a decrease in cell Cl^- has been suggested to play a permissive role in the activation of the cotransporter as well as the exchanger. It is important to note, however, that RVI was actually recently demonstrated in Ehrlich cells without a preconditioning hypotonic challenge. Provided the driving force(s) for Na$^+$, K$^+$, and/or Cl$^-$ are large enough during the hypertonic challenge, an RVI response proceeds without a prior RVD (Jensen *et al.*, 1993), though the response is slower than observed following the "RVI-after-RVD protocol." Therefore several factors seem to be involved in the difference between the two RVI responses.

Figure 1 shows the major changes during the RVD and the RVI responses using Ehrlich ascites tumor cells as an example. For further details, see earlier reviews (Hoffmann *et al.*, 1988, 1993; Hoffmann and Simonsen, 1989). Hypotonically swollen Ehrlich cells (Fig. 1, left) recover their volume by net loss of K$^+$ and Cl$^-$ via conductive transport pathways (Hoffmann *et al.*, 1986), by net loss of taurine via a leak pathway (Hoffmann and Lambert, 1983), and under some circumstances, e.g., low external Ca^{2+}, by K$^+$, Cl$^-$ cotransport (Thornhill and Laris, 1984). Kramhøft *et al.* (1986) demonstrated and distinguished both K$^+$, Cl$^-$ cotransport and K$^+$ and Cl$^-$ conductance channels.

The conductive transport pathways activated during RVD in Ehrlich cells are channels for K$^+$ and for Cl$^-$ (Christensen and Hoffmann, 1992). This mechanism is common to many other cell types (see later discussion). The volume-sensitive K$^+$ channels in Ehrlich cells can be blocked by Ba^{2+}

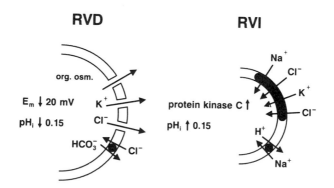

FIG. 1 Some features of regulatory volume decrease and regulatory volume increase in Ehrlich ascites tumor cells. During RVD, CTX-insensitive K$^+$ channels and mini Cl$^-$ channels are activated. The increase in Cl$^-$ conductance is much larger than that in K$^+$, and therefore the membrane potential (E$_m$) depolarizes by 20 mV (the downward-pointing arrow). The K$^+$ loss is partially balanced electrically by HCO$_3^-$ loss through the Cl$^-$/HCO$_3^-$ exchanger, resulting in acidification of the cells (decrease in pH$_i$). During RVI, the Na$^+$, K$^+$, 2Cl$^-$ cotransporter as well as the Na$^+$/H$^+$ exchanger are activated. The latter results in an alkalinization of the cells (increase in pH$_i$). The activation of protein kinase C seems to be involved in activation of both transport pathways.

(Lambert *et al.*, 1984) and by quinine (Hoffman *et al.*, 1984), but not by charybdotoxin ($CT\chi$) or clotrimazole (Harbak and Simonsen, 1994). The volume-sensitive Cl^- channels are inhibited by diphenylamine-2-carboxylate (DPC) and indacrinone (MK-196) (Aabin and Hoffmann, 1986) as well as by unsaturated fatty acids (e.g., arachidonic acid, oleic acid, linoleic acid, linolenic acid, and palmitoleic acid) (Lambert, 1987; Lambert and Hoffmann, 1994). The volume-sensitive Cl^- channel is only slightly sensitive to 4,4'-disothiocyano-2,2'-stilbene-disulfonic acid (DIDS) (Lambert and Hoffmann, 1994).

About 70% of the total cellular osmolytes lost during RVD are KCl; the remaining 30% are amino acids, predominantly taurine (Hoffmann and Hendil, 1976). The volume-induced taurine efflux is by a pH-dependent leak pathway (an organic osmolyte channel, see later discussion) which is independent of the Na^+-, and Cl^--dependent saturable taurine transport system (Lambert, 1984, 1985; Lambert and Hoffmann, 1991).

The Cl^- conductance is increased to a larger extent (68-fold) than the K^+ conductance (twofold). Therefore a depolarization of $\sim$20 mV in a hypotonic solution, 50% of isotonic osmolality, is observed during RVD in Ehrlich cells, measured by a fluorescent probe (Lambert *et al.*, 1989) or by conventional microelectrodes (Lang *et al.*, 1987). A depolarization during the RVD response has also been demonstrated in human lymphocytes (Grinstein *et al.*, 1982d; Sarkadi *et al.*, 1984a), astrocytes (Kimelberg and O'Connor, 1988), cultured opossum kidney cells (Ubl *et al.*, 1988), and neuroblastoma cells (Falke and Misler, 1989). In a line of cultured human epithelial cells, intestinal 407 cells, biphasic membrane potential changes during RVD have been demonstrated with an initial hyperpolarization (caused by opening of K^+ channels) followed by a transient depolarization (caused by opening of Cl^- channels) (Hazama and Okada, 1988, 1990).

During RVD in Ehrlich ascites cells, K^+ loss exceeds Cl^- loss by about 1.6-fold (Hendil and Hoffmann, 1974). Acidification of the cytoplasm during RVD, demonstrated using an intracellular pH-sensitive fluorescent probe (Livne and Hoffmann, 1990), was explained by a reuptake of Cl^- in exchange for HCO_3^- via the Cl^-/HCO_3^- exchanger (Livne and Hoffmann, 1990), producing a proton uptake via the Jacobs–Stewart cycle (Jacobs and Stewart, 1942). This explains the greater loss of K^+ during RVD. Acidification by this mechanism is predicted by the analysis of Lew and Bookchin (1986), according to which the K^+ loss will exceed the Cl^- loss to an extent dependent upon the cellular pH buffering capacity (Freeman *et al.*, 1987; Lew and Bookchin, 1986). The Ehrlich cell shows clear examples of these predicted types of distributions of ions during RVD. In other cell types, different types of transports have been implicated in RVD.

The RVI response in Ehrlich cells (see Fig. 1, right) is mediated by a normally quiescent, electroneutral, Na^+ and Cl^- dependent cotransport system (Hoffman *et al.*, 1983), later shown to be the Na$^+$, K$^+$, 2Cl$^-$ cotransporter (Levinson, 1991; Jensen *et al.*, 1993). The cotransporter can be blocked by bumetanide (Hoffmann *et al.*, 1983) and by various other loop diuretics (Kramhøft *et al.*, 1984). A role for Na$^+$, Cl$^-$ or Na$^+$, K$^+$, 2Cl$^-$ cotransport in cell volume regulation has also been established in many other cell types (see later discussion). In other cell types, transport pathways other than these cotransporters have been implicated in RVI.

A. Volume-Sensitive K$^+$ and Cl$^-$ Channels Involved in Regulatory Volume Decrease

Ion channels sensitive to changes in cell volume have been reported in a number of different cell types in recent years. They include Cl$^-$ channels, K$^+$ channels, and nonselective channels. Volume-sensitive Cl channels have been described in human airway epithelial cells (McCann *et al.*, 1989a; Kunzelmann *et al.*, 1989; Chan *et al.*, 1992; Solc and Wine, 1991), a colonic tumor cell line (T84) (Worrell *et al.*, 1989), Ehrlich ascites tumor cells (Christensen and Hoffmann, 1992), human small intestinal epithelial cells (Kubo and Okada, 1992), ciliary ocular epithelial cells (Yantorno *et al.*, 1992), cardiac myocytes (Tseng, 1992; Sorota, 1992), lymphocytes (Cahalan and Lewis, 1988; Lewis *et al.*, 1993), bovine chromaffin cells (Doroshenko and Neher, 1992), and *Xenopus* oocytes (Ackerman *et al.*, 1994). The widely expressed ClC-2 channel protein cloned from rat brain (Gründer *et al.*, 1992) is a volume-sensitive Cl$^-$ channel and the P-glycoprotein has also been proposed to be a swelling-induced Cl$^-$ channel (Valverde *et al.*, 1992; Han *et al.*, 1994). That the P-glycoprotein is a swelling-activated Cl$^-$ channel has been disputed in several cell lines (Rasola *et al.*, 1994), in lymphocytes (Cahalan *et al.*, 1994) as well as in Ehrlich cells (T. Litmann and E. K. Hoffmann, unpublished results). In these latter two studies, the volume-activated Cl$^-$ efflux was independent of whether the cells express the P-glycoprotein.

Volume-sensitive K$^+$ channels have, for example, been described in Ehrlich cells (Christensen and Hoffmann, 1992), tracheal epithelial cells (Butt *et al.*, 1990), and lymphocytes and thymocytes (see Rabin *et al.*, 1991). The nonselective cation channels will be discussed later.

Activation by cell swelling of separate, conductive fluxes of both K$^+$ and Cl$^-$ was first proposed for Ehrlich ascites tumor cells (Hoffmann, 1978) although a swelling-induced increase in the conductive Cl$^-$ permeability of the basolateral membrane of frog skin epithelial cells had already been suggested in 1961 (MacRobbie and Ussing, 1961; Ussing, 1982a, 1985). The steady-state membrane conductances in Ehrlich cells are estimated at 10.4 μS/cm^2 for K$^+$, 3.0 μS/cm^2 for Na$^+$, and 0.6 μS/cm^2 for Cl$^-$ (Lambert *et al.*, 1989). During the RVD response, Na$^+$ conductance is reduced (Hoffmann, 1978) whereas K^+ conductance is increased about twofold, and Cl^- conductance is increased about 60-fold (Lambert et al., 1989).

Swelling-induced increase in both K^+ and Cl^- conductive pathways was later reported in a number of cell types, including the basolateral membrane of rabbit proximal tubule (Welling and O'Neil, 1990), Madin-Darby canine kidney (MDCK) cells (Bandarali and Roy, 1992), human intestinal epithelial cells (Hazama and Okada, 1988), human lymphocytes (Cheung et al., 1982; Grinstein et al., 1982a,b,c, 1984; Sarkadi et al., 1984a,b, 1985), human platelets (Livne et al., 1987), human fibroblasts (Rugolo et al., 1989), rat astrocytes (Kimelberg and O'Connor, 1988), jejunal villus cells (MacLeod and Hamilton, 1991), and Ehrlich cells (Hoffman, 1978, 1985; Hoffmann et al., 1986).

Single-channel studies of swelling-activated channels have been performed in Ehrlich ascites tumor cells by Hudson and Schultz (1988) and by Christensen and Hoffmann (1992). Single-channel currents were recorded in excised patches and in cell-attached patches with the cell exposed to isotonic or hypotonic medium. Three types of channels have been identified in the cell-attached mode: a small Cl^- channel, a small K^+ channel, and a 23-pS nonselective cation channel. They are all activated by swelling, with a delay of about 1 min (Christensen and Hoffmann, 1992). Figure 2 shows clear examples of the opening of two types of swelling-activated channels: K^+ channels (upward deflections) and nonselective cation channels (downward deflections).

1. Properties of the Volume-Sensitive K^+ Channels

A small Ca^{2+}-dependent, inwardly rectifying K^+ channel was demonstrated in isolated inside-out patches of Ehrlich ascites cells (Christensen and Hoffmann 1992). This K^+ channel resembles the small Ca^{2+}-activated K^+ channel previously found in several cell types (Hoffmann et al., 1993), such as human red blood cells (Grygorzyk et al., 1984; Christophersen, 1991), HeLa cells (Sauvé et al., 1986), aortic endothelial cells (Sauvé et al., 1988), MDCK cells (Friedrich et al., 1988), and rabbit proximal tubule cells (Parent et al., 1988).

A K^+ channel with spontaneous activity and with characteristics similar to those of the K^+ channel seen in excised patches was found in patch-clamp studies in the cell-attached mode. The single-channel conductance at 5 mM external K^+ was ~7 pS (Christensen and Hoffmann, 1992). No evidence was found of voltage gating of the K^+ channels, or channel activation by membrane stretch. The channel could be activated by a Ca^{2+} ionophore. It appeared that swelling-induced conductive K^+ transport was mediated predominantly by this type of Ca^{2+}-activated channel (Christensen and Hoffmann, 1992).

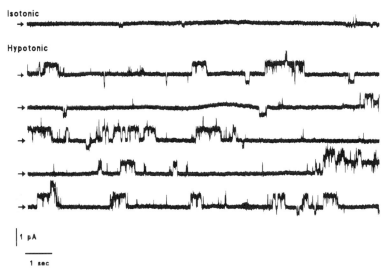

Isotonic

Hypotonic

1 pA

1 sec

FIG. 2   Cell-attached patch-clamp recording of channel activity in Ehrlich ascites cells induced by cell swelling. The hypotonic medium was 1/2 isotonic. The arrows to the left of each trace indicate zero current. The electrode contained isotonic medium. The bath was nominally at zero voltage and the interior of the patch electrode in this experiment was held at 0 mV. The upward deflections (possitive current, from the cell into the electrode) are K$^+$ channels and the downward deflections are nonselective cation channels. (F. Jørgensen, unpublished results. Similar results were published by Christensen and Hoffmann, 1992.)

However, more recent evidence suggests that this is not the case. In Ehrlich cells, we have compared the sensitivities to K$^+$ channel blockers of the K$^+$ loss induced by hypotonic swelling and by thrombin stimulation. The K$^+$ loss observed during RVD was unaffected by charybdotoxin (100 nM), an inhibitor of some classes of Ca^{2+}-activated K$^+$ channels, whereas the thrombin-induced K$^+$ loss was inhibited by about 70%. Similarly, K$^+$ loss during RVD was unaffected by clotrimazole (5 μM), a potent inhibitor of Ca^{2+}-dependent K$^+$ channels (Alvarez *et al.*, 1992), whereas the thrombin-induced K$^+$ loss was inhibited by about 50% (Harbak and Simonsen, 1995). Finally, the thrombin-induced K$^+$ loss was unaffected by Ba^{2+}. Therefore the K$^+$ channels activated during RVD in Ehrlich cells were concluded to be different from those activated by a rise in cytoplasmic Ca^{2+} (Harbak and Simonsen, 1995), which is consistent with the findings of Thomas-Young *et al.* (1993).

In "inside-out" vesicles made from human red cell membranes, a high K$^+$ permeability with a large Ca^{2+}-independent component was induced by osmotic swelling (Rossi and Schatzmann, 1982). In frog red blood cells, two classes of K$^+$ channels were demonstrated in patch clamp studies.

Swelling of frog red cells caused an increase in K^+ permeability, mediated primarily by a small conductance channel which lacked the striking sensitivity to Ca^{2+} of the larger conductance channel (Hamill, 1983). In human peripheral blood lymphocytes, RVD also seems to be mediated by Ca^{2+}-independent K^+ channels. Two types of Ca^{2+}-independent, voltage-sensitive channels were found to be involved in RVD in lymphocytes, one CTX-sensitive (n-type K^+ channel) and the other CTX-insensitive, whereas only the CTX-insensitive channel was involved in rat thymocytes (Grinstein and Smith, 1990; Rotin et al., 1991). In this regard, Ehrlich ascites tumor cells resemble the thymocytes.

Recently Deutsch and Chen (1993) have shown that transient transfection of a lymphocyte cell line which is unable to regulate volume with the K^+ channel isoform KV1.3 reconstitutes the cells' ability to regulate volume, whereas transfection with KV3.1 did not. KV1.3 encodes the charybdotoxin-sensitive, voltage-sensitive K^+ channel (the n-type channel), which in these cells is the volume-regulating channel.

It seems fair to conclude that there are no K^+ channels which are directly activated by swelling. When the cells depolarize enough to open the voltage-sensitive K^+ channels, they function in volume regulation; when Ca^{2+} increases enough to open Ca^{2+}-activated K^+ channels, they are used. That both types of K^+ channels are functioning in volume regulation under various conditions is suggested for Ehrlich cells (see also Section IV,G,2) and for lymphocytes (Schlichter and Sakellarapoulus, 1994).

2. Properties of the Volume-Sensitive Cl⁻ Channels

Intermediate conductance (25–75 pS), volume-sensitive Cl^- channels have been described in airway epithelia (Welsh, 1987; Solc and Wine, 1991), lobster walking leg (Lukacs and Moczydlowski, 1990), T84 colonic cell line (Worrell et al., 1989), and rat colonic epithelium (Diener et al., 1992). The volume-sensitive organic osmolyte-anion channel (VSOAC) is an intermediate conductance volume-sensitive anion channel, which is widely distributed (Jackson and Strange, 1993). Small (2–7 pS) volume-sensitive Cl^- channels have been described in Necturus choroid plexus epithelial cells (Christensen et al., 1989), in Ehrlich cells (Christensen and Hoffmann, 1992), lymphocytes (Lewis et al., 1993), and chromaffin cells (Doroshenko, 1991; Doroshenko and Neher, 1992). They all seem to have an anion selectivity and a voltage dependence different from the earlier-mentioned small volume-sensitive CIC-2 channel cloned by Günder et al. (1992). Large (400 ps) volume-activated Cl^- channels are found in cardiac myocytes (Coulombe and Coraboeuf, 1992).

A "mini Cl^- channel" with low spontaneous activity was observed in excised inside-out patches from Ehrlich cells (Christensen and Hoffmann,

1992). The channel activity was insensitive to Ca^{2+}_i in excised patches, and could not be activated by membrane stretch exerted by suction. Recordings from cell-attached patches showed single-channel currents with characteristics similar to those seen in isolated patches. The single-channel conductance of the fully activated channel was estimated at 3–7 pS. In the cell-attached mode, the channel in most patches was activated by osmotic swelling. There was a delay of about 1 min in its activation (Christensen and Hoffmann, 1992). In intact cells, the channel was activated by a Ca^{2+} ionophore. Therefore the swelling-activated Cl^- channel may be activated by intracellular mediators released by Ca^{2+} (see later discussion). The "mini Cl^- channel" in Ehrlich cells is a nonselective anion channel; it is permeable to Cl^-, Br^-, NO_3^-, and SCN^- (Hoffmann et al., 1986).

From a single-channel conductance at 7 pS and the total chloride conductance at 40 $\mu S/cm^2$ during RVD (Lambert et al., 1989), the open channel density was estimated at 6×10^6 channels per cm^2 or 80 open channels per cell, and at this density the small Cl^- channel accounts for the major part of the volume-activated Cl^- efflux (Christensen and Hoffmann, 1992).

3. Other Cl⁻ Channels

Two other types of Cl^- channels have been recorded in patch-clamp studies on Ehrlich cells, a 400 pS and a 34 pS chloride channel (the "maxi Cl^- channel" and the "medium Cl^- channel," respectively) (Christensen and Hoffmann, 1992). The maxi Cl^- channel has properties similar to the "maxi-chloride channel" reported in many cell types: cultured rat muscle (Blatz and Magleby, 1983), macrophages (Schwarze and Kolb, 1984; Kolb and Ubl, 1987), and a mouse B-lymphocyte hybridoma cell line (Bosma, 1989). The "maxi Cl^- channel" was voltage dependent (activated by depolarization of the cell membrane) but not directly activated by Ca^{2+} (Christensen and Hoffmann, 1992). The 34-pS "medium Cl^- channel" was an inward rectifier with properties similar to the Cl^- channel originally described in the apical membrane in human airway epithelial cells (Frizzell et al., 1986; Welsh, 1986), which is activated by cAMP-dependent protein kinase (PKA) and protein kinase C (PKC) (Li et al., 1989). There is no direct evidence for a role of either the maxi Cl^- channel or the medium Cl^- channel in the RVD response in Ehrlich cells (Christensen and Hoffmann, 1992). These two types of chloride channels have been observed in human B- and T-cell lines, again without evidence of the involvement of either channel in volume regulation (Rotin et al., 1991).

4. Cl⁻ Channel Inhibitors—The Gramicidin Trick

When the K^+ channel is bypassed by addition of gramicidin, a K^+ and Na^+ ionophore, the rate of cell shrinkage in an Na^+-free medium can be used to evaluate the effect of inhibitors on the swelling-increased Cl^- permeability (Hoffmann *et al.*, 1986). Figure 3 shows that the RVD in Ehrlich cells in hypotonic Na^+-free medium (with choline$^+$ substituted for Na^+) is inhibited by MK196, which is a Cl^- channel blocker (DiStefano *et al.*, 1985; Dürr and Larsen, 1986) and by arachidonic acid. In contrast, DIDS, an inhibitor of anion exchange and some anion channels (Cabantchik and Greger, 1992; Gunn, 1992), had only a marginal effect.

The $^{36}Cl^-$ efflux from Ehrlich cells in hypotonic Cl^- free medium is almost completely inhibited by MK196 (2 mM) and by arachidonic acid (200 μM) (Lambert and Hoffmann, 1994). The experiments were performed in Cl^--free medium (with gluconate substituted for Cl^-) and in the presence of bumetanide in order to block $^{36}Cl^-$ efflux via the anion exchanger and the Cl^--dependent cotransport systems, respectively.

B. K⁺, Cl⁻ Cotransport Involved in Regulatory Volume Decrease

Activation of K^+, Cl^- cotransport can play a role in RVD by promoting efflux of K^+ and Cl^- and osmotically obliged water. K^+, Cl^- cotransport

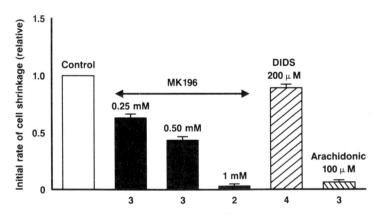

FIG. 3   Effects of MK196, DIDS, and arachidonic acid, Cl^- channel blockers, on regulatory volume decrease in Ehrlich ascites tumor cells. Cells were swollen in a hypotonic choline chloride medium (150 mosmol) with 500 μM gramicidin, and cell volumes were measured in a Coulter counter. DIDS (200 μM), MK196 (0.25 to 1 mM), or arachidonic acid (100 μM) were added at the time of hypotonic exposure. Gramicidin was added at the time of maximal cell swelling, $t = 0.9$ min, to ensure a high K^+ permeability. The initial rates of volume recovery (measured in fl/min) are the rates of cell shrinkage within 1.5 min after swelling. The rates of cell shrinkage are given relative to the rates measured in control cells in hypotonic medium + gramicidin. Values are means ± S.E.M.s, with the numbers of experiments indicated below the bars. (Data with 1 mM MK196 are from Hoffmann *et al.*, 1988; data with arachidonic acid are from Lambert, 1987; all other data are from Lambert and Hoffmann, 1994.)

was first noted as a swelling-activated K^+ efflux in duck red blood cells (Kregenow, 1971). Subsequently, swelling-activated K^+ efflux observed in sheep red blood cells of the LK phenotype was shown to require Cl^-, with only Br^- serving as an alternative substrate anion (Dunham and Ellory, 1981). K^+, Cl^- cotransport has now been observed in the red blood cells of a number of other mammalian species, including humans, rabbits, pigs, horses, and dogs, as well as in nucleated red cells of teleosts (Lauf et al., 1992). As stated earlier, K^+, Cl^- cotransport can be demonstrated under certain conditions in Ehrlich ascites cells (Thornhill and Laris, 1984; Kramhøft et al., 1986). Recently, direct evidence has been presented for the direct coupling of K^+ and Cl^- during K^+, Cl^- cotransport (Brugnara et al., 1989) and for a 1:1 stoichiometry of the coupling (Kaji, 1993).

K^+, Cl^- cotransport has also been reported in various epithelia, including gall bladder (Reuss, 1983; Corcia and Armstrong, 1983) and choroid plexus (Zeuthen, 1991) of Necturus, early distal tubule of Amphiuma kidney (Guggino, 1986), and cortical thick ascending limb (Greger and Schlatter, 1983) and proximal tubule of rabbit kidney (Eveloff and Warnock, 1987; Sasaki et al., 1988). Some of these reports are subject to controversy. For example, KCl efflux from Necturus gall bladder was proposed to be by separate conductive channels rather than K^+, Cl^- cotransport (Furlong and Spring, 1988), and K^+, Cl cotransport was reported to be absent in mammalian proximal tubule (Chen and Verkman, 1988; Grassl et al., 1987). Where it is present in epithelial cells, both secretory and reabsortive, K^+, Cl^- cotransport participates in transcellular efflux of salt and water, as well as in the maintenance of constant volume of the cells.

K^+, Cl^- cotransport is not measurable in mature human erythrocytes (Duhm, 1987) unless they are osmotically swollen (Kaji, 1986). However, the cotransporter is functional in immature human red cells (Hall and Ellory, 1986; Canessa et al., 1987), in which it has been demonstrated to promote RVD (O'Neill, 1989). Because of its absence in mature red cells, it follows that K^+, Cl^- cotransport promotes a decrease in cell volume during maturation of circulating human red cells, and red cells of other mammals as well. Regulation of cell volume by K^+, Cl^- cotransport has been demonstrated in young red blood cells of humans (O'Neill, 1989) and sheep (Lauf and Bauer, 1987).

K^+, Cl^- cotransport may play a role in the heritable pathological condition, sickle cell anemia. The primary defect in the disease is one of several mutations in the β-chain of hemoglobin (HbS). With this mutation, there is polymerization of hemoglobin in response to deoxygenation and a characteristic sickling of the cells. The rate and extent of polymerization of HbS are greatly enhanced by small decreases in red cell volume (Eaton and Hofrichter, 1987). Volume-sensitive K^+, Cl^- cotransport is very active in red cells from patients with sickle cell disease (Brugnara et al., 1986; Canessa *et al.,* 1987), and may contribute to the lower mean cell volume of red cells in these patients and the resulting enhanced susceptibility to polymerization of their hemoglobin.

A direct interaction between the mutant hemoglobin and the K^+, Cl^- cotransporter was suggested by results of studies on K^+, Cl^- cotransport in red cells with various mutations of the β-chain of hemoglobin. With mutations at the β_6 or β_7 other than the β_6 Glu $\rightarrow$ Val of HbS, the red cells had elevated K^+, Cl^- cotransport as in red cells with HbS. Mutations at other positions along the β-chain (e.g., β_{26} Glu $\rightarrow$ Lys and β_{121} Glu $\rightarrow$ Lys) with similar changes in net charge caused by the HbS mutation did not have elevated K^+, Cl^- cotransport (Olivieri *et al.,* 1992). Therefore the N-terminal of the β-chain of hemoglobin with a modified charge may interact with the membrane and affect K^+, Cl^- cotransport.

The Ca^{2+}-activated channel of sickle cells has also been implicated in the reduced volume of deoxygenated red cells with HbS (Bookchin *et al.,* 1987), though this conclusion is subject to controversy (Joiner, 1993). For a review of volume regulation in red blood cells with HbS, see Joiner (1993).

Some features of the regulation of volume-sensitive K^+, Cl^- cotransport have been worked out in some detail in mammalian red cells, particularly from sheep and rabbits. Swelling of sheep red cells increases both the J_{max} of the cotransporter and its apparent affinity for K^+ (Bergh *et al.,* 1990). These changes in the kinetic parameters are separate events in that they could be induced independently: reducing cell Mg^{2+} concentration caused an increase in J_{max} without changing the $K_{1/2}$ for K^+ and swelling of low Mg^{2+} cells reduced the $K_{1/2}$ without further affecting J_{max} (Bergh *et al.,* 1990).

Studies of presteady-state kinetics of swelling activation of K^+, Cl^- cotransport of rabbit red cells led to the development of a two-state model. $A \leftrightarrow B$, in which most cotransporters in cells at physiological volume are in the A state with low (or zero) flux, and swelling increases the fraction of transporters in the B state, which has a higher flux. The kinetic analysis showed that swelling activation and shift of the equilibrium distribution of transporters toward the B state was a consequence of an inhibition of the reverse reaction (the conversion of B to A), and not stimulation of the forward reaction, A to B (Jennings and Al-Rohil, 1990). The evidence included the demonstration of a substantial delay in swelling activation, with a time constant of ~10 min, and a rapid shrinkage inactivation, with no measurable delay. As will be discussed later, there is good evidence that the forward reaction is promoted by a protein phosphatase and the reverse reaction by a protein kinase. Therefore swelling activation is a consequence of inhibition of a volume-sensitive kinase, leading to reduced phosphorylation of the transporter or an associated regulatory protein.

Regulation of swelling activation of K^+, Cl^- cotransport in sheep red cells is more complex than in rabbit red cells (Dunham *et al.,* 1993). Normally there was the same delay in swelling activation as in rabbit red cells. However, there was no delay in the swelling activation of low Mg^{2+} cells, in which cotransport was already activated, as shown by Delpire and Lauf (1991). This led to the development of a three-state model for swelling activation in sheep red blood cells, in which the first step, A → B, is rate limiting, dependent on Mg^{2+}, entails an increase in J_{max} and is controlled by phosphorylation and dephosphorylation reactions (Dunham et al., 1993). The second step, B → C, is rapid and entails an increase in apparent affinity for K^+. The mechanism of the control of this step has not been established, but recent results suggest it is likely to involve mechanical changes in the cytoskeleton (S. J. Kelley and P. B. Dunham, unpublished results). The three-state model is shown in Fig. 4. By the mechanism just outlined, 10% swelling provokes an increase of three- to fourfold in K^+, Cl^- cotransport. From the minimum cotransport flux in cells shrunk 40% to cells swollen 50%, near hemolytic volume, the increase in K^+, Cl^- cotransport is more than 70-fold, a consequence of ~10-fold increases in both J_{max} and the apparent affinity for K^+ (Dunham et al., 1993). Though similar regulatory mechanisms may exist for control of K^+, Cl^- cotransport in epithelial cells, they have not been worked out.

C. Na^+, K^+, $2Cl^-$ Cotransport Involved in Regulatory Volume Increase

Electroneutral Na^+, K^+, $2Cl^-$ cotransport systems are present in a wide variety of cells and tissues. The involvement of cotransport systems in RVI was first described in avian red blood cells (Kregenow, 1973, 1981) and

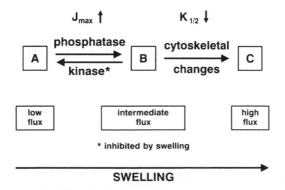

FIG. 4   Three-state model for the swelling activation of K^+, Cl^- cotransport in sheep red blood cells (Dunham et al., 1993).

was later established both in epithelia such as frog skin epithelial cells (Ussing, 1986), retinal pigment epithelial cells (Adorante and Miller, 1990; Kennedy, 1994), the apical membrane of rabbit kidney medullary thick ascending limb of Henle's loop (MTAL) (Eveloff and Warnock, 1987; Sun et al., 1991), jejunal villus cells (MacLeod and Hamilton, 1990), tracheal epithelial cells (Shorofsky et al., 1984; Musch and Field, 1989), and salivary gland cells (Robertson and Foskett, 1994), and in other nonepithelial cells such as vascular endothelial cells (O'Neill and Klein, 1992; O'Donnell, 1993), Ehrlich ascites tumor cells (Hoffmann et al., 1983; Levinson, 1990; Jensen et al., 1993), red blood cells from rat (Duhm and Göbel, 1984) and ferret (Mercer and Hoffman, 1985), L cells (Gargus and Slayman, 1980), simian virus-transformed 3T3 cells (Bakker-Grunwald et al., 1982), squid giant axon (Breitwieser et al., 1990), chick cardiac cells (Frelin et al., 1986), rabbit cardiac cells (Drewnowska and Baumgarten, 1991), and astrocytes from rat cerebral cortex (Kimelberg and Frangakis, 1986). For reviews on the Na^+, K^+, $2Cl^-$ cotransport system, see Chipperfield (1986), Geck and Heinz (1986), Lauf et al. (1987), O'Grady et al. (1987), Haas (1989, 1994), Parker and Dunham (1989), Hoffmann and Simonsen (1989), Hoffmann and Ussing (1992), and Palfrey and O'Donnell (1992).

In many cell types at normal, steady-state volume, the Na^+, K^+, $2Cl^-$ cotransport is virtually inactive. A hyperosmotic challenge and cell shrinkage activate the cotransporter, with the resultant influx of salt and the osmotically obliged water. The restoration of cell volume to near the initial level may be complete in a few minutes, at which time the cotransporter becomes inactive again. In Ehrlich ascites tumor cells, the bumetanide-inhibitable influxes of K^+ and Cl^- are near zero at physiological steady state, and increase to near 30 μmol $\times$ g dry wt $\times$ min with cell shrinkage (Hoffman et al., 1983). The transport of all three substrate ions required the simultaneous presence of the other two (Hoffmann et al., 1983; Jensen et al., 1993). In the presence of ouabain and Ca^{2+} the stoichiometry of the bumetanide-sensitive net fluxes was 1.0 Na^+, 0.8 K^+, 2.0 Cl^-, close to 1:Na, 1:K, 2:Cl. The K^+ and Cl^- flux ratios (influx/efflux) for the bumetanide-sensitive components were estimated at 1.34 $\pm$ 0.8 and 1.82 $\pm$ 0.15. These ratios should be close to the combined concentration ratio for the Na^+, K^+, $2Cl^-$ cotransport system calculated from the external and intracellular concentrations of the ions (see later discussion). The chemical driving force is a function of this ratio. The experimentally determined combined concentration ratio, 1.75 $\pm$ 0.24, is close to the value predicted by the flux ratios (Jensen et al., 1993), and the driving force for a Na^+, K^+, $2Cl^-$ cotransporter in Ehrlich cells can account for the measured flux ratios during RVI for K^+ and Cl^-.

In the medullary thick ascending limb of rabbit kidney, both Na^+, Cl^- and Na^+, K^+, $2Cl^-$ cotransport systems have been identified; Na^+, K^+, $2Cl^-$

was proposed to predominate during RVI (Eveloff and Warnock, 1987). In tracheal epithelial cells, a K⁺-independent Na⁺, Cl⁻ cotransport is activated by osmotic shrinkage (Musch and Field, 1989). In Ehrlich ascites tumor cells, the RVI response was originally reported to involve a K^+-independent Na^+, Cl^- cotransport system (Hoffmann et al., 1983), but recent evidence demonstrates that activation of Na^+, K^+, $2Cl^-$ cotransport predominates during RVI in Ehrlich cells (Levinson, 1991; Jensen et al., 1993). Indeed, it is not clear whether there is K^+-independent Na^+, Cl^- cotransport in Ehrlich cells. In cells where both transporters are present, the precise relation between them during RVI is uncertain. The two transport systems are distinguishable, but may represent alternative modes of operation of the same transporter. Sun et al. (1991) reported that the addition of antidiuretic hormone (ADH) to mouse medullary thick ascending limb cells converts the K^+-independent Na^+, Cl^- cotransporter to a K^+-dependent Na^+, K^+, $2Cl^-$ cotransporter. This Na^+, Cl^- cotransporter is not the same as the thiazide-sensitive Na^+, Cl^- cotransporter (see later discussion).

1. Mechanism and Primary Structure

The stoichiometry of Na^+, K^+, $2Cl^-$ cotransporter is $1Na:1K:2Cl$ in all cells examined except in the squid giant axon, where the stoichiometry is $2Na:1K:3Cl$ (Russell, 1983). Variable stoichiometries reported for certain other cells can probably be accounted for by K^+/K^+ and Na^+/Na^+ exchanges mediated by the cotransporter (Lauf et al., 1987; and Duhm, 1987). Net uptake by the cotransporter is driven by the inwardly directed concentration gradients for Na and Cl. The driving force is given by: $RT\ln([Na]_o[K]_o[Cl]_o^2/[Na]_c[K]_c[Cl]_c^2)$. The driving force is independent of voltage because transport is electroneutral. Because of the stoichiometry, the inwardly directed Cl^- gradient is the main contributor to driving force. A combined concentration ratio, the term in parentheses, of greater than one gives an inwardly directed driving force. We find that in Ehrlich cells, the combined concentration ratio after the RVD/RVI protocol, as well as after increasing Na^+, K^+, and Cl^- in the medium, is above one and the cells can perform RVI. In contrast, the ratio after shrinking the cells by addition of sucrose is 0.13 and the cells cannot perform RVI (Jensen et al., 1993).

Studies by McManus and his associates (Lytle and McManus, 1986; McManus, in Lauf et al., 1987) led to a model for binding and release of ions by the Na^+, K^+, $2Cl^-$ cotransporter (Parker and Dunham, 1989; Haas, 1994). According to this model presented in Fig. 5, binding of the four substrate ions on the outside of the cell and their release on the inside are both ordered, and the order of binding and release are the same: Na^+, Cl^-, K^+,

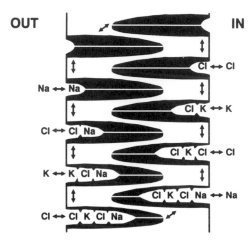

FIG. 5   Model for the Na^+, K^+, $2Cl^-$ cotransporter showing "first on, first off" binding and release of the substrate ions (Lytle and McManus, 1986; McManus, in Lauf et al., 1987).

Cl^-. This "first on, first off" binding and release is known as "glide symmetry" (Stein, 1986).

Bumetanide, the diagnostic inhibitor of the cotransporter, was suggested to compete for the second Cl^- site (Haas and McManus, 1983), although it is also possible that bumetanide binds to a separate site whose affinity for bumetanide is Cl^- dependent (Hedge and Palfrey, 1992; Haas, 1994). Hoffmann et al. (1994) recently demonstrated an increased affinity for bumetanide in Ehrlich cells caused by disruption of actin filaments by cytochalasin B, which also activates cotransport in these cells (Jessen and Hoffmann, 1992). This raises the possibility of a regulatory function for the site where bumetanide binds.

Identification and purification of the cotransporter has been attempted in several studies with different approaches: (1) photoafinity labelling studies using a photosensitive, ^3H-labelled bumetanide analog; (2) purification of detergent-solubilized cell membranes on a bumetanide-Sepharose affinity gel; and (3) immunoprecipitation of the photoaffinity-labelled protein with monoclonal antibodies against the partially purified protein. Method (1) identified a 150-kDa protein in membranes from duck red blood cells and mouse and dog kidney, and a 195-kDa protein in membranes from shark rectal glands (Haas, 1994). Method (2) was used to isolate two proteins from detergent-solubilized Ehrlich cell membranes with molecular weights of 85 kDa and 39 kDa (Feit et al., 1988). Antibodies raised against these purified proteins inhibited Na^+, K^+, $2Cl^-$ cotransport in Ehrlich cells (Fig. 6) (Dunham et al., 1990). Method (3) was used to purify further the 195-

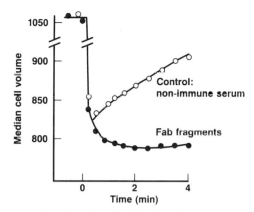

FIG. 6   Effect of antibodies against purified bumetanide-binding proteins on regulatory volume increase in Ehrlich ascites cells. Fab fragments were made from the antibodies by papain cleavage. Cells were preincubated with the Fab fragments for 20 min. Control cells were preincubated with nonimmune rabbit serum. Mean cell volumes were measured using a Coulter counter. Cells were shrunken by preincubation with hypotonic medium and then transfer to isotonic medium at time zero. (Redrawn from Dunham *et al.*, 1990.)

kDa protein from shark rectal gland labelled with an ^{3}H-analog of bumetanide (Lytle *et al.*, 1992). A series of monoclonal antibodies was made against the partially purified protein. Four of the antibodies immunoprecipitated the ^{3}H-labelled protein. A cDNA library from shark rectal gland was screened with two of these antibodies, which led to the isolation of a cDNA clone encoding a protein of 1191 amino acids with a molecular weight of 130 kDa, subsequently verified by expression to be the Na^+, K^+, $2Cl^-$ cotransporter (Xu *et al.*, 1994) (see later discussion).

There is a wide range of bumetanide affinities of the Na^+, K^+, $2Cl^-$ cotransporter in different cells. High-affinity bumetanide binding is found in duck and ferret red blood cells and in apical membranes of rabbit kidney thick ascending limb (TAL), and low affinity in bumetanide is found in Ehrlich ascites tumor cells and basolateral membranes of secretory epithelia like shark rectal gland and dog trachea (Haas, 1994). This suggested that there may be several different isoforms of the Na^+, K^+, $2Cl^-$ cotransporter. Several isoforms have now been identified by cDNA cloning and sequencing as well as by Northern blotting analysis. A cDNA clone was first isolated from a shark rectal gland cDNA library using antibodies against a 195-kDa protein thought to be the shark rectal gland cotransporter (Xu *et al.*, 1994). The clone was expressed in HEK-293 cells, a human embryonic kidney cell line, and a 10-fold increase in the bumetanide-sensitive Na^+- and Cl^--dependent ^{86}Rb influx was found. The transporter had a low affinity for bumetanide typical of the rectal gland (Xu *et al.*, 1994). The corresponding protein is called NKCC-1 and hydropathy analysis yielded a scheme for the association of the protein with the membrane with 12 transmembrane helices. The primary structure has two consensus phosphorylation sites.

Subsequently Payne and Forbush (1994) cloned a rabbit kidney cDNA (NKCC-2) with a 61% amino acid sequence identity with the shark rectal gland Na^+, K^+, $2Cl^-$ cotransporter sequence, and Gamba et al. (1994) cloned a rat kidney cDNA (rBSC1) encoding a Na^+, K^+, $2Cl^-$ cotransporter (a protein of ~115 kDa) as shown by expression in *Xenopus* oocytes (Gamba et al., 1994). Also, rBSC1 is a membrane protein with 12 membrane-spanning helices. It is predominantly expressed in epithelial cells of the thick ascending limb, where the Na^+, K^+, $2Cl^-$ cotransporter is in the apical membrane. In contrast, the recently cloned isoform BSC2 is expressed in the basolateral membrane of secretory epithelia and in symmetric cells (Delpire et al., 1994). Additional isoforms will probably soon be added. It will then be possible to explore directly the relationships among the different cotransport systems.

2. Occlusion of Ions by the Na^+, K^+, $2Cl^-$ Cotransporter

The Na^+, K^+, $2Cl^-$ cotransporter is permanently activated in the large membrane vesicles called cytoplasts, prepared from Ehrlich cell membranes as described in Section IV,A. Therefore cytoplasts provided an advantagous system to test for occlusion of ions by the cotransporter. The washed cytoplasts were solubilized in HEPES–Tris buffer containing *N*-octylglucoside and ouabain to inhibit the Na^+, K^+ pump. ^{86}Rb occlusion was tested by incubating aliquots of the cytoplasts in media with K^+ concentrations from zero to 10 m*M*. Nonoccluded ^{86}Rb was separated from the cytoplast suspensions by cation-exchange chromatography. Free ^{86}Rb was retained by the columns, and the occluded ^{86}Rb was eluted with the membranes. Occluded K^+ (expressed in μmoles/g protein) plotted against K^+ concentrations yields a hyperbolic function with a $K_{1/2}$ of 500 ± 80 μM K^+ (mean $\pm$ S.E.M., $n = 4$). K^+ binding to the cytoplasts requires both Cl^- (Krarup et al., 1994) and Na^+ (T. Krarup and E. K. Hoffmann, unpublished results), and bumetanide enhanced the K^+ binding.

D. Na^+H^+, K^+H^+, and $Cl^-HCO_3^-$ Exchangers in Volume Regulation

1. Parallel Na^+H^+, and $Cl^-HCO_3^-$ Exchangers in RVI

Amiloride-sensitive Na^+/H^+ exchange functionally coupled to Cl^-, HCO_3^- exchange has been shown to mediate RVI in many cell types, including

Amphiuma red blood cells (Cala, 1980, 1983b), lymphocytes (Grinstein *et al.,* 1984), and many epithelial cell types. For extensive reviews on the functions and regulation of the Na^+, H^+ exchanger, see Mahnensmith and Aronson (1985), Grinstein and Rothstein (1986), Wakabayashi *et al.* (1992a), M. Tse *et al.* (1993), and C.-H. Tse *et al.* (1994).

Activation of the Na^+/H^+ exchanger seems to be the primary event in the volume increase in these cells, leading to increase in intracellular pH that secondarily stimulates the Cl^-/HCO_3^- exchanger (Cala, 1980; Grinstein and Foskett, 1990). The loss of H^+ is compensated for by release of protons from intracelluar buffers until their capacity is depleted. Subsequently, intracellular pH increases, resulting in an increase in intracellular concentration of HCO_3^- ($[HCO_3^-]_i$) owing to the action of carbonic anhydrase. The increase in $[HCO_3^-]_i$ drives the influx of Cl^- via the Cl^-, HCO_3^- exchanger, and the net result is uptake of NaCl and water (Grinstein *et al.,* 1992a).

For most cell types examined, the existence of highly pH-sensitive sites on both exchangers restricts their activities to separate ranges of pH_i at normal cell volume (Mason *et al.,* 1989). The volume-dependent activation of the Na^+/H^+ exchanger is largely a response to an upward shift in its sensitivity to pH_i (Grinstein *et al.,* 1985), determined by an allosteric modifier site (Aronson, 1985), and resulting in its activation at normal pH_i and concomitant intracellular alkalinization. If pH_i rises above the "set point" for the Cl^-/HCO_3^- exchanger, both exchangers can operate simultaneously. (The set point is the threshold volume at which a transporter is activated.) This results in volume gain and restoration of pH_i, with concomitant cessation of the activity of both exchangers.

The relative contributions of these exchangers to regulation of cell volume and pH_i vary considerably among cell types and are determined by the capacities and pH_i set points of the exchangers, as well as by the intracellular buffering capacities. When the set points of the two exchangers are sufficiently close to each other to allow the exchangers to be active simultaneously after cell shrinkage, the following will apply: In cells with a large buffer capacity and a large capacity for Cl^-/HCO_3^- exchange, activation of the Na^+/H^+ exchanger will result in little change in pH_i, but substantial cell swelling. Conversely, if the buffer capacity and Cl^-, HCO_3^- exchange are limited, activation of Na^+/H^+ exchange will result in intracellular alkalinization without an increase in cell volume (Cala and Grinstein, 1988).

In the nominal absence of HCO_3^-, RVI in Ehrlich cells is mediated by the Na^+, K^+, $2Cl^-$ cotransporter (Hoffmann *et al.,* 1993). When HCO_3^- is available, however, both the Na^+, K^+, $2Cl^-$ cotransporter and a Na^+/H^+ exchanger contribute to RVI (Pedersen *et al.,* 1995). When the Na^+/H^+ exchanger is activated by shrinkage of Ehrlich cells in nominally HCO_3^--free medium, the result is alkalinization and not RVI (Pedersen *et al.,* 1995).

At normal pH_i in HCO_3^--free medium, the Cl^-/HCO_3^- exchanger is quiescent in Ehrlich cells (Kramhøft $et\ al.$, 1994). DIDS did not affect the rate of change in pH_i after cell shrinkage, so the exchanger is also inactive in shrunken Ehrlich cells (Pedersen $et\ al.$, 1995). In media containing 25 mM HCO_3^-, pH_i is considerably more alkaline than in HCO_3^--free media (Kramhøft $et\ al.$, 1994). The increased HCO_3^- concentration and the alkaline pH_i are both likely to promote activation of Cl^-/HCO_3^- exchange following shrinkage activation of Na^+/H^+ exchange. Therefore the Na^+/H^+ exchanger is unlikely to mediate RVI in Ehrlich cells in the absence of HCO_3^- because it is "turned off" at a pH_i not sufficiently alkaline for the Cl^-/HCO_3^- exchanger to be "turned on." In contrast, in HCO_3^- medium, the pH_i will promote the activity of both exchangers, and they will promote RVI following cell shrinkage. As pointed out by Cala and Maldonado (1994), the increased buffering in cells in the presence of HCO_3^- will retard the inactivation of the Na^+/H^+ exchanger because the shrinkage-induced intracellular alkalinization is slower when the buffer capacity is higher.

Four distinct mammalian isoforms of the Na^+/H^+ exchanger have been cloned. The NHE-1 isoform appears to be present in the plasma membrane of nearly all mammalian cells; NHE-2, NHE-3, and NHE-4 are mainly found in epithelia (M. Tse $et\ al.$, 1993; C.-H. Tse $et\ al.$, 1994; Orlowski $et\ al.$, 1992; Fliegel and Frøhlich, 1993). Functional characterization of the four isoforms is still incomplete. The "housekeeping" NHE-1 isoform is thought to be responsible for pH regulation (Sardet $et\ al.$, 1989) and is also involved in cell volume regulation (Grinstein $et\ al.$, 1992b; Sarkadi and Parker, 1991). NHE-2 and NHE-3 are assumed to be associated with Na^+ absorption (Fliegel and Frøhlich, 1993). Recently, however, Kapus $et\ al.$ (1994) showed that NHE-2 as well as NHE-1 was stimulated by hypertonicity and inhibited in hypotonic media, whereas NHE-3 was markedly inhibited by hypertonicity and unaffected by hypotonicity. The three isoforms were tested after transfection into antiporter-deficient Chinese hamster ovary cells. It is likely that NHE-1 as well as NHE-2 are volume regulatory transporters. The role of NHE-3 in the physiology of epithelia is still uncertain (Kapus $et\ al.$, 1994).

The NHE molecules contain 10–12 membrane-spanning regions. Deletion studies indicated that the N-terminal contains the Na^+, H^+ transport site, an amiloride binding site, and an H^+ modifier site, while the C-terminal contains most of the molecule's consensus phosphorylation sites and also putative cytoskeleton binding sites (Fliegel and Frøhlich, 1993). Recently NHE-1 was shown to be colocalized with focal adhesion sites in adherent fibroblasts, and a direct physical interaction of the exchanger with the components of the cytoskeleton was suggested to be involved in the activation of Na^+/H^+ exchange after cell adhesion as well as after osmotic shrinkage (Grinstein $et\ al.$, 1993; Wakabayashi $et\ al.$, 1992b).

2. K$^+$/H$^+$ Exchange in RVD

The K$^+$/H$^+$ exchanger, in parallel with Cl$^-$/HCO$_3^-$ exchange, was shown to promote volume regulation after cell swelling in *Amphiuma* red blood cells (Cala, 1980). The K$^+$/H$^+$ exchanger had also been shown to promote K$^+$ loss from mitochondria following swelling (Garlid, 1978). While the distribution of the Na$^+$/H$^+$ exchanger is essentially ubiquitous, the K$^+$/H$^+$ exchanger has been found in a limited number of cell types. These include rat ileum brush border (Binder and Murer, 1986), turtle bladder epithelium (Youmans and Barry, 1991), and corneal epithelium (Bonano, 1991). It has not been established if the K$^+$/H$^+$ exchanger plays a role in volume regulation in these tissues. In corneal epithelium, its role appears to be in cell acidification (Bonano, 1991).

The function of the K$^+$/H$^+$ exchanger is best known in *Amphiuma* red cells, but even in these cells, its control and even its identity is uncertain. For example, there is evidence that the same membrane protein may mediate both K$^+$/H$^+$ exchange and Na$^+$/H$^+$ exchange (Cala, 1983b, 1986). On the other hand, there is a recent report of specific labeling of the K$^+$/H$^+$ exchanger in *Amphiuma* red cells with [^{3}H$_2$]-DIDS (Maldonado and Cala, 1994). There is no evidence that DIDS inhibits Na$^+$/H$^+$ exchange, so the uncertainty of the identity of the K$^+$/H$^+$ exchanger remains.

Ca^{2+} is found to be a modulator of the K$^+$/H$^+$ exchanger in *Amphiuma* red blood cells (Cala, 1983b).

III. Regulation of Amino Acids as Cellular Osmolytes

The immediate response to cell swelling or shrinkage (in seconds to minutes) is mainly loss or uptake of electrolytes by the cells (short-term RVD and RVI respectively). On a longer time scale (hours to days), volume regulation depends mainly upon loss or accumulation within the cells of nonperturbing, "compatible" organic osmolytes (long-term RVD and RVI, respectively). Organic osmolytes have been implicated in volume regulation in cells of animals, plants, bacteria, and fungi (Chamberlin and Strange, 1989; Law, 1991). For additional reviews on the volume-regulatory role of nonelectrolytes, see Fugelli and Rohrs (1980), Gilles (1983, 1987), Goldstein and Kleinzeller (1987), Law and Burg (1991), and Garcia-Pérez and Burg (1991).

The type of organic osmolytes used are restricted to five classes of solutes: free amino acids and their derivatives, certain methylated compounds, sugars, polyols, and urea (Yancey, 1994). The renal medulla will tolerate high extracellular osmolality by accumulation of five principal organic osmolytes: sorbitol, inositol, betaine glycerophosphorylcholine, and taurine (Garcia-Pérez and Burg, 1991).

Only a few amino acids (alanine, glycine, GABA, taurine, glutamate, β-alanine, and proline) are quantitatively important in cell volume regulation. In general, the amino acids serving as osmolytes require less energy for their production *de novo* than other amino acids. Taurine was the first amino acid shown to be involved in volume regulation in fish erythrocytes (Fugelli and Zachariasen, 1976), in mammalian cells, and in Ehrlich ascites tumor cells (Hoffmann and Hendil, 1976).

A. Regulation of Taurine

The aminosulfonic acid taurine (2-aminoethanesulfonic acid) is a naturally occurring β-aminosulfonic acid. Taurine is the end product of the metabolism of the sulfur-containing amino acids methionine and cysteine, and is present at high concentrations in most invertebrate and vertebrate cells and tissues (Jacobsen and Smith, 1968). It is not used in protein synthesis, with the exception of some small peptides in the brain (Marnela *et al.*, 1984), but is nevertheless considered to be a conditional essential amino acid (Huxtable, 1992). Taurine is involved in several physiological processes, including modulation of Ca^{2+} flux, neuronal excitability, and regulation of cell volume (Huxtable, 1992). Its inert biochemical nature and the low permeability of membranes to taurine make it well suited as an osmolyte. More than 95% of the taurine molecules are zwitterionic at physiological pH, giving a very low diffusion across the lipid bilayer.

In addition to Ehrlich cells (Hoffmann and Hendil, 1976), taurine effluxes play a role in RVD in cells from several species of fish: red cells, heart muscle, and brain of the stenohaline little skate (King and Goldstein, 1983; Goldstein and Brill, 1991), and red cells of different marine species (Fugelli and Thoroed, 1986; Thoroed and Fugelli, 1994b). Taurine has also been found to play a role in RVD in other mammalian cells: heart (Thurston *et al.*, 1981), cultured MDCK cells (Sánchez-Olea *et al.*, 1991); mouse astrocytes (Pasantes-Morales *et al.*, 1990); mouse neurons (Schousbo *et al.*, 1991), and rabbit lymphocytes (J. J. Garcia *et al.*, 1991). Solis *et al.* (1990) suggested that taurine plays an important role as a volume-protective osmolyte in the brain during hypoxic ischemia.

For amino acids to promote RVD, their cellular concentrations must be relatively high. The taurine concentration is in general high in teleost red cells—20–75 mmol/liter of cell water (Thoroed and Fugelli, 1994b), and in Ehrlich ascites tumor cells—53 mmol/liter of cell water (Hoffmann and Lambert, 1983). In contrast, the taurine concentration in the ascites fluid is only 0.01 mM (Christensen *et al.*, 1954). The high taurine concentration is maintained in Ehrlich cells, as in other types of cells, by a high-affinity, pH-dependent transport system, designated the β-system (Kromphardt, 1963, 1965), a $2Na^+$, $1Cl^-$, 1 taurine cotransporter (Lambert, 1984; Lambert and Hoffmann, 1993). Taurine efflux in Ehrlich cells and in many other cells is predominantly mediated by an Na^+-independent leak pathway (Lambert and Hoffmann, 1993).

B. Swelling-Activated Taurine Channel

The taurine efflux from Ehrlich cells is increased fourfold by swelling in hypotonic media with half isotonic osmolality, and the cellular taurine concentration declines from 53 mM to 7 mM within 40 min after swelling (Hoffmann and Lambert, 1983). The swelling-activated taurine efflux pathway is different from the taurine transporting systems operative under isotonic conditions, both with respect to their voltage and pH dependence (Hoffmann and Lambert, 1983; Lambert, 1985; Lambert and Hoffmann, 1993).

A volume-sensitive taurine efflux pathway different from the high-affinity β-system has also been demonstrated in rat astrocytes (Kimelberg et al., 1990), cerebellar granule cells (Schousboe et al., 1991), MDCK cells (Sánchez-Olea et al., 1991), and flounder erythrocytes (Thoroed and Fugelli, 1993, 1994a).

In most cells investigated, the volume-sensitive taurine efflux is sensitive to DIDS and other anion transport blockers (Ballatori and Boyer, 1992; Garcia-Romeu et al., 1991; Goldstein et al., 1990; Kimelberg et al., 1990; Kirk et al., 1992; Kirk and Kirk, 1993, 1994; Pasantes-Morales et al., 1990; Roy and Malo, 1992; Sánchez-Olea et al., 1991; Schousboe et al., 1991; Thoroed and Fugelli, 1994a; Lambert and Hoffmann, 1994). This made several of the authors suggest that an anion channel is involved in the swelling-activated loss of taurine. DIDS inhibits anion exchange as well as anion channels (Cabantchik and Greger, 1992; Gunn, 1992). Thus Goldstein and co-workers have suggested that taurine leaves the cell via the anion exchanger in skate erythrocytes (Goldstein and Brill, 1991). However, the swelling-activated taurine efflux from Ehrlich cells is shown not to be an exchange of cellular taurine with extracellular Cl^- via the anion-exchange system because the taurine efflux from Ehrlich cells is similar in hypotonic Cl^--containing and Cl^--free media (Lambert and Hoffmann, 1994).

Depolarization of the cell membrane decreases the rate constant for taurine efflux in osmotically swollen Ehrlich cells (Lambert and Hoffmann, 1994). This is in agreement with the view that during RVD taurine leaves the cells either as an anion or as an uncharged zwitterion via a DIDS- and MK196-sensitive channel which is potential dependent, i.e., inhibited by depolarization of the cell membrane. If taurine behaves as an anion, a substantial acidification of the cells would be expected as a consequence of the large net efflux of taurine during RVD and of the fact that 95% of cellular taurine is present in the cells as an uncharged zwitterion (Lambert and Hoffmann, 1993). However, Livne and Hoffmann (1990) demonstrated that only a minor acidification takes place during RVD, and that the acidification was a result of a redistribution of Cl^- and HCO_3^- via the Cl^-/HCO_3^- exchanger. Therefore it is likely that taurine leaves osmotically swollen Ehrlich cells as zwitterion. To account for this, it was suggested that the taurine channel in Ehrlich cells is a nonselective, voltage-dependent channel, which is permeable to some neutral solutes as well as to anions.

In other cell types, some selectivity of the channel has been demonstrated; for example, in flounder erythrocytes the sequence of selectivity is taurine $= \beta$ Ala $>$ GABA $>>>$ choline (Thoroed and Fugelli, 1994a,b); in mouse erythrocytes it is taurine $= \beta$ Ala $>$ GABA $> \alpha$ amino acids (Pasantes-Morales et al., 1994) and in rat mammary tissue it is taurine $>>$ glycine $>>>$ AIB (Shennan et al., 1994). It should be noted that the channel apparently can be permeable even to a positive ion like choline (Thoroed and Fugelli, 1994a,b; Joyner and Kirk, 1994). The swelling-activated anion channel in MDCK cells has relatively high permeabilities to taurine, glutamate, and aspartate with $P_{tau}/P_{Cl} = 0.75$ and $P_{glut}/P_{Cl} = 0.20$ (Bandarali and Roy, 1992). The volume-sensitive, nonselective anion channel which Jackson and Strange (1993) have termed volume-sensitive organic osmolyte/anion channel (VSOAC) (Jackson et al., 1994), is found in many cells, including brain glial cells, and it seems that many organic osmolytes such as myoinositol, taurine, and sorbitol leave the cell through this channel.

Pasantes-Morales et al. (1994) were able to prevent RVD in astrocytes by partly replacing NaCl in the external medium with anionic or neutral amino acids. The VSOAC channel has been demonstrated in C6 glioma cells to show the same depolarization-induced inactivation (Jackson and Strange, 1994) as the volume-activated taurine channel in Ehrlich cells (Lambert and Hoffmann, 1994). However, arachidonic acid and oleic acid inhibited the channel (Strange et al., 1993), which is at variance with results in the Ehrlich cells, where arachidonic acid and oleic acid stimulate taurine efflux in osmotically swollen cells, as well as in cells at normal steady-state volume (Lambert and Hoffmann, 1994).

Activation of organic osmolyte channels by cell swelling may play a role in the response to ischemia or anoxia because cytotoxic cell swelling takes place following ischemic stroke (Friedman and Haddad, 1993; Section I). In the brain, where taurine and glutamate both have been shown to function as osmolytes (Kimelberg, 1991), on one hand, the activation of taurine and glutamate efflux by cell swelling would reduce damage to brain cells from swelling and on the other, the release of large amounts of glutamate would contribute to the rapid glutamate toxicity observed after the onset of ischemia (see Section I). Anion channel blockers specific for the VSOAC channel might thus be protective during ischemia.

As described in Section IV,F, swelling in Ehrlich cells stimulates a phospholipase A_2 and/or a 5-lipoxygenase and thereby initiates enhanced synthesis of leukotrienes. Leukotriene D_4 (LTD_4) then acts as a second messenger for the activation of the taurine channel as well as the mini Cl^- channel (Hoffmann et al., 1993). The question arose of whether the swelling-activated taurine efflux in Ehrlich cells is actually mediated by the swelling-activated small Cl^- channel or whether it is a separate anion channel with an identical activation mechanism. The latter was shown to be the case (Lambert and Hoffmann, 1994).

C. Comparison of the Swelling-Activated Effluxes of Cl^- and Taurine

The mini Cl^- channel and the taurine channel have been found to have different characteristics with respect to two effectors—DIDS and arachidonic acid. Figure 7 shows that arachidonic acid (25 μM) strongly inhibits swelling-activated Cl^- efflux (left panel) and activates swelling-activated taurine efflux (right panel). DIDS was a strong inhibitor of the taurine channel and a weak inhibitor of the Cl^- channel. Taken together, these

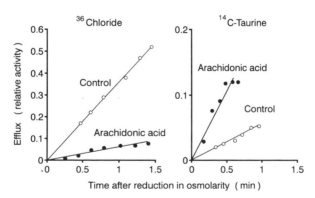

FIG. 7 Comparison of the effects of arachidonic acid on the swelling-activated effluxes of [$^{36}Cl^-$] and [^{14}C] taurine from Ehrlich ascites cells. The effluxes were measured in a hypotonic Cl^--free gluconate medium. At time zero, arachidonic acid (25 μM) was added to cells loaded with [^{36}Cl] or [^{14}C]taurine and the osmolarity of the medium was reduced by 50%. The efflux is expressed as the ratio of the activity of the tracer in the medium at the times indicated to the initial activity of the tracer in the cells. (Redrawn from Lambert and Hoffmann, 1994.)

results indicate that the swelling-activated "mini Cl⁻ channel" and the swelling-activated taurine channel are separate pathways.

The elevated Cl⁻ conductance after cell swelling represents fluxes through both the mini Cl⁻ channel and the taurine channel. The mini Cl⁻ channel accounts for the majority of the Cl⁻ conductance activated by cell swelling, and the taurine channel for very little of it. This is different from the VSOAC channel in C6 glioma cells, which accounts for the majority of the whole-cell anion current activated by swelling (Jackson and Strange, 1994). Thus, unless there are only a few copies of the taurine channels per Ehrlich cell, their Cl⁻ conductance must be quite low. The VSOAC channel was concluded to be a 35–40 pS channel, which resembles the medium Cl⁻ channel in Ehrlich cells described earlier. Therefore the VSOAC of C6 glioma is unlikely to be identical with the taurine channel in Ehrlich cells because:

1. The Cl⁻ conductance of VSOAC seems to be higher than the Cl⁻ conductance of the taurine channel.

2. Unsaturated fatty acids have the opposite effect on the two channels. As in Ehrlich cells, the volume-activated taurine efflux in rat mammary tissues occurs through a pathway different from the dominating volume regulatory anion channel (Shennan *et al.*, 1994).

IV. Volume Sensing and Signal Transduction

The changes in volume-sensitive transport pathways in response to cell volume changes that have been described require the following mechanisms: a signal of volume change, a sensor of the signal of the change, transduction of the signal to the effector, i.e., the transporter, and changes in the function of the transporter. These mechanisms are of course necessary for all transport pathways involved in the regulation of cell volume.

Basically two kinds of mechanisms can be envisioned as signals of volume change: mechanical changes and changes in concentration of cellular solutes. Mechanical changes include membrane tension or deformation and changes in the relation between the membrane and the cytoskeleton. Changes in concentration of cytoplasmic proteins can signal volume change, that is, a change in macromolecular crowding (see Section IV,C). Sensors of the signals can be mechanosensitive channels, and membrane or cytoplasmic enzymes. The cytoskeleton may sense swelling signals as well as generating them. The sensors of volume change signals initiate the signal transduction pathway. Examples include an increase in cytosolic free Ca^{2+}, polyphosphoinositide metabolism, activation of protein kinases and phosphatases, changes in cAMP concentration, and production of eicosanoids. The signal transduction pathway causes changes in the function of the effectors, the volume-sensitive transporters.

A. Role of the Cytoskeleton

The cytoskeleton consists of three distinct filamentous components: actin filaments, intermediate filaments, and microtubules. In recent years, the cytoskeleton has been found to be a modulator of ion transport and cell volume regulation. Many studies have shown that ion channels, ion exchangers, and ion pumps are directly associated with the cytoskeleton (Luna and Hitt, 1992) and several studies have suggested that the cytoskeleton contributes to volume regulation in a variety of different cell types (Mills et al., 1994).

1. RVD-Ion Channels

Cytochalasin B, which inhibits actin assembly by inhibiting the addition of actin monomers to microfilaments (MacLean-Fletcher and Pollard, 1980; Stossel, 1989), inhibits RVD in rat liver slices (Van Rossum and Russo, 1981), in *Necturus* gall bladder epithelial cells (Foskett and Spring, 1985), in Ehrlich cells (Hoffmann and Kolb, 1991; Cornet et al., 1993a) (see Fig. 8), in mouse peritoneal macrophages (Galkin and Khodorov, 1988), and

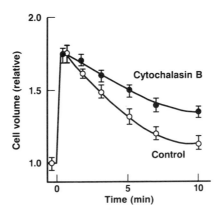

FIG. 8 Inhibition of regulatory volume decrease in Ehrlich ascites cells by cytochalasin B. Cells were preincubated with cytochalasin at 50 μM for 30 min. At time zero, the cells were swollen by transfer to hypotonic medium (osmolality 1/2 of iosotonic). Mean cell volumes were measured using a Coulter counter and are expressed on a relative scale. Shown are means ± S.E.M.s, six separate experiments. (Redrawn from Cornet et al., 1993a.)

in various cultured mammalian cells (Cornet *et al.*, 1987, 1988, 1993b). These studies suggest that an intact microfilament network is necessary for a normal RVD response. It should be noted that treating astrocytes with cytochalasin B does not affect swelling-induced taurine release (Kimelberg *et al.*, 1990), suggesting that mirofilament polymerization is not involved in the activation of the taurine leak pathway in these cells.

That an intact cytoarchitecture is required for the RVD response is also evident from the observation that the expected increase in K^+ permeability in response to either cytoplasmic Ca^{2+} or hypotonic swelling was absent in isolated membrane vesicles from lymphocytes (rabbit thymocytes) (Grinstein *et al.*, 1983). Mills and Lubin (1986) and Mills (1987) showed that an increase in the levels of cAMP caused a rearrangement of F-actin, a net loss of KCl, and cell shrinkage in MDCK cells. Thus, they proposed that the state of organization of F-actin has an effect on membrane elements that play a role in volume control processes.

In shark rectal gland (Ziyadeh *et al.*, 1992; Mills *et al.*, 1994) and Ehrlich ascites tumor cells (Cornet *et al.*, 1993a), rearrangement and depolymerization of F-actin takes place during RVD in parallel with the loss of taurine in the shark rectal gland and the loss of taurine and KCl in Ehrlich cells. Changes in the organization of F-actin induced by hypotonic stress have also been demonstrated in a cultured mammalian cell (Cornet *et al.*, 1987). In Ehrlich cells, the volume-induced depolymerization of F-actin is observed only in the presence of external Ca^{2+}, in contrast to the RVD response, which is essentially unaltered in Ca^{2+} media (Hoffmann *et al.*, 1986). The disappearance of F-actin proceeded over a time interval of 1 to 10 min after hypotonic exposure (Cornet *et al.*, 1993a), whereas the ion channels are fully activated after about 1 min (Christensen and Hoffmann, 1992). The above results suggest that an intact microfilament network is a prerequisite for activation of K^+ and Cl^- channels during RVD, but that the rearrangement of F-actin during the process may play other roles than being involved in the channel activation.

On the other hand, if Ca^{2+} is added 1 min after Ca^{2+}-depleted cells are transferred to Ca^{2+}-free hypotonic medium, there is a rapid microfilament reorganization and a simultaneous twofold acceleration of the RVD response. Ca^{2+} plus the Ca^{2+}-ionophore A23187 causes an eightfold acceleration of the RVD response and a more rapid reorganization of the microfilaments (Cornet *et al.*, 1993a).

Calmodulin has been shown to play an important role in microfilament disruption as well as in the activation of the ion channels (Cornet *et al.*, 1993a). It is involved in the regulation of microfilament formation and organization in various cells and binds with a high affinity to the β-subunit of spectrin, a central element of the cytoskeleton (Glenney *et al.*, 1980; Pritchard and Moody, 1986). These results are consistent with an earlier observation on osmotically swollen sea urchin coelomocytes in which disruption of microfilaments was observed and was shown to be regulated by the calcium–calmodulin complex (Henson and Schatten, 1983).

Profilin is one of several actin-binding proteins which inhibits actin assembly by binding to actin monomers (Stossel, 1989). Gelsolin is a Ca^{2+}-binding protein which binds to the filaments and blocks further assembly, resulting in short, stable F-actin filaments. Removal of gelsolin with a Ca^{2+} chelator promotes filament assembly (Weber and Osborne, 1982; Stossel, 1989). Therefore intracellular redistribution of Ca^{2+} can result in either a shortening or even depolymerization of F-actin at high local Ca^{2+} or an elongation and polymerization at low local Ca^{2+}. It follows that Ca^{2+} influx through stretch-activated nonselective cation channels promoted by osmotic swelling can be involved in reorganization of actin filaments.

Phosphatidylinositol 4,5-bisphosphate (PIP_2) promotes dissociation of actin-profilin complexes in vitro (Lassing and Lindberg, 1985), so breakdown of PIP_2 should promote the association between profilin and actin and disassembly of actin filaments. Thus, the disruption of the microfilament system during RVD which occurs, e.g., in Ehrlich cells, might be caused by PIP_2 breakdown (S. Christensen et al., 1988) as well as by a local increase in intracellular free Ca^{2+}, which in turn results from Ca^{2+} entry through stretch-activated nonselective cation channels (O. Christensen and Hoffmann, 1992).

It is likely that the actin reorganization observed in Ehrlich cells and other cells after osmotic cell swelling results in the immediate closing of the swelling-activated K^+ and Cl^- channels (the opening of which is dependent on an intact microfilament network). However, the fragmentation and depolymerization of F-actin seen during RVD might also result in the opening of other channels. In fact, it has recently been demonstrated that cell swelling and membrane stretch activate a 305-pS mechanosensitive anion channel in renal cortical collecting duct cells (RCCT-28A) by a mechanism involving the fragmentation and depolymerization of F-actin (Mills et al., 1994).

2. RVI-Na^+, K^+, Cl^--Cotransport System

Disassembly of the actin filaments, either by hypotonic treatment in the presence of Ca^{2+} or by addition of cytochalasin B, also results in activation of Na^+, K^+, $2Cl^-$ cotransport, as shown in Fig. 9. Membrane vesicles made from Ehrlich ascites cells by treatment with cytochalasin B contain the Na^+, K^+, $2Cl^-$ cotransporter in a permanently activated state (Hoffman et al., 1994). Incubation of the cells with cytochalasin B causes the formation of blebs in the surface membrane, which are removed by gentle homogenization. These large vesicles, called cytoplasts, contain neither mitochondria

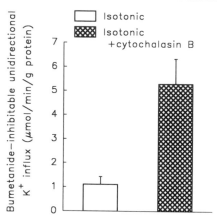

FIG. 9   Effect of cytochalasin B (42 μM) on the bumetanide-inhibitable unidirectional K$^+$ influx. The cells were preincubated in an isotonic standard medium for 15 min, and the experimental group for 1 min with cytochalasin B. The bumetanide-inhibitable K$^+$ influx was measured as the difference between the influx in the absence and the presence of bumetanide (30 μM). ^{86}Rb was used as tracer for K$^+$. The vertical bars represent S.E. of seven and five experiments in the isotonic and isotonic plus cytochalasin B groups, respectively. (From Jessen and Hoffmann, 1992, with permission.)

nor a nucleus nor other membranous organelles. Pretreatment of intact cells with cytochalasin B for 1 min stimulates the bumetanide-inhibitable K$^+$ influx ~ fivefold. The influx into purified cytoplasts expressed per gram of protein is three- to fourfold higher than the influx into cytochalasin B-treated intact cells. Several standard protocols confirmed that the bumetanide-inhibitable K$^+$ influx is Na$^+$, K$^+$, 2Cl$^-$ cotransport. SDS gels of the cytoplast proteins and immunoblots of these gels stained with the antiserum against the purified Ehrlich cell cotransporter proteins (Dunham et al., 1990) showed that the cotransporter protein is a major protein in the cytoplast membranes (Hoffmann et al., 1994).

3. Possible Mechanisms Involved in the Regulation of Ion Transporters by the Cytoskeleton

Microfilaments can regulate ion channels and other ion transporters by several mechanisms.

1. They may act as the mechanical transducer by which membrane stretch causes channel activation (see later discussion). Spectrin may be the cytoskeletal element involved in stretch activation of ion channels (Sachs, 1991).

2. Microfilaments may promote insertion of channels from a cytoplasmic store into the membrane, as proposed by Foskett and Spring (1985) and Lewis and de Moura (1982). Although the insertion of vesicles containing ion and water transport proteins into the plasma membrane is a well-described phenomenon, especially after agonist stimulation (Al-Awqati, 1989; Lewis and Donaldson, 1990; Mills *et al.*, 1994), almost nothing is known about the role of this process in cell volume regulation.

3. Microfilaments may activate a second messenger cascade leading to the synthesis of products which either activate or modulate ion transporting systems. Mechanical deformation has been shown to activate protein kinase C in the heart atrium (von Harsdorfet *et al.*, 1988), in rat skeletal muscle (Richter *et al.*, 1987), and in endothelial cells (Rosales and Sumpio, 1992). Cell swelling has been found to directly stimulate adenylate cyclase in mouse lymphoma cells (Watson, 1991). Since many Cl^- channels are activated by cAMP, this might be relevant to volume regulation. Very recently, phospholipase A_2 was found to be stimulated directly by cell swelling in thrombocytes (Margalit *et al.*, 1993b) and in Ehrlich ascites tumor cells (Thoroed *et al.*, 1994). In both systems this was suggested to constitute a novel "indirect" form of channel stretch activation, although it is also possible that the activation of phospholipase A_2 might be a consequence of a decrease in macromolecular crowding (see Section IV,C, and G and Fig. 13).

Microtubules do not appear to be implicated in the volume-regulatory responses of Ehrlich ascites tumor cells. Reduction of the extracellular osmolality does not induce any significant change in their structure or density, and high doses of colchicine, which disrupt microtubules, have no significant effect on RVD (Cornet *et al.*, 1993a). The same is true in *Necturus* gall bladder epithelium (Foskett and Spring, 1985), whereas disruption of microtubules in macrophages resulted in activation of anion channels (Melmed *et al.*, 1981). Obviously the picture is rather complicated and different channels and transporters involved in volume regulation may have different associations with the cytoskeleton.

B. Stretch-Activated Channels

The first stretch-activated channels were described in chick embryonic skeletal muscle (Guharay and Sachs, 1984) and embryonic *Xenopus* muscle (Brehm *et al.*, 1984). The mechanical and electrical properties of these channels were investigated using patch-clamp techniques. More recently, ion channels activated by membrane stretch have been detected in many vertebrate cells, including cells in which their function seems readily apparent, such as auditory cells, stretch receptors, muscle spindles, and vascular endothelium (Guharay and Sachs, 1984; Lansman *et al.*, 1987; Sackin, 1994), but also in cells in which their function is less obvious, such as *Necturus* renal proximal tubule cells (Sackin, 1987; Filipović and Sackin, 1991a), choroid plexus from salamander (Christensen, 1987), an opossum kidney cell line (Ubl *et al.*, 1988), *Xenopus* oocytes (Yang and Sachs, 1989), corneal ephithelium (Cooper *et al.*, 1986), neuroblastoma cells (Falke and Misler, 1989), osteoblasts (Duncan and Misler, 1989), a human fibroblast cell line (Stockbridge and French, 1988), frog proximal tubule cells (Hunter, 1990), and Ehrlich ascites tumor cells (Christensen and Hoffmann, 1992). Stretch-activated channels might be a part of volume-sensing mechanisms in general (Sachs, 1988; Rugolo *et al.*, 1989; Morris, 1990; Hoffmann and Kolb, 1991; Sackin, 1994).

Sackin (1994) suggested that "A reliable feed-back system for short time volume regulation ultimately requires some type of mechanical sensor to convey information about cell size." An alternative volume-sensing mechanism would detect changes in the concentration of intracelluar solutes following cell swelling or shrinkage. Parker and co-workers (Colclasure and Parker, 1992; Parker, 1993b) presented evidence that the reduction of cytoplasmic proteins with swelling of dog red blood cells is the signal resulting in activation of volume-sensitive K^+, Cl^- cotransport. This was interpreted in terms of excluded volume effects, i.e., changes in macromolecular crowding (Zimmerman and Minton, 1993; See Section IV,C for further discussion). In the authors' opinion, both mechanical changes in the membrane and changes in macromolecular crowding can serve as signals of changes in cell volume. Recent results on the effects of urea on K^+, Cl^- cotransport in sheep red cells indicated that both types of signals operate, and in sequence (Dunham, 1995).

In general, the stretch-activated channels can be activated in patches with negative pressures of as little as 5 mm Hg. Furthermore, several recent studies have demonstrated that stretch-activated channels are activated during volume regulatory processes (Christensen, 1987; Ubl *et al.*, 1988, 1989; Christensen and Hoffmann, 1992). It was therefore suggested that these stretch-activated channels could be involved in volume regulation (Christensen, 1987; Ubl *et al.*, 1988, 1989; Christensen and Hoffmann, 1992; Cornet *et al.*, 1993b). There are two categories of the selectivity of these channels.

1. Nonselective Stretch-Activated Cation Channels

Nonselective stretch-activated cation channels (SA-cat) are permeable to Ca^{2+} and may allow sufficient Ca^{2+} to enter to serve as a second messenger (Christensen, 1987; Lansman *et al.*, 1987). Most of the examples of SA channels are in this category. Their single channel conductances range between 25 and 35 pS. The trivalent lanthanide, gadolinium (Gd^{3+}), blocks this type of channel (Bennett, 1985). SA channels with similar conductance and ion selectivity have been demonstrated in frog oocytes (Tanglietti and Tesclli, 1988), in *Xenopus* oocytes (Yang and Sachs, 1989), and in renal proximal tubule cells (Filipović and Sackin, 1991a,b).

Christensen (1987) and Lansman *et al.* (1987) proposed that SA channels were activated by the swelling-induced increase in membrane tension and that influx of Ca^{2+} through these channels resulted in an increase in cytoplasmic Ca^{2+}, which in turn opened the Ca^{2+}-activated K^+ channels. Single-channel records have now demonstrated swelling activation of cation (SA-cat) channels in several cell types, including:

1. Opossum kidney cells (Ubl *et al.*, 1988), in which a nonselective, low-conductive channel was found. The swelling- and suction-activated channels had identical characteristics (Ubl *et al.*, 1988). The SA-cat channel may be functionally coupled to a small-conductance (15 pS) Ca^{2+}-dependent K^+ channel (Ubl *et al.*, 1988).

2. In rat liver cells, cell swelling activated Ca^{2+}-permeable SA-cat channels (Bear, 1990b).

3. In neuroblastoma cells, SA-cat channels were rapidly activated by swelling (Falke and Misler, 1989). Channel activity subsided during regulatory volume decrease, as expected if the signal of activation was a consequence of swelling.

4. SA-cat channels are activated by swelling in Ehrlich ascites tumor cells, measured with a cell-attached patch electrode (Christensen and Hoffmann, 1992). A typical result is shown in Fig. 2, the downward deflecting channels. Channels with identical characteristics were activated by suction (Christensen and Hoffmann, 1992) and were permeable to cations, including Ba^{2+}, demonstrating that the channels activated by swelling are SA-cat channels.

It was shown earlier that channels are activated in Ehrlich cells by only 5% swelling (Hudson and Schultz, 1988). In rabbit proximal tubule cells, Lohr and Grantham (1986) showed that there must be a sensor that responds to cell volume changes that are both small and slow. It is not clear if SA channels are involved; if they are, they are coupled with high gain to a sensor with exquisite sensitivity. The unphysiologically extensive cell swelling employed in most studies on SA channels raises the question of whether slow, small, changes in physiological volume actually activate SA channels.

The role of SA-cat channels in volume regulation is probably a consequence of Ca^{2+} currents through them and not K^+ or Na^+ currents. Swelling-activated Ca^{2+} currents in gadolinium-sensitive channels have been demonstrated in intestine 407 cells (Okada *et al.*, 1990). Gadolinium also blocks the response to swelling in these cells, so the Ca^{2+} entry probably is responsible for the major part of the swelling-induced increase in cytosolic free Ca^{2+} in intestine 407 cells (Okada *et al.,* 1990).

However, there seems to be a quantitative discrepancy between the amount of Ca^{2+} that would enter the cells and the amount necessary to turn on Ca^{2+}-sensitive K^+ channels. Fluorescence measurements of cytosolic Ca^{2+} in rabbit proximal tubule (Beck *et al.,* 1991) as well as in Ehrlich ascites tumor cells (Jørgensen *et al.,* 1994) during cell swelling also suggested that the change in free cellular Ca^{2+} during swelling was too small to activate Ca^{2+}-dependent K^+ channels. At physiological Ca^{2+} concentrations, the estimated Ca^{2+} current through the open channel will be only 20 fA (Christensen and Hoffmann, 1992). It is possible that SA-cat channels could establish local regions of elevated Ca^{2+} concentration adjacent to the interior of the cell membrane. Activation of the K^+ channels by this Ca^{2+} would require either that the K^+ channels be adjacent to the SA channels, or that both channels have access to the same intracellular compartment with elevated Ca^{2+}.

In experiments with Ca^{2+}-containing NaCl solutions (3 mM Ca^{2+}) in a pipet, concurrent activation of SA-cat channels and K^+ channels was consistently observed (Christensen and Hoffmann, 1992). Activation of an SA-cat channel, seen as a downward current deflection, is followed after about 20 sec by activation of a K^+ channel (upward current deflection), which is taken to represent the Ca^{2+}-activated inward-rectifying K^+ channel. It should be noted, however, that this type of K^+ channel accounts for only a minor part of the volume-activated K^+ efflux through volume-activated K channels in Ehrlich cells, since it is reduced by only 20% in the absence of external Ca^{2+} (Kramhøft *et al.,* 1986). The major part of the K^+ efflux during RVD is through channels insensitive to Ca^{2+} and charybdotoxin (see Section II,A,1). In many epithelial cells Ca^{2+} influx after cell swelling is likely to have greater importance than in Ehrlich cells (see later discussion).

A significant cell depolarization during RVD has been observed for several cell systems, including Ehrlich cells (Lambert *et al.,* 1989) and may play a role in K^+ channel activation (see Section II,A,1). Such depolarization is caused partly by activation of SA-cat channels and partly by activation of Cl^- channels.

2. Selective Stretch-Activated Channels

Stretch-activated channels selective for Ca^{2+}, K^+, and Cl^- have been reported. Two types of stretch-activated, volume-sensitive K^+ channels can be demonstrated in the basolateral membrane of *Necturus* proximal tubules (Sackin, 1989; Filopović and Sackin, 1992b). Efflux of K^+ through these channels may play a role in regulation of renal cell volume (Filipović and

Sackin, 1992b). Stretch-activated Cl⁻ channels are not as common as SA-cat or SA-K channels. Most mechanosensitive anion channels are large, >300pS (Sackin, 1994; Mills *et al.*, 1994; Section IV,A,1). The swelling-activated K⁺ and Cl⁻ channels described for Ehrlich cells *cannot* be activated by stretch administered by suction (Christensen and Hoffmann, 1992).

3. Stretch-Inactivated Channels

These channels are still relatively rare and probably are not involved in volume regulation (Morris, 1990).

4. Cytoskeleton and Stretch-Activated Channels

Are the stretch forces coupled to SA channels by the cytoskeletal network? For a detailed discussion of this question, see Sackin (1994). The well-known links between the cytoskeleton and volume regulation in a variety of cells were described earlier. However, there are few direct studies at the channel level. Actin filaments were reported to regulate Na⁺ channels in A6 cells (Cantiello *et al.*, 1991), and altered properties of the mechanosensitive ion channel were demonstrated in a mutant strain of *E. coli* where a lipoprotein is lacking that normally links the cell wall to the membrane which contains stretch-sensitive channels (Kubalski *et al.*, 1993). On the other hand, in a quantitative video microscopic study of patch-clamped membranes, it was demonstrated that agents which disrupt tubulin and actin did not block SA channel activity (Sokabe *et al.*, 1991). These authors suggested that spectrin may be the cytoskeletal element involved in stretch activation of some types of channels.

C. Changes in Macromolecular Crowding as a Signal of Cell Volume Change

There have been several suggestions that the change in the concentration of a cytoplasmic solute accompanying cell shrinkage or swelling is the signal of volume change. In duck red blood cells, changes in Mg^{2+} concentration were proposed to regulate K⁺, Cl⁻ and Na⁺, K⁺, 2Cl⁻ cotransport (Starke and McManus, 1990). However, the high sensitivity of both cotransporters to small changes in volume makes it unlikely that changes in Mg^{2+} concentration are the signals (Dunham *et al.*, 1993). Cl⁻ modulates Na⁺, K⁺, 2Cl⁻ cotransport (Levinson, 1990; Lytle and Forbush, 1992a; Breitwieser *et al.*, 1990), but a shrinkage-induced increase in Cl⁻ concentration is unlikely to be the signal of volume change; the clearest evidence is that Cl⁻ actually inhibits the Na⁺, K⁺, 2Cl⁻ cotransporter (Breitwieser *et al.*, 1990).

There is a compelling proposal that changes in the concentration of total cytoplasmic protein can be the signal of changes in volume. The first evidence came from studies on K^+, Cl^- cotransport and Na^+/H^+ exchange in resealed ghosts made from dog red blood cells (Colclasure and Parker, 1991, 1992; Parker, 1993b). Ghosts were made with the same hemoglobin concentration as intact cells, but with one third the volume of intact cells. The threshold volumes, or set points, at which K^+, Cl^- cotransport and Na^+/H^+ exchange were activated were compared for intact cells and resealed ghosts. The set points for both transport pathways were exactly the same for cells and ghosts when they were expressed as percent dry weight, but they were very different when expressed per cell or ghost volume (Colclasure and Parker, 1991, 1992). Therefore it was not changes in cell or ghost volume that activated changes in transport, but changes in the concentration of something inside. When half of the hemoglobin in the ghosts was replaced with serum albumin, the set points were about the same as when essentially all of the soluble protein was hemoglobin. The signalling substance was not a small molecular weight substance, because these were all diluted by replacing hemoglobin with albumin. The signalling substance was also obviously not hemoglobin; it was total protein, and independent of the specific protein.

These results were interpreted in terms of excluded volume effects or macromolecular crowding (Minton, 1983, 1994; Minton *et al.*, 1992; Zimmerman and Minton, 1993). The thermodynamic activity of all soluble proteins depends on total protein concentration, and this dependence is strong at the protein concentration typical of cells, ~300 g/liter. For example, the activity coefficient of hemoglobin is reduced tenfold by a twofold dilution of the hemoglobin from its physiological concentration (Minton, 1983). This striking dependence of thermodynamic activity on concentration is due to the occupancy by the protein molecules of a large fraction of the solution volume (Zimmerman and Minton, 1993). It applies to all macromolecules in solution, is a function of their molecular weights, and results in large deviations of the expected behavior of proteins derived from simple mass action relations valid for dilute solutions (Zimmerman and Minton, 1993).

It was speculated earlier that cells use changes in macromolecular crowding to sense changes in their volumes (Zimmerman and Harrison, 1987). The work on the dog resealed ghosts provided evidence that supported this suggestion (Colclasure and Parker, 1991, 1992). Based on this work, a formal proposal was developed by which changes in macromolecular crowding could serve as signals of volume change and promote regulation of cell volume (Minton *et al.*, 1992; Minton, 1994). Specifically, it was proposed that the volume sensitivity of transport pathways could be attributed to the reversible association of proteins of the transporters and soluble regulatory proteins.

As discussed later (Section IV,I), a volume-sensitive kinase has been proposed to be involved in the regulation of K^+, Cl^- cotransport in mammalian red blood cells. Swelling inhibits this kinase, thereby activating cotransport. This kinase was envisioned to be the soluble regulatory protein in the scheme of Minton et al. (1992). The enzymatic activity of the kinase is strongly dependent on total protein concentration, i.e., macromolecular crowding, and a slight swelling and slight decrease in crowding would greatly inhibit the kinase. There must also be a phosphatase which activates cotransport that is volume-insensitive. This phosphatase must be insensitive to macromolecular crowding as a result of some kind of compartmentation, e.g., tight binding to the transporter (Minton et al., 1992; Minton, 1994).

There is additional support for this view of a volume-sensing mechanism from several different studies. Parker (1993a) showed that urea strongly activates K^+, Cl^- cotransport without affecting cell volume. The same phenomenon has been reported for sheep red blood cells (Dunham, 1995). This action of urea can be interpreted in terms of an effect on macromolecular crowding as follows. There are attractive interactions between urea and cytoplasmic proteins which reduce the thermodynamic activity of the proteins (Collins and Washabaugh, 1985; Minton, 1983; Prakash et al., 1981) and thereby reduce macromolecular crowding without changing cell volume. This reduction in crowding will affect volume-sensitive enzymes as if the cells were swollen.

More recently, evidence of a different nature was obtained for macromolecular crowding serving as a signal of cell volume. Intact dog red blood cells were prepared in which cell volumes and total intracellular salt concentrations varied independently (Parker et al., 1995). Five sets of cells were prepared with approximately the same ranges of cell volumes, ~60 to ~70% cell water. Each of the five sets of cells had a different range in intracellular ionic strengths, owing primarily to varying Na^+ and Cl^- concentrations. The set of cells with the lowest range of ionic strengths contained $Na^+ + Cl^-$ concentrations from 60 to 100 mM; the highest range was 200 to 260 mM $Na^+ + Cl^-$. On these five sets of cells, the set points, i.e., the threshold cell volumes of activation, were determined for both K^+, Cl^- cotransport and Na^+/H^+ exchange as functions of cell volume and therefore of ionic strength. Increasing ionic strength shifted the set points for both transporters to smaller cell volumes. At the highest ionic strength, the set points were 60–62% cell water and at the lowest they were 70–72% cell water. Figure 10 shows the results of one of these experiments. The set points for influxes of Na^+ and K^+ are shown for two sets of cells with the same ranges of cell volumes but with the cells represented by the open circles having ionic strengths ~ 30% greater than the cells represented by the solid circles. The

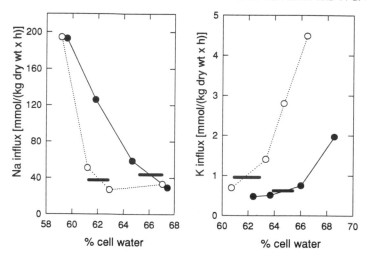

FIG. 10   Unidirectional influxes of Na$^+$ (left panel) and K$^+$ (right panel) in dog red blood cells as functions of cell water content and intracellular ionic strength. The filled symbols represent cells in a medium containing 100 mM NaCl, 4 mM KCl, 5 mM glucose, and buffers, pH 7.4. The open symbols represent cells in a medium containing 100 mM NaCl + 75 mM N-methyl-D-glucamine NO$_3$ plus KCl, glucose, and buffers as above. The cells in 100 mM NaCl + 75 mM NMDG NO$_3$ (open symbols) had ionic strengths approximately 30% greater than the cells in 100 mM NaCl at each percent cell water. The horizontal bars indicate the ranges of estimates of the four set points— two for the two sets of Na$^+$ influxes and two for the two sets of K$^+$ influxes. (From Parker *et al.*, 1995. Reproduced from *J. Gen. Physiol.* by permission of The Rockefeller University Press.)

increase in ionic strength shifted the set points for both volume-sensitive transporters to smaller cell volumes.

The results of a series of these experiments were interpreted in terms of nonspecific electrostatic interactions between small electrolytes and intracellular proteins. It was demonstrated that the observed dependence of set points on ionic strength was expected in terms of the attractive interactions between the small electrolytes and the proteins, the resultant reduction in thermodynamic activity of the proteins, and reduced macromolecular crowding (Parker *et al.*, 1995).

D. Ca²⁺–Calmodulin in Signal Transduction

In most cells investigated, Ca^{2+} has been implicated as having either a direct or participatory role in the RVD response following swelling, but the nature of the role played by Ca^{2+} is still not clear. The understanding of intracellular Ca^{2+} homeostasis is therefore critical to an evaluation of cell volume regulatory phenomena. For reviews of the role of Ca^{2+} in cell volume regulation, see Pierce and Politis (1990), McCarty and O'Neil (1992), Hoffmann *et al.* (1993), and Foskett (1994). Several questions can be considered in reviewing the evidence for the involvement of Ca^{2+} in regulating cell volume.

1. Ca⁺ Influx and the RVD Response

In several cell types, mainly epithelial cells, entry of extracellular Ca^{2+} and Ca^{2+} across the cell membrane are prerequisites for the RVD response. Examples include proximal tubule cells (McCarty and O'Neil, 1991a,b), inner medullary duct cells (Bevan *et al.*, 1990), MDCK cells (Rothstein and Mack, 1990, 1992), intestine 407 cells (Okada *et al.*, 1990), frog urinary bladder (Davis and Finn, 1987), and jejunal enterocytes from guinea pigs (MacLeod *et al.*, 1992b). In these cells, volume recovery is blocked by removal of external Ca^{2+} and by Ca^{2+} channel blockers. In isolated toad bladder cells, the cellular $^{45}Ca^{2+}$ uptake increased threefold during RVD (Wong and Chase, 1986).

In other cell types, e.g., Ehrlich cells (Hendil and Hoffmann, 1974; Hoffmann *et al.*, 1984), opossum kidney cells (Ubl *et al.*, 1988, 1989; Hoffmann and Kolb, 1991), cultured inner medullary duct cells (Montrose-Rafizadeh and Guggino, 1991), rat hepatocytes (Bear, 1990; Corasanti *et al.*, 1990), rat salivary gland acinar cells (Foskett *et al.*, 1994), lymphocytes (Grinstein *et al.*, 1982a, 1984; Grinstein and Smith, 1990), and platelets (Szirmai *et al.*, 1988), the volume response is essentially unaffected by removal of external Ca^{2+}.

2. Intracellular Free Ca²⁺ and the RVD Response

In Ehrlich cells, prolonged preincubation in Ca^{2+}-free media containing EGTA reduces the RVD response only slightly (Hoffmann *et al.*, 1984). It can be demonstrated, however, that the activation of K^+ and Cl^- channels is reduced by about 25%, but that this is partly compensated for by an activation of K^+, Cl^- cotransport (Kramhøft *et al.*, 1986) and by an increased taurine efflux (Lambert, 1994). There is nearly a complete block of the activation of K^+ and Cl^- channels after chelating Ca^{2+}_i by BAPTA (Jørgensen *et al.*, 1994), whereas the RVD response in lymphocytes is unaffected by chelation of cytosolic Ca^{2+} (Grinstein and Smith, 1989), or by depletion of intracellular Ca^{2+} (Grinstein and Smith, 1990).

In contrast, increasing the cellular Ca^{2+} concentration either by releasing Ca^{2+} from intracellular stores (induced by agonists such as bradykinin, thrombin, or histamine), or by promoting Ca^{2+} influx (by adding Ca^{2+} to Ca^{2+}-depleted cells or by adding the Ca^{2+}-ionophore A23187) significantly stimulates the RVD response (see Fig. 10 in Hoffmann *et al.*, 1993). Therefore Ca^{2+}, released from internal stores or entering from outside, definitely can stimulate RVD in the Ehrlich cells, but is nevertheless not necessary for it.

3. Calmodulin Involvement

Several studies have implicated Ca^{2+}–calmodulin as a mediator of RVD. The evidence in most cases is the sensitivity of RVD to inhibitors of calmodulin. Several drugs which inactivate calmodulin inhibit the loss of KCl induced by cell swelling or by A23187 in lymphocytes (Grinstein *et al.*, 1982a,b, 1983) and in Ehrlich cells (Hoffmann *et al.*, 1984, 1986). There is also evidence of a role for calmodulin in the RVD response in *Amphiuma* red cells (Cala *et al.*, 1986), *Necturus* gall bladder (Foskett and Spring, 1985), molluscan (*Noetia*) red blood cells (Pierce *et al.*, 1989), and guinea pig jejunal enterocytes (MacLeod *et al.*, 1992b). Calmodulin is present in Ehrlich cells, and the cell membranes bind calmodulin in a Ca^{2+}-dependent manner (Aabin and Kristensen, 1987). The anticalmodulin drug pimozide (diphenyl piperdine) blocks the swelling-activated conductance of K^+ as well as swelling-activated conductance of Cl^- in Ehrlich cells. It was suggested that calmodulin is involved in swelling-induced channel activation (Hoffmann *et al.*, 1986), though pimozide is not a highly specific anticalmodulin agent. In mouse fibroblasts (Okada *et al.*, 1987) and guinea pig jejunal enterocytes (MacLeod *et al.*, 1992b), calmodulin appeared to be involved only in the regulation of K^+ conductance and not of Cl^- conductance.

For all of these studies, the lack of specificity of the anticalmodulin drugs, primarily phenothiazines, such as pimozide, make the conclusions uncertain. In human lymphocytes (Grinstein *et al.*, 1982a) and in *Noetia* red blood cells (Pierce *et al.*, 1989), however, it has been shown that the sulfoxide derivatives of the phenothiazines, which still have the anesthetic properties of the drug, but which do not block calmodulin, had no effect on RVD. In a more definitive result, Ca^{2+}-activated Cl^- current measured by whole cell patch-clamp in airway epithelial cells was inhibited by a specific Ca^{2+}-calmodulin kinase inhibitor, but the swelling-activated Cl^- current was not (Chan *et al.*, 1992).

The role of Ca^{2+}–calmodulin in the activation of the taurine leak pathway following osmotic cell swelling also differs among cell types. An increase in cellular Ca^{2+} seems to be required for initiation of the swelling-induced activation of the taurine leak in the nucleated erythrocyte of the blood clam (Pierce and Rowland-Faux, 1992). The swelling-activated taurine efflux can be mimicked by the Ca^{2+}-ionophore A23187 in skate red cells, but calmodulin seems not to be involved in the response because anticalmodulin drugs failed to inhibit the taurine efflux (Leite and Goldstein, 1987). Taurine efflux from Ehrlich cells in hypotonic media is of similar magnitude in the presence and absence of extracellular Ca^{2+}; indeed, it may be higher in the absence of Ca^{2+} (Lambert and Hoffmann, 1993). However, taurine release from osmotically swollen cells is inhibited by pimozide (Lambert and Hoffmann, 1993), suggesting that Ca^{2+} and calmodulin might be involved in swelling activation of the taurine leak pathway. Thus Ca^{2+}–calmodulin might be involved in one or more steps in activation of the conductive K^+ and Cl^- pathways as well as in the activation of the taurine leak pathway in osmotically swollen Ehrlich cells.

4. Volume Response and Change in Intracellular Free Ca^{2+}

Several studies have shown a rise in Ca^{2+}_i during RVD. In most of these, Ca^{2+} influx from the extracellular medium was necessary (Foskett, 1994). A sizable increase in Ca^{2+}_i during RVD, which was completely dependent upon the presence of extracellular Ca^{2+}, was reported for isolated toad bladder cells (Wong and Chase, 1986; Wong et al., 1990), cultured intestinal epithelial cells (Hazama and Okada, 1988, 1990; Okada et al., 1990), rabbit medullary thick ascending limb cells (Montrose-Rafizadek and Guggino, 1991), MDCK cells (Rothstein and Mack, 1992), rabbit proximal tubule cells (Suzuki et al., 1990), and in a osteosarcoma cell line UMR-106-01 (Yamaguchi et al., 1989). A biphasic Ca^{2+} increase is induced by hypotonicity in intestine 407 cells (Hazama and Okada, 1990). Both of the increases were dependent on extracellular Ca^{2+}, but the second rise was due to Ca^{2+} release from internal stores, a Ca^{2+}-induced Ca^{2+} release resulting from swelling-activated influx of Ca^{2+}. All of these examples are epithelial cells. It was demonstrated recently that swelling of hepatocytes induced by Na^+-dependent amino acid accumulation also increases the concentration of free cytosolic Ca^{2+} (Baquet et al., 1991).

However, no change in cytosolic, free Ca^{2+} concentration could be detected in human lymphocytes during RVD (Rink et al., 1983; Grinstein and Smith, 1989) and no significant increase in the Ca^{2+} concentration has been detected in the Ehrlich cell within the first few minutes following hypotonic challenge during RVD, using the Ca^{2+} indicator Fura-2 (Jensen et al., 1993; Thomas-Young et al., 1993; Harbak and Simonsen, 1994). When Ca^{2+} was measured in single Ehrlich cells using fluorescence microscopy, a remarkable heterogeneity in the response was observed (Jørgensen et al., 1994). With normal external Ca^{2+} (1 mM), a transient increase in Ca^{2+}_i was observed in only 40% of the experiments, and in less than 10% of the cells. In only half of these experiments was the increase observed within the first 2 min after hypotonic exposure. In media with low (100 μM) or zero extracellular Ca^{2+} concentration, a transient increase in Ca^{2+}_i was almost never observed.

Since the RVD response via K^+ and Cl^- channels at low extracellular Ca^{2+} is reduced only 25% (Kramhøft et al., 1986), it is now clear that a measurable increase in Ca^{2+}_i is not necessary for an RVD response in

Ehrlich cells, although the response is probably faster in the cells where a rise in Ca^{2+}_i occurs (see Section IV,G,2). Elevated Ca^+_i is also not necessary for RVD in either lymphocytes (Grinstein and Smith 1990; Grinstein and Foskett, 1990), salivary acinar cells (Foskett et al., 1994), or rabbit proximal tubule cells (Breton et al., 1992). The increase in Ca^{2+}_i always seen during RVD in salivary acinar cells and proximal tubule cells and sometimes in Ehrlich cells therefore is probably an epiphenomenon. In Ehrlich cells it can be explained as an effect of the increased synthesis and excretion of leukotrienes (see later discussion). Recently Schlichter and Sakellaropoulos (1994) have also recorded a rise in Ca^{2+}_i in human T lymphocytes during RVD. It should be noted, as discussed in some detail by Lambert (1994), that the dramatic decrease in cellular taurine concentration during RVD (Hoffmann and Hendil, 1976) might result in very local increases in Ca^{2+} stemming from a decrease in the binding affinity of Ca^{2+} to some membrane sites (Lambert, 1994).

5. Agonist-Induced Ca^{2+}-Dependent Cell Volume Changes

Several growth factors, hormones, neurotransmitters, and other agonists induce a transient rise in Ca^{2+}_i by receptor-mediated activation of the phosphoinositide-specific phospholipase C, resulting in a release of inositol 1,4,5-trisphosphate [Ins(1,4,5)P$_3$ or IP$_3$] and diacylglycerol (DAG). IP$_3$ binds to the IP$_3$ receptor (Taylor and Richardson, 1991), releases Ca^{2+} from internal stores, and produces a transient increase in Ca^{2+}_i (Berridge and Irvine, 1989).

Since Ca^{2+} modulates the activation of several transport pathways, it is likely that such stimuli will alter cell volume. The relation between Ca^{2+}_i and cell volume has been elegantly studied in single salivary gland acinar cells during activation of fluid secretion (Foskett and Melvin, 1989; Foskett et al., 1994; Foskett, 1994). According to a simplified model for epithelial secretion discussed by Hoffmann and Ussing (1992) (see also Section I), the simultaneous opening of basolateral K^+ channels and apical Cl^- channels will result in loss of KCl and water and thus in cell shrinkage. The shrinkage, or rather the decrease in Cl^- concentration (Robertson and Foskett, 1994), will in turn activate Na^+, K^+, $2Cl^-$ cotransport and the coupled Na^+/H^+ and Cl^-/HCO_3^- exchangers.

A similar sequence of events can be demonstrated in Ehrlich ascites tumor cells. We have studied the release of Ins(1,4,5)P$_3$, the increase in Ca^{2+}_i, and the relation between Ca^{2+}_i and cell volume after agonist stimulation in these cells (Hoffmann et al., 1993). Bradykinin, thrombin, and histamine all induce a transient increase in Ins(1,4,5) and in Ca^{2+}_i in Ehrlich cells (Hoffmann et al., 1993). This is illustrated in Fig. 11, left panel, for thrombin. In isotonic medium, the agonists induce a fast cell shrinkage, as illustrated

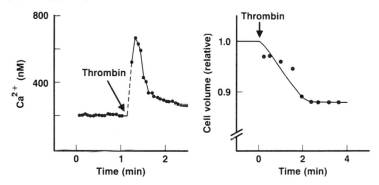

FIG. 11   Effect of thrombin (1 IU/ml) on free cytoplasmic Ca^{2+} (left panel) and on cell volume (right panel) in Ehrlich ascites tumor cells. Free cytoplasmic Ca^{2+} was measured using the fluorescent probe Fura-2 and the "ratio method" based on dual wavelength excitation. Cell volume was followed with time using a Coulter counter. Bumetanide (30 μM) was added to inhibit an RVI response resulting from activation of the Na^+, K^+, $2Cl^-$ cotransport system. (Redrawn from Hoffmann and Kolb, 1991; results from Simonsen et al., 1990.)

in the right panel of Fig. 11. Figure 11 shows the response in Na^+-free medium. In Na^+-containing medium, the shrinkage is transient owing to salt uptake by the cotransporter followed by water. The salt uptake is inhibited by bumetanide and by substituting NO_3^- for Cl^- or $NMDG^+$ for Na^+ (Hoffmann et al., 1993). The initial KCl loss and cell shrinkage is mediated by activation of K^+ channels and nonselective anion channels (Hoffmann et al., 1993). As mentioned in Section II,A, the K^+ channel activated by agonist stimulation is different from the K^+ channel that is dominant during RVD (see Section II,A,1). A bradykinin-induced increase in inositol 1,4,5-triphosphate (IP_3) and Ca^{2+} as well as activation of K^+ channels was also demonstrated in MDCK cells (Lang et al., 1991).

E. Inositol Phosphate Cycle

Swelling causes a transient increase in IP_3 in rat hepatocytes and this results in activation of K^+ channels and an associated RVD (Dahl et al., 1991). In cultured rabbit proximal tubule cells, the levels of IP_3 and inositol 1,3,4,5-tetrakisphosphate (IP_4) increase ~threefold after cell swelling (Suzuki et al., 1990). Finally, swelling of hepatocytes by Na^+-dependent amino acid accumulation resulted in a sustained elevation of IP_3 (Baquet et al., 1991).

It was suggested previously that in Ehrlich cells cell swelling activates phospholipase C (Hoffmann et al., 1993). The RVD response in Ehrlich cells had been reported to be associated with a rapid decrease in membrane PIP_2 and a substantial increase in cytoplasmic IP_3 (Christensen et al., 1988), suggesting that IP_3 plays a role in Ca^{2+} mobilization during the RVD response in these cells. In more recent experiments, however, a swelling-induced IP_3 response was not found consistently. The IP_3 response was observed in only a fraction of the experiments in the presence of external Ca^{2+} and was never observed in its absence (N. K. Jørgensen, H. Havbak, S. Christensen, L. O. Simonsen, and E. K. Hoffmann, unpublished results). Therefore, the increase in IP_3 during RVD, like the Ca^{2+} increase described earlier, appears to be an epiphenomenon, perhaps resulting from the volume-induced release of leukotrienes (see later discussion). It remains to be explored if direct swelling-induced activation of phospholipase C occurs in other cell types.

F. Eicosanoids

Eicosanoids have their name because of their common origin from C_{20} polyunsaturated fatty acids, the eicosaenoic acids. The most abundant of these acids in most lipids is arachidonic acid; others are dihomo-γ-linolenic acid and eicosapentaenoic acid (EPA, the well-known "fish oil"). The eicosanoids are distinguished by their potent physiological effects, their low levels in tissues, and their rapid metabolic turnover. The most important classes of this extensive family of compounds are the prostanoids (prostaglandins, prostacyclin, thromboxanes), the leukotrienes, and the lipoxins. Eicosanoids act as intracellular messengers or as paracrine messengers to neighboring cells, serving as locally acting hormones. Evidently they act through binding to specific cellular receptors. The effects of the eicosanoids include modification of the activation of membrane channels and protein kinases, and release of Ca^{2+} from internal stores.

Arachidonic acid is released from membrane phospholipids by such stimuli as hormones (e.g., bradykinin and epinephrine), neurotransmitters, growth factors, and certain antigens (Berridge, 1987; Smith, 1989; Axelrod, 1990). Stimulation leads to the release of arachidonic acid by several different pathways, either through the action of a specific phospholipase A_2 (PLA_2) on the acyl bond at the sn-2 position of phosphatidylcholine (PC) or phosphatidylethanolamine (PE) (Van den Bosch, 1980; Force and Bonventura, 1994) or through the action of a phospholipase C on phosphatidylinositol, yielding a diacylglycerol, which in turn is cleaved to give arachidonic acid (Irvine, 1982; Neufeld and Majerus, 1983).

Alternatively, diglyceride lipase releases arachidonic acid from phosphatidic acid produced by phospholipase-D (PLD)-mediated hydrolysis of phospholipids. Because intracellular levels of arachidonic acid are extremely low (Kunze and Vogt, 1971), the liberation of arachidonic acid is likely to be a rate-limiting step in whatever processes follow. Several factors are involved in the regulation af of intracellular phospholipases, including Ca^{2+}, protein kinases, and G-proteins (Van den Bosch, 1980; Feinstein and Sha'afi, 1983; Axelrod, 1990; Force and Bonventura, 1994).

There are three primary pathways for arachidonic acid metabolism (Force and Bonventura, 1994; Lambert, 1994):

1. The cyclooxygenase (CO) pathway (promoted by cyclooxygenase and peroxidase), which forms prostaglandins and thromboxanes. The cyclooxygenase step is rate limiting in prostanoid synthesis (Yamamoto, 1989). It is the CO activity that is inhibited by aspirin and indomethacin. Arachidonic acid is the preferred substrate for CO, but dihomo-γ-linolenic acid and to a lesser extent EPA are also converted to prostaglandins (Van Dorp, 1967; Samuelsson, 1983; Yamamoto, 1989).

2. The lipoxygenase pathways, which form leukotrienes and certain mono-, di- and trihydroxy acids. There are three mammalian lipoxygenases—5-LIP, 12-LIP, and 15-LIP—which add O_2 to C-5, C-12, or C-15, respectively, forming the hydroperoxy eicosatetraenoic acids (HPETEs) (Yamamoto, 1989). The HPETE products can be reduced to the hydroxy eicosatetraenoic acids (HETEs). 5-HPETE can also be converted by 5-lipoxygenase to an unstable epoxide intermediate LTA_4, which is converted by a cytosolic LTA_4 hydrolase to LTB_4 or conjugated with glutathione by the enzyme leukotriene C_4 synthase to form LTC_4. Subsequent modifications of the peptide chain yield the related compounds LTD_4 and LTE_4. Both arachidonic acid and EPA are substrates in the 5-LIP pathway and therefore both LTB_4 and LTB_5 as well as LTC_4 and LTC_5 can be produced.

Synthesis of the leukotrienes via 5-HPETE involves translocation of the 5-LIP from the cytosol to an integral membrane protein, the 5-LIP activating protein (FLAP). This translocation can be blocked by inhibitors such as MK-866 (Ford-Hutchinson, 1994). The translocation is Ca^{2+} dependent and is reversed by Ca^{2+} removal (Ford-Hutchinson, 1991; Samuelsson et al., 1991; Sigal, 1991; Ford-Hutchinson, 1994). It has been suggested that FLAP acts as a "docking" protein for 5-LIP in the membrane and recently it has been clear that FLAP is located in the cisternum of the nuclear envelope and the associated ER and that 5-LIP moves there after activation (Ford-Hutchinson, 1994).

Leukotriene-induced effects are generally mediated through interaction with stereospecific, high-affinity, G-protein-coupled plasma membrane receptors (Sjölander and Grönroos, 1994). Ehrlich ascites cells seem to have an LTD_4 receptor coupled to a pertussis toxin-sensitive G-protein with high specificity: neither LTB_4, LTC_4, nor LTE_4 have an effect (Lambert et al., 1987; Lambert, 1987, 1989). LTD_4 receptors, distinct from those for LTC_4, have been demonstrated, for example, in lung tissues from guinea pig (Mong et al., 1984; Watanabe et al., 1990), in a human monocytic leukemic cell line THP-1 (Rochette *et al.*, 1993), and in human intestine cells (Sjö-lander *et al.*, 1990). The human THP-1 cell LTD_4 receptor is a 65-kDa *N*-glycosylated membrane protein (Rochette *et al.*, 1993) whereas the LTD_4 receptor in lung tissues from guinea pig was suggested to be at least a dimer, with the molecular mass of one of the subunits around 70 kDa (Watanabe *et al.*, 1990). Among other products, 12-HPETE can be metabolized to hepoxilin A_3 (Pace-Asciak and Asorta, 1989), which was shown to play a role in cell volume regulation in platelets (see later discussion), and to stimulate calcium mobilization in association with neutrophil activation (Dho *et al.*, 1990).

3. Cytochrome P-450, the "epoxygenase" pathway, produces epoxyeicosatrienoic acids (EETs), which can then be metabolized to dihydroxy eicosatriienoic acids (DHETs) by hydrolases.

A number of the metabolites of arachidonic acid mentioned earlier have properties suggesting roles in cell volume regulation. Cyclooxygenase products inhibit arterial muscarinic K^+ channels (Scherer and Breitwieser, 1990). Prostaglandin E_2 (PGE_2) is suggested to increase Na^+ permeability in Ehrlich cells, but has no effect on the swelling-induced Cl^- permeability or the K^+ permeability (Lambert *et al.*, 1987). Therefore, the reduced synthesis of PGE_2 in Ehrlich cells after cell swelling (see later discussion) may account for the resultant reduction in the passive conductance of Na^+ (Hoffmann, 1978). PGE_2 stimulates Ca^{2+}-sensitive K^+ channels in Madin-Darby canine kidney cells, but this activation is secondary to a PGE_2-induced release of intracellular Ca^{2+} (Steidl *et al.*, 1991). In cultured ciliary epithelial cells, PGE_2 was suggested to mediate the swelling activation of the K^+ channels, probably by stimulation of IP_3 receptors (Civan *et al.*, 1994). An unidentified epoxygenase product was proposed to regulate the volume-sensitive Cl^- channel in these cells.

1. Role of Lipoxygenase Products in Cell Volume Regulation

The RVD response of human platelets is controlled by the 15-lipoxygenase product hepoxilin A_3, which elevates exclusively the K^+ current and not the Cl^- current (Margalit and Livne, 1991, 1992; Margalit *et al.*, 1993a,b). The RVD response in Ehrlich cells is controlled by the 5-lipoxygenase product LTD_4, which activates both K^+, Cl^-, and organic osmolyte channels (Lambert *et al.*, 1987; Lambert, 1994; Lambert and Hoffmann, 1993; Hoffmann *et al.*, 1993). Ehrlich cells and the platelets in this regard will be dealt with in more detail in Section IV,G.

In Ehrlich cells, hypotonic cell swelling results in a stimulation of leukotriene synthesis (Fig. 12, left) and in a reduction in prostaglandin synthesis (Fig. 12, right). Thus the metabolism of arachidonic acid via the 5-LIP pathway is favored during RVD and the resulting increased synthesis of leukotrienes occurring after osmotic cell swelling (Fig. 12, left) seems to

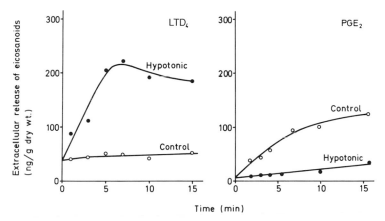

Time (min)

FIG. 12   Synthesis and release of the eicosanoids LTC$_4$ and PGE$_2$ from Ehrlich ascites cells following transfer to isotonic medium (open circles) or to hypotonic medium (normal osmolarity reduced by 50%) (closed circles). The release of the eicosanoids was measured by radioimmune assay and the release was calculated as described by Lambert *et al.* (1987). (From Lambert *et al.*, 1987, with permission.)

be essential for activation of the volume-sensitive K$^+$ and Cl$^-$ channels. This conclusion was based on the following findings:

1. LTD$_4$ stimulates separate K$^+$ and Cl$^-$ channels in Ehrlich cells (Lambert *et al.*, 1987), just as the Ca^{2+}-ionophore A23187 (Hoffmann *et al.*, 1984), the receptor agonists bradykinin and thrombin, and osmotic cell swelling do (Hoffmann *et al.*, 1993).

2. Inhibition of the PLA$_2$, either indirectly by addition of pimozide, or directly by addition of the PLA$_2$ inhibitor RO 31-4638, leads to an inhibition of the RVD response. In both cases LTD$_4$ is able to overcome the inhibition by the PLA$_2$ inhibitors (Hoffmann *et al.*, 1993).

3. Prevention of the attachment of 5-LIP to the FLAP protein by MK886 or inhibition of the 5-LIP by addition of ETH 615-139 also blocks the RVD response in Ehrlich cells (Hoffmann *et al.*, 1993; Lambert, 1994). It is thus concluded that LTD$_4$ is the messenger which activates the volume regulatory channels in Ehrlich cells. LTD$_4$ has also been suggested to act as a mediator for the activation of basolateral Cl$^-$ channels in the isolated crypts of rat colonic epithelium during the RVD response (Diener and Scharrer, 1993). 5-Lipoxygenase is also required for RVD in human fibroblasts (Mastrocola *et al.*, 1993).

2. Role of LTD$_4$ in Activation of Taurine Leak Pathway

Addition of LTD$_4$ to Ehrlich cells in isotonic medium strongly stimulates the taurine efflux, an effect similar to stimulation by cell swelling (Lambert and Hoffmann, 1993). Swelling activation of the taurine channel is inhibited by inhibition of PLA_2 by RO 31-4639, and the inhibition is overcome by addition of LTD_4 (Lambert and Hoffmann, 1993). The 5-LIP inhibitors nordihydroguaiaredic acid (NDGA) and ETH 615-139 also block the taurine efflux activated by cell swelling (Lambert and Hoffmann, 1993). Thus, LTD_4 apparently serves as a second messenger for the activation of the taurine channel as well as for the K^+ and Cl^- channels. Thoroed and Fugelli (1994a) recently demonstrated that Ca^{2+}–calmodulin as well as leukotrienes contribute to the control of taurine loss from flounder erythrocytes following cell swelling.

3. Effect of Polysaturated Fatty Acids on Volume Regulation

The effects of dietary fish oil supplement on cell volume regulation were tested in Ehrlich cells (Lauritzen *et al.*, 1993). The object of these studies was to learn if modification of the amount of eicosanoid precursors in the phospholipid pool can affect eicosanoid-dependent cellular functions. A dietary supplement of MaxEPA (7.5% wt/wt) was given the mouse hosts of the Ehrlich cells for 2 wks; control mice received olive oil. In the Ehrlich cells grown in mice receiving MaxEPA, there was a significant increase in the relative content of eicosapentaenoic acid ($C20:5$, n-3) and a significant decrease in arachidonic acid ($C20:4$, n-6) in both phosphatidylcholine and phosphatidylethanolamine. In cells from MaxEPA-fed mice, there was an accelerated RVD response and net Cl^- loss following hypotonic exposure owing to an increased Cl^- conductance during RVD (Lauritzen *et al.*, 1993). The RVD response was inhibited by pimozide, but the inhibition could be overcome by addition of leukotrienes (see Section IV,F,1). It was demonstrated that LTD_4 and LTD_5 are equipotent in lifting the inhibition by pimozide. Since cells enriched with EPA are better volume regulators even though LTD_4 and LTD_5 were equipotent in activating the volume response, it was suggested that the total leukotriene synthesis (LTD_4 plus LTD_5) in cells enriched with EPA must be larger than the total leukotriene synthesis (predominantly LTD_4) in control cells. In support of this idea, there is evidence that EPA may be a better substrate than arachidonic acid for leukotriene synthesis (Careaga-Houck and Sprecher, 1990). In addition, EPA is a poor substrate for the cyclooxygenase pathway as well as a competitive inhibitor of arachidonic acid in this pathway (Needleman *et al.*, 1979). The swelling-induced shift from prostaglandin to leukotriene synthesis (see Fig. 12) could thus be augmented in cells enriched with EPA and thereby accelerate volume recovery (Lauritzen *et al.*, 1993). Whether this effect has any significance for humans on a fish oil-rich diet is unknown.

Unsaturated fatty acids are also shown to have a direct effect on ion channels. Arachidonic acid activates a mechanosensitive K^+ channel in heart cells (Kim, 1992), and an organic osmolyte channel in Ehrlich cells (Lambert and Hoffmann, 1994), whereas arachidonic acid and other polyunsaturated fatty acids inhibit the volume-sensitive mini Cl^- channel in Ehrlich cells and the VSOAC channel in C6 glioma cells (see Section III,C). The apical membrane Cl^- channel in airway epithelia is likewise inhibited by arachidonic acid (Anderson and Welsh, 1990).

G. Model for the Swelling-Initiated Signal Transduction Pathway That Provokes the RVD Response

Two related models for the activation mechanism of the RVD response are given in the following section—one for activation in human platelets and the other for activation in Ehrlich ascites tumor cells.

1. Human Platelets

In human platelets it has been demonstrated that cell swelling activates the Ca^{2+}-insensitive PLA_2 by a PTX-sensitive G-protein (Margalit et al., 1993a). There is an increased production of arachidonic acid which, through stimulation of the 12-lipoxygenase pathway, leads to an increased production of 12-HPETE and subsequently of hepoxilin A_3 (Margalit et al., 1993b). The conclusions are based on the following arguments:

1. RVD is inhibited by PLA_2 inhibitors.
2. Formation of hepoxilin A_3 has been measured during RVD; this is not inhibited in Ca^{2+}-free media or by depletion of intracellular Ca^{2+} with BAPTA.
3. The RVD response is inhibited by GDP β S and PTX (Margalit et al., 1993a,b).

The hepoxilin A_3 produced (at that time only identified as a lipoxygenase product, LP) is excreted and activates K^+ channels, which then results in an RVD response (Margalit and Livne, 1991). Platelet eluates from cells undergoing RVD can reconstitute RVD in other cells when lipoxygenase is blocked. Centrifugation of the cells, or shear stress from flow through a 1-mm tube, is sufficient to express LP activity (Margalit and Livne, 1992).

2. Ehrlich Ascites Tumor Cells

For Ehrlich cells, the following model is suggested (see Fig. 13): Cell swelling activates PLA_2 through a G-protein as in the platelets. Two kinds of mechanisms are envisioned as signals of volume change:

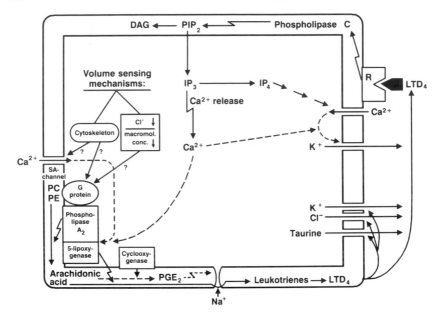

FIG. 13 Model of swelling-induced intracellular signalling in Ehrlich ascites tumor cells. Cell swelling is proposed to induce G-protein activation, which might be mediated by the cytoskeleton and/or macromolecular dilution. The G-protein activates phospholipase A_2, resulting in release of arachidonic acid (AA). The AA is metabolized into LTC_4 and LTD_4 by the 5-lipoxygenase pathway. Prostaglandin synthesis is reduced. LTD_4 activates K^+ channels, Cl^- channels, and organic osmolyte channels. In higher concentrations LTD_4 reacts with a receptor and stimulates release of Ca^{2+} via release of IP_3. The concomitant synthesis of DAG stimulates protein kinase C, which might be involved in reclosing of the channels.

1. Mechanical changes, including membrane tension or deformation and changes in the relation between the membrane and the cytoskeleton
2. Changes in concentrations of cytoplasmic solutes, e.g., proteins; that is, a change in macromolecular crowding (see Section IV,C) or small ions, e.g., Cl^-

Sensors of the signals of volume change are the nonselective stretch-activated channels, or the G-protein controlling PLA_2. The cytoskeleton may sense swelling signals as well as generating them. The sensors of signals of volume change (the stretch-activated channels and the G-protein–PLA_2 complex) initiate the signal transduction pathway with a small local increase in cytosolic free Ca^{2+} and increased arachidonic acid production (see Fig. 14) and by stimulation of 5-lipoxygenase and increased production of 5-HPETE and subsequently of LTC_4 and LTD_4. This is followed by a decrease in synthesis of prostaglandins. LTD_4 is excreted. The signal transduction

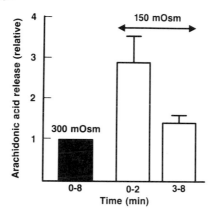

FIG. 14 The effect of reduction of osmolality from 300 to 150 mosmol kg^{-1} on the release of [^{3}H]arachidonic acid (cpm 10^6 cells^{-1} min^{-1}) from Ehrlich ascites tumor cells The cells were loaded with [^{3}H]arachidonic acid (58 μCi, 7.2 ml cytocrit 0.5%) at 300 mosmol kg^{-1} for 2 hr. The incubation media contained 0.5% (w/v) bovine serum albumin (BSA) to trap the arachidonic acid released from the cells. Mean values $\pm$ S.D. are shown (n = 6). (Results from Thoroed et al., 1994.)

pathway causes changes in the function of the effectors, the volume-sensitive channels for Na$^+$, Cl$^-$, K$^+$, and organic osmolytes. A decrease in prostaglandins closes Na$^+$ channels and an increase in LTD$_4$ opens Cl$^-$, K$^+$, and organic osmolyte channels. Provided the increase in extracellular LTD$_4$ becomes more pronounced, the LTD$_4$ receptor is engaged.

Binding of LTD$_4$ to the receptor is known to activate a G-protein-regulated phospholipase C (PLC) (Crooke et al., 1989; Cristol et al., 1989; Sjölander et al., 1990), which hydrolyzes PIP$_2$ into IP$_3$ and DAG. The latter stimulates PKC, while IP$_3$ stimulates the release of Ca^{2+} from intracellular stores followed by the formation of the Ca^{2+}–calmodulin complex. A rise in free Ca^{2+} induced by LTD$_4$ (shown in Fig. 15 for Ehrlich cells) will activate the charybdototoxin-sensitive, Ca^{2+}-gated K$^+$ channel (Thomas-Young et al., 1993; Harbak and Simonsen, 1995) and, moreover, enhance activation of PLA$_2$ and 5-LIP. All this constitutes a positive feedback loop for activation of the effectors. As mentioned in Section IV,D,4, a remarkable heterogeneity in the Ca^{2+} response during RVD was observed, probably reflecting the fact that the extracellular LTD$_4$ concentration does not usually rise high enough to trigger the Ca^{2+} response.

The G-protein which activates PLA$_2$ is both the sensor of the signal of volume change and the initiating step in the signal transduction pathway. The signal is likely to involve changes in either the cytoskeleton or macromolecular crowding, with a change in Cl$^-$ concentration being a possible additional modulator. The nonselective stretch-activated cation channel,

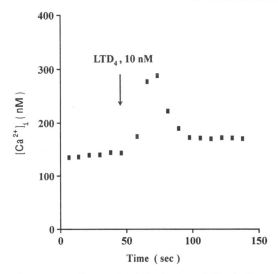

FIG. 15   Ehrlich ascites tumor cells were loaded with Fura-2 (2 μM Fura-2-AM in standard NaCl medium with 0.2% BSA, 0.4% cytocrit, 20 min at 37°C). The cells were washed to remove extracellular Fura-2. [Ca^{2+}]$_i$ was followed in a Perkin–Elmer LS-5 luminescence spectrometer (excitation, 340 and 380 nm; emission, 510 nm). The 340/380 ratios were converted to Ca^{2+} concentrations using an *in vitro* calibration. LTD$_4$ was added at the time indicated to 10 nM. (Data from Jørgensen *et al.*, 1994.)

which permits Ca^{2+} entry, is probably also a sensor of volume change and an initiator of signal transduction. The activation of nonselective cation channels will depolarize the membrane potential; this is important for the opening of the CTX-insensitive, voltage-sensitive K channels, which are suggested to play a central role in volume regulation (see Section II,A,1). Thus, changes in the membrane potential can also be part of the signal transduction pathway. Following the swelling of Ehrlich cells, the Cl⁻ conductance increases more than the K⁺ conductance, which also results in depolarization of the membrane potential. As mentioned in Section II,A,1, it is suggested, as a working model, that Ehrlich cells use charybtotoxin-insensitive voltage-gated channels when Ca^{2+} does not increase during RVD, and small Ca^{2+}-activated channels when LTD$_4$ increases high enough in the medium to give a spike in $Ca^{2+}$$_i$. A similar involvement of two types of K⁺ channels (a Ca^{2+}-dependent and a voltage-sensitive K⁺ channel) has been suggested in human lymphocytes (Schlichter and Sakellaropoulus, 1994).

This sequence of events during the RVD response results in a net loss of KCl and taurine, and an inhibition of the Na⁺ uptake. In this way, Ehrlich ascites cells maximize the net loss of cell osmolytes by the concomitant stimulation of loss and inhibition of gain.

The experimental evidence for the working model is found in Sections II,A; II,B; IV,A; IV,B,1; IV,D,4; and IV,F,1.

a. Termination of the Swelling Signal An important requirement of transduction of the swelling signal is its termination when volume is restored. Since PLA_2 activity is linked to a G-protein, termination may occur if GTP is hydrolyzed to GDP by the GTP-ase associated with the α-subunit of the G-protein as soon as cell volume is restored or, alternatively, after a certain period. In addition, LTD_4 receptors are known to be downregulated after extended exposure (Robertson, 1986). As an example, desensitization of LTD_4 receptors followed by downregulation of LTD_4 receptor-mediated second messenger systems such as Ca^{2+} mobilization is seen after 5 min of stimulation of RBL-1 cells with LTD_4 (Winkler *et al.*, 1987; Chan *et al.*, 1994).

Similar downregulation of the LTD_4-mediated Ca^{2+} response as well of the volume response is found in Ehrlich cells (Jørgensen *et al.*, 1995). The signal can also be shut off by activation of PKC. PKC has been shown to be involved in desensitization of LTD_4 receptors (Vegesna *et al.*, 1988). An increase in PKC was recently demonstrated in the late phase of RVD (Larsen *et al.*, 1994). This late increase could mean that PKC, through a downregulation of the LTD_4 receptor(s), is involved in the closing of the K^+ and Cl^- channels. It is still not clear whether separate LTD_4 receptors are involved in K^+ and Cl^- channel activation and in the Ca^{2+}_i response, or whether we are dealing with one LTD_4 receptor coupled to different G-proteins. That LTD_4 can activate two G-proteins has been described in other cells types (Sjölander and Grönross, 1994).

Finally, as discussed in Section IV,A,1, it is likely that the actin reorganization observed during RVD in some cells is involved in the closing of the swelling-activated channels.

3. Dependence of Signal Transduction Pathway on Mode of Cell Swelling

Results of studies with jejunal enterocytes from guinea pigs suggest that the events taking place during RVD depend on the mode of cell swelling (MacLeod *et al.*, 1992b). RVD in the enterocytes following cell swelling in a hypotonic medium requires extracellular Ca^{2+}, is insensitive to the K^+ channel blocker charybdotoxin, and is inhibited by calmodulin antagonists. In contrast, RVD in the enterocytes following cell swelling induced by Na^+-coupled intake of glucose or alanine is not dependent on extracellular Ca^{2+}, shows CTX sensitivity of the involved K^+ conductive pathway, involves release of Ca^{2+} from thapsigargin-sensitive stores, and involves a PKC-activated Cl^- conductive pathway (MacLeod *et al.*, 1992a). The explanation of the different behaviors is not clear. MacLeod (1994) suggested that the Na$^+$-nutrient swelling stimulates phospholipase C activity and inositol lipid hydrolysis, thereby causing a release of intracellular Ca^{2+} and stimulation of protein kinase C, whereas hypotonic cell swelling does not have this effect.

Ehrlich cells can also accomplish RVD by different mechanisms, depending on cellular Ca^{2+} and pH; in low pH medium, swelling activates K$^+$, Cl$^-$ cotransport rather than conductive channels as an important pathway for osmolyte loss (Kramhøft et al., 1986).

H. Role of Protein Phosphorylation and Dephosphorylation in Control of Volume-Sensitive Transport Pathways

This section focuses on the regulation of the Na$^+$, K$^+$, 2Cl$^-$ and K$^+$, Cl$^-$ cotransporters and the Na$^+$/H$^+$ exchanger. For these three systems, much of the evidence on the role of dephosphorylation/phosphorylation reactions comes from studies employing modifiers of these reactions. In addition, there are recent studies on the direct phosphorylation of the Na$^+$, K$^+$, 2Cl$^-$ cotransporter protein in response to volume changes.

1. Na$^+$, K$^+$, 2Cl$^-$ Cotransporter

The earlier evidence that Na$^+$, K$^+$, 2Cl$^-$ cotransport is regulated by protein kinases and phosphatases was reviewed by Haas (1989). Much of the evidence was indirect, and the control mechanisms appeared to differ in different tissues. For example, there was evidence for promotion of Na$^+$, K$^+$, 2Cl$^-$ cotransport by activation of protein kinase C with a phorbol ester in hamster fibroblasts (Paris and Pouysségur, 1986) and inhibition of cotransport in vascular smooth muscle by phorbol ester (see O'Donnell, 1993). Furthermore, cyclic nucleotides activated cotransport in some tissues and inhibited in others (Haas, 1989). With the availability of potent inhibitors of these protein kinases and phosphatases, a clearer picture has emerged. It is now clear that Na$^+$, K$^+$, 2Cl$^-$ cotransport is activated by protein kinases and inhibited by phosphatases. Furthermore, in several different tissues, more than one type of kinase seems to be capable of activation of cotransport; perhaps only one of these kinases is volume sensitive, and other kinases may respond to other regulatory inputs such as hormones and other extracellular chemical mediators.

Studies on two representative systems—avian erythrocytes and Ehrlich ascites cells—will be presented as illustrations. In duck erythrocytes, a cAMP analog activated Na$^+$, K$^+$, 2Cl$^-$ cotransport, and the potent broad specificity protein kinase inhibitors—K252a and H-9—rapidly reversed this activation (Pewitt et al., 1990). In the same study, it was shown that the protein kinase inhibitors also block activation of Na^+, K^+, $2Cl^-$ cotransport by cAMP-independent stimuli such as osmotic shrinkage. However, higher concentrations of the kinase inhibitors were required. A serine–threonine phosphatase inhibitor, okadaic acid, stimulated Na^+, K^+, $2Cl^-$ cotransport and did so to a greater extent than the cAMP analog. K252a prevented activation by okadaic acid. The concentration dependence of the inhibition by K252a was like that of its inhibition of the effects of osmotic shrinkage, and not of cAMP. Several lines of evidence indicate that activation of Na^+, K^+, $2Cl^-$ cotransport by hypertonicity does not involve PKA, the cAMP-dependent kinase. Shrinkage raises neither cell cAMP concentration (Kregenow et al., 1976) nor cAMP-dependent protein phosphorylation (Alper et al., 1980b). One interpretation is that two protein kinases activate Na^+, K^+, $2Cl^-$ cotransport in duck red blood cells—PKA and an unknown kinase. PKA responds to stimulation of the cells by catecholamines (Alper et al., 1980a). Cell shrinkage activates the unknown kinase or perhaps inhibits a phosphatase (Palfrey and Pewitt, 1993).

In Ehrlich ascites cells, the level of cAMP does not seem to be involved in the volume sensitivity of Na^+, K^+, $2Cl^-$ cotransport (Geck and Pfeiffer, 1985). However, there is ample evidence for roles of protein kinases and phosphatases in the regulation of cotransport. Bradykinin activates Na^+, K^+, $2Cl^-$ cotransport in Ehrlich cells (Jensen et al., 1993; Hoffmann et al., 1993). This agent activates phospholipase C and thereby activates PKC both by mobilizing intracellular Ca^{2+} by releasing IP_3, and by the release of DAG. There was also a small stimulation of cotransport by a phorbol ester, an activator of PKC. It was demonstrated subsequently that PKC activity is stimulated in Ehrlich cells by hypertonic shock, as shown in Fig. 16, and that the activated PKC activity is primarily membrane associated (Larsen et al., 1994). The PKC inhibitors H-7 and chelerythrine greatly inhibited the shrinkage-induced PKC activity, but inhibited shrinkage-induced cotransport only 20–30% (Larsen et al., 1994). Staurosporine, the most potent of the PKC inhibitors, had no effect on shrinkage-induced cotransport (Jensen et al., 1993).

Pimozide was a more effective inhibitor of shrinkage-induced Na^+, K^+, $2Cl^-$ cotransport than any of the PKC inhibitors (Jensen et al., 1993), suggesting the involvement of another kinase, the Ca^{2+}–calmodulin-dependent kinase. To investigate a possible role for a Ca^{2+}–calmodulin-dependent kinase, a specific inhibitor of Ca^{2+}–calmodulin-dependent kinase II (Ca^{2+}–CaM kinase II) was tested on both bradykinin-activated and shrinkage-activated Na^+, K^+, $2Cl^-$ cotransport. This inhibitor, KN-62, had no significant effect on cotransport after either activating strategy (L. Jakobsen and E. K. Hoffmann, unpublished results). Furthermore, deltamethrin, a potent inhibitor of protein kinase 2B, which is dependent on Ca^{2+}–CaM kinase II, also had no effect on cotransport, either in isotonic or hypertonic

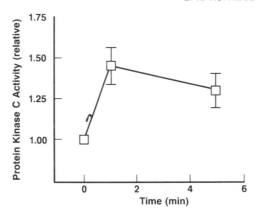

FIG. 16   Effect of shrinkage of Ehrlich ascites cells on protein kinase C activity. Cells were preincubated in isotonic medium (300 mosmol) and then subjected to a hypertonic challenge (600 mosmol) at time zero. Control cells were kept in isotonic medium, and PKC activity was measured on both sets of cells at times shown (Larsen *et al.*, 1994). Results are ratios of activities in cells in hypertonic to isotonic medium. (Redrawn from Larsen *et al.*, 1994.)

media (L. Jakobsen and E. K. Hoffmann, unpublished results). Therefore a role for a Ca^{2+}–CaM kinase in activating Na^+, K^+, $2Cl^-$ cotransport, suggested by the results with bradykinin and pimozide, was not supported by experiments employing KN-62 or deltamethrin. Perhaps a Ca^{2+}-dependent kinase other than PKC or Ca^{2+}–CaM II is involved; alternatively, the effects of pimozide and/or bradykinin may not have been as they were interpreted. If the shrinkage induced by bradykinin was prevented, there was negligible activation of the Na^+, K^+, $2Cl^-$ cotransporter (Hoffmann, 1993) and when thrombin-induced cell shrinkage was blocked by CTX (100 μM), the activation of the cotransporter was also blocked (Harbak and Simonsen, 1993), suggesting the involvement of an additional shrinkage-activated kinase. One result confirming the role of a serine–threonine protein kinase was the marked activation of Na^+, K^+, $2Cl^-$ cotransport by calyculin A, a potent inhibitor of phosphatases 1 and 2A (Jakobsen *et al.*, 1994).

The studies on avian red cells and Ehrlich cells demonstrate that phosphorylation activates Na^+, K^+, $2Cl^-$ cotransport; that more than one class of kinase is involved, but perhaps only one responds to cell shrinkage. For a discussion of similar results in other systems, see Haas (1994).

The phosphorylation of Na^+, K^+, $2Cl^-$ cotransporters has been demonstrated directly in two cell types—shark rectal gland (Lytle and Forbush, 1992b) and duck salt gland epithelia (Torchia *et al.*, 1992). In both studies, phosphorylation was demonstrated by immunoprecipitation using the monoclonal antibodies raised against the shark gland cotransporter. Both osmotic shrinkage and forskolin promoted phosphorylation of the shark gland cotransporter, the latter result implicating PKA. Phosphorylation was demonstrated on both serine and threonine residues, but not on tyrosines (Lytle and Forbush, 1992b). The duck salt gland transporter was phosphorylated in response to vasoactive intestinal peptide, cAMP analogs, and carbachol, all of which also promote cotransport (Torchia et al., 1992). Therefore phosphorylation is both cAMP- and Ca^{2+}-dependent, and once again, in a recurrent pattern, more than one protein kinase is involved. What is not known about phosphorylation of the $Na^+, K^+, 2Cl^-$ cotransporter is whether activation is by stimulation of a kinase, inhibition of a phosphatase, or both.

2. K^+, Cl^- Cotransporter

Whereas $Na^+, K^+, 2Cl^-$ cotransport is activated by phosphorylation, K^+, Cl^- cotransport is activated by dephosphorylation. Most of the evidence comes from studies employing modifiers of kinase and phosphatase activities in mammalian red cells. The first clear evidence for the control of K^+, Cl^- cotransport by phosphorylation was the inhibition of cotransport in rabbit red cells by okadaic acid, an inhibitor of the serine–threonine protein phosphatases 1 and 2A (Jennings and Schulz, 1991). A comparison of the inhibitory effects on transport of okadaic acid and calyculin A, another phosphatase inhibitor, implicated protein phosphatase 1 as the enzyme involved (Starke and Jennings, 1993). Similar observations were made for K^+, Cl^- cotransport in red cells of humans (Kaji and Tsukitani, 1991) and sheep (Bize and Dunham, 1994). Therefore activation of K^+, Cl^- cotransport requires dephosphorylation of the transporter or of an associated regulatory protein. Direct demonstration of this awaits the identification of the protein of the K^+, Cl^- cotransporter.

If a phosphatase activates K^+, Cl^- cotransport, then a protein kinase must inhibit transport. Measurements of presteady-state kinetics of swelling activation and shrinkage inactivation of K^+, Cl^- cotransport in rabbit red cells showed that it is the reverse reaction of activation that is volume sensitive, not the forward reaction promoted by a phosphatase (Jennings and Al-Rohil, 1990). These observations have been confirmed for red blood cells of sheep (Dunham et al., 1993) and humans (Kaji and Tsukitani, 1991) (see Fig. 4). This volume-sensitive protein kinase appears to be neither PKC nor PKA (Jennings and Shulz, 1991); other classes of kinase have not been ruled out. Staurosporine, a highly potent kinase inhibitor with broad specificity, stimulated K^+, Cl^- cotransport in sheep red cells (Bize and Dunham, 1994). However, this was shown not to be an effect on the volume-sensitive kinase. Rather, staurosporine activates the phosphatase indirectly, apparently by inhibiting a kinase which, in turn inhibits the phosphatase (Bize and Dunham, 1994). These results show another level of complexity of the regulation of the cotransporter. The regulatory input to the kinase inhibiting the phosphatase has not been identified and the volume-sensitive kinase remains elusive.

3. Na$^+$/H$^+$ Exchanger

All known isoforms of the Na$^+$/H$^+$ exchanger contain several consensus phosphorylation sites, apparently all serine residues (Tse *et al.*, 1993). NHE-1 (the mammalian "housekeeping" Na$^+$/H$^+$ exchanger) has sites for Ca^{2+}–calmodulin kinase II and for most isoforms of the MAP kinase family, but not for PKA (Fliegel and Frøhlich, 1993). It was proposed that phosphorylation of NHE-1 is essential for the activation of cells by growth factors (Sardet *et al.*, 1990). However, it is not clear that volume sensitivity of the exchanger is controlled by phosphorylation; shrinkage-activation of the exchanger in two tissues did not result in an increase in its level of phosphorylation (Grinstein *et al.*, 1992b). There may be a dual control of Na$^+$/H$^+$ exchange by phosphorylation-dependent and -independent mechanisms, and cell shrinkage activates the latter type of mechanism.

Nevertheless shrinkage-induced activation of the Na$^+$/H$^+$ exchanger is dependent on intracellular ATP in many cell types (Grinstein *et al.*, 1992a; Bianchini *et al.*, 1991), including Ehrlich cells, as shown in Fig. 17. One possible explanation is the participation of G-proteins and the depletion of cellular GTP concomitant with the loss of ATP. Recent evidence suggests a role for a G-protein in the osmotic activation of the Na$^+$, H$^+$ exchanger in barnacle muscle fibers (Davis *et al.*, 1992; Boron *et al.*, 1994). Another possible explanation for the requirement for ATP is phosphorylation of polypeptides other than the exchanger. Grinstein *et al.* (1986) reported swelling-induced phosphorylation of a number of proteins in lymphocytes. Bianchini *et al.* (1991) observed that treatment with the phosphatase inhibitor okadaic acid restored the response to hypertonicity in ATP-depleted cells, which is consistent with the involvement of phosphorylation events in volume sensitivity. Several other tentative explanations for the ATP dependence have been offered by Grinstein *et al.* (1994). ATP may bind to, but not phosphorylate, the exchanger or an associated protein. There might be similar ATP-dependent interactions with the elements of cytoskeleton or with phospholipids involved in volume regulation.

There are conflicting results on whether PKC and/or other protein kinases are involved in the shrinkage activation of the Na$^+$/H$^+$ exchanger. Shrinkage-induced activation of the Na$^+$/H$^+$ exchanger in Ehrlich cells is inhibited by chelerythrine (Pedersen *et al.*, 1994; see Fig. 17), which is reported to be a specific inhibitor of PKC (Herbert *et al.*, 1990) and cell shrinkage stimulates PKC in Ehrlich cells (see Fig. 16 and Larsen *et al.*, 1994). In addition, the shrinkage-induced activity of the Na$^+$/H$^+$ exchanger is stimu-

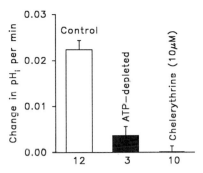

FIG. 17   Cells were loaded with 2',7'-bis-(2-carboxyethyl)-5,6 carboxyfluorescein (BCECF) and changes in pH_i were measured spectrophotometrically at 25°C. Control cells were given a hypertonic challenge by the RVI protocol (for ATP-depleted cells) or by the RVD-RVI protocol (chelerythrine-treated cells). Cells were depleted of ATP by preincubation in standard medium containing 10 mM azide and 10 mM deoxyglucose for 30 min (during BCECF loading). ATP, measured by high-performance liquid chromatography, was 13.6 nmoles/mg dry wt in control cells and 0.12 nmole/mg dry wt in depleted cells. Cell viability was not impaired by ATP depletion. Other aliquots of cells were preincubated with chelerythrine (10 μM) in standard medium for 14 min. These cells were exposed to a hypertonic challenge using the RVD-RVI protocol. The rates of change in pH_i were calculated from fitted slopes of the measurements obtained 2–4 min after the hypertonic challenge. (Redrawn from Pedersen *et al.*, 1994). ATP-depleted cells are more acidic than control cells by 0.22 pH units. When the shrinkage-induced changes in pH_i are expressed in units of H^+ concentration, the inhibition of proton efflux by ATP depletion is to 30%, not 14% as shown in the figure.

lated by calyculin A (Pedersen *et al.*, 1994), a potent inhibitor of the phosphatases PP1 and PP2A. This is consistent with findings in dog red blood cells where okadaic acid, another phosphatase inhibitor, activated the volume-sensitive Na^+/H^+ exchanger (Parker *et al.*, 1991).

In lymphocytes (Grinstein *et al.*, 1986; Grinstein and Foskett, 1990), in bone cell lines (Dascalu *et al.*, 1992), and in barnacle muscle fibers (Boron *et al.*, 1994), a PKC-dependent pathway does not seem to account for the activation of the Na^+/H^+ exchanger after cell swelling.

The effect of cAMP on the shrinkage-induced activation of the exchanger is also variable. In Ehrlich cells (Pedersen *et al.*, 1994; see Fig. 17), and in pig red cells (Sargeant *et al.*, 1989) cAMP was inhibitory, whereas in bone cell lines (Dascalu *et al.*, 1992) and in barnacle muscle fibers (Boron *et al.*, 1994) it seemed to have no effect. In several cell types cAMP has been found to inhibit the exchanger under isotonic conditions (Tse *et al.*, 1993).

It is relevant to note that phosphatases have important roles in regulation of cytoskeleton structure (Hosoya *et al.*, 1993) and that Grinstein *et al.* (1993) have suggested that interaction between the Na^+/H^+ exchanger and the cytoskeleton may underlie the osmotic activation of the exchanger.

Thus the observed effects of calyculin may be exerted on cytoskeleton as well as directly on transport pathways.

4. Channels

Modulation of the activity of ion channels by phosphorylation and dephosphorylation reactions is a general phenomenon. Ligand-gated channels, voltage-gated channels, and Ca^{2+}-activated K^+ channels are all subject to this type of control (Levitan, 1994). It is reasonable to suppose that mechano-sensitive channels are as well. However, there is no direct evidence for control of a volume-sensitive channel by the level of phosphorylation of the channel or of a protein which regulates it.

I. Coordinated Control of Volume-Regulatory Transport Processes

As described earlier, $Na^+, K^+, 2Cl^-$ cotransport is promoted by phosphorylation and K^+, Cl^- cotransport by dephosphorylation. It was proposed that pairs of transporters in a cell which promote RVD and RVI, respectively, can be regulated in a coordinated fashion as follows: if a transport pathway which promotes RVI is activated by a kinase, the same kinase will inhibit the transport pathway which promotes RVD (Parker *et al.*, 1991; Grinstein *et al.*, 1992a; Parker, 1994). Likewise, a phosphatase which activates the transport pathway promoting RVD inhibits the transport pathway promoting RVI.

Much of the evidence for coordinated regulation of pairs of transporters comes from changes in set points of the transporters. When a manipulation of a cell changes the set points of two transporters in the same direction and to the same extent, this is evidence for, but not proof of, coordinated control of the two transporters. In duck red blood cells, as in many cell types, $Na^+, K^+, 2Cl^-$ cotransport is activated by shrinkage and K^+, Cl^- cotransport is activated by swelling (Haas and McManus, 1985). It was subsequently proposed that the regulation of two transporters is coordinated (Starke and McManus, 1990), but the evidence was limited.

The clearest evidence for coordinated control comes from dog red blood cells, and that evidence is the focus of this discussion. The advantage of these cells in studying coordinated control is that the swelling-activated transporter, K^+, Cl^- cotransport, and the shrinkage-activated transporter, Na^+/H^+ exchange, have no substrate ions in common and the two transporters can be distinguished with no uncertainty. Dog red cells lack Na^+, K^+ pumps, their intracellular Na^+ and K^+ concentrations approximate those of plasma, and activation of the two volume-sensitive transporters by cell shrinkage or swelling cannot regulate cell volume at physiological cellular cation concentrations. Nevertheless, these cells provide a convenient model system for the study of volume-sensitive transporters which participate in volume regulation in other cell types.

A series of treatments affected the set points of the two transport pathways in the same way. Loading dog red blood cells with Mg^{2+} or Li^+ stimulated Na^+/H^+ exchange, inhibited K^+, Cl^- cotransport, and shifted the set points of the two transporters in the same direction and to the same extents to higher cell volumes (Parker *et al.*, 1990). Urea stimulated K^+, Cl^- cotransport, inhibited Na^+/H^+ exchange, and also shifted the set points the same direction and to the same extents to lower cell volumes (Parker, 1993a). Figure 18 shows that okadaic acid, a phosphatase inhibitor, had the same effect as Mg^{2+} and Li^+: inhibition of K^+, Cl^- cotransport, stimulation of Na^+/H^+ exchange, and a shift of both of their set points to higher cell volumes (Parker *et al.*, 1991). This result implicates a phosphatase in the control of both transport pathways, and leaves open the possibility that it is the same phosphatase.

More recently, other modifiers of protein kinases and phosphatases have been tested. All of these agents either stimulated one of the transporters and inhibited the other, or had no effect on either. Calyculin A, another phosphatase inhibitor, had the same effects as okadaic acid on both trans-

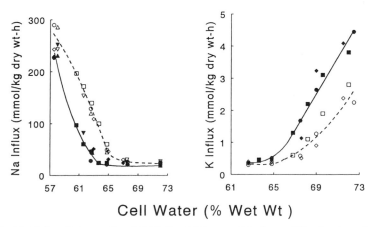

Cell Water (% Wet Wt)

FIG. 18 Na^+ influx (left panel) and K^+ influx (right panel) in dog red blood cells as functions of cell water content. (Shrinkage-activated Na^+ influx and swelling-activated K^+ influx are entirely Na^+/H^+ exchange and K^+, Cl^- cotransport, respectively.) Aliquots of cells were preincubated with 50 nM okadaic acid (open symbols). Results are shown from five experiments. Note that okadaic acid stimulated Na^+ influx and inhibited K^+ influx and shifted the set points for both fluxes to higher volumes. (From Parker, 1994, with permission.)

porters, but with higher affinity. Vanadate, a tyrosine phosphatase inhibitor, had slight but consistent effects: inhibition of K^+, Cl^- cotransport and stimulation of Na^+/H^+ exchange. Staurosporine, a protein kinase inhibitor, stimulated K^+, Cl^- cotransport and inhibited Na^+/H^+ exchange. The following agents had no effect on either pathway: phorbol myristate acetate, an activator of protein kinase C, K252a, and H-7; inhibitors of PKC, PKA, and cGMP-dependent kinases; and genistein and tyrphostin 25, tyrosine kinase inhibitors (R. P. Thomas, H. Gitelman, and P. B. Dunham, unpublished results). These results are consistent with coordinated control: the same enzymes, a kinase and a phosphatase, regulating both transport pathways. Direct demonstration of coordinated control will require the identification, isolation, and characterization of the enzymes.

V. Induction of Organic Osmolyte Transporters

As mentioned earlier, regulation of constant cell volume over long periods of extracellular hypertonicity depends in part on accumulation of "compatible" organic osmolytes. This osmolyte maintenance can result from increased synthesis and/or inhibited degradation of the organic osmolytes, or from regulation of the osmolyte transporter (Garcia-Pérez and Burg, 1991; Kwon, 1994). The renal medulla maintains the following osmolytes to aid in maintenance of constant volume: sorbitol, inositol, betaine, glycerophosphocholine (GPC), and taurine. The degradation of glycerophosphocholine by GPC:choline phosphodiesterase is inhibited by hypertonicity (Garcia-Pérez and Burg, 1991). Moreover, during antidiuresis, aldose reductase activity increases, resulting in increased synthesis of sorbitol from glucose, and increased cellular osmolyte concentration (Cowley et al., 1990; Smardo et al., 1992). This is a consequence of increased transcription of the genes encoding for aldose reductase in response to the extracellular hypertonicity (Smardo et al., 1992).

Transcription of the genes encoding for the Na^+-dependent transporters for betaine and myoinositol is also stimulated within several hours in MDCK cells in response to hypertonicity, resulting in cellular accumulation of these osmolytes (Yamauchi et al., 1992, 1993; Uchida et al., 1993). When the hypertonic cells were returned to isotonic medium, the mRNA levels for betaine and myoinositol transporters returned to their basal levels within 8 hr (Yamauchi et al., 1993; Uchida et al., 1993). For a recent review of molecular regulation of these mammalian osmolyte transporters, see Kwon (1994).

Cohen and Gullans (1994) suggested that "hyperosmotic stress in renal epithelial cells increases expression of a subset of immediate-early genes (i.e., primary response genes), which in turn activates expression of a series of effector genes comprising an adaptive response to hyperosmotic stress." In this way the synthesis of proteins essential for adaptation to anisotonic media, including enzymes and membrane transport proteins, can be induced.

Also, an increased number of Na^+-dependent amino acid transporters can be induced in response to hypertonicity. Cultured human fibroblasts shrunken in hypertonic media can restore their initial volumes when the medium contains a substrate for the Na^+-dependent amino acid transport system A. In contrast, substrates for transport system L are unable to promote the RVI response. The explanation for the difference is that system A can generate high concentration gradients and system L cannot. Two points of evidence indicate that the response of system A to shrinkage involves *de novo* protein synthesis. The protein-synthesis inhibitor cycloheximide blocks shrinkage activation and the time course for the activation is slow (Gazzola *et al.*, 1991). Isotonic RVI in human fibroblasts involves the same regulatory mechanisms. Replacement of the amino acid-containing culture medium with an isotonic inorganic salt solution results in cell shrinkage caused by a loss of intracellular organic osmolytes. Following replacement of this isotonic medium with the complete medium, there is a substantial increase in amino acid influx through system A, indicating that the stress of isotonic cell shrinkage, as well as hypertonic cell shrinkage, promotes synthesis of new system A transporters. This latter amino acid influx promotes cell swelling, which in turn initiates RVD by the release of K^+, Cl^-, and organic osmolytes. In time, the level of system A returns to its original, basal level (Gazzola *et al.*, 1991).

VI. Concluding Remarks

Cell volume homeostasis is crucial for the integrated function of cells. Thus it has recently been shown that volume-sensitive transport pathways play a primary role in cellular growth and proliferation:

1. Mitogens activate the Na^+, H^+ antitransporter in a number of cell types, resulting in cell swelling and an increase in cellular Na^+, which leads to stimulation of the Na^+, K^+ ATPase, which in turn increases the intracellular K^+ concentration (Bianchini and Grinstein, 1993).

2. In 3T3 fibroblasts transfected with the *Ha ras* oncogene, the activity of the Na^+, K^+, $2Cl^-$ cotransporter was increased fourfold. This was associated with a parallel increase in cell volume and an increased rate of cellular proliferation; furosemide inhibited both phenomena (Meyer *et al.*, 1991).

The role played by the increase in K^+ and/or in cell volume in controlling growth and proliferation is, however, still not clear.

In addition, cell volume regulatory processes appear to be an integral part of the function of secreting and absorbing epithelia. For a detailed discussion of the role of cell volume regulatory processes in secreting and absorbing epithelia, with a special emphasis on the regulation of isotonic secretion, see Hoffmann and Ussing (1992). For a discussion of the major role played by minor cell volume changes in the cross-talk in epithelia, see Reuss and Cotton (1994).

In the liver it has been proposed that cell volume changes are the basis for metabolic control (Hässinger and Lang, 1991; Häussinger et al., 1993). Transporters involved in volume regulation are modulated by the primary hormones of metabolism—insulin and glucagon—such that exposure to insulin produces swelling (by stimulation of Na^+, K^+, $2Cl^-$ cotransport or Na^+/H^+ exchange) and exposure to glucagon causes shrinkage (by activating K^+ and Cl^- channels). Häussinger and Lang (1991) suggested that, because osmotic swelling and shrinkage affect metabolism in a way similar to that of insulin and glucagon, respectively, volume perturbation is actually the "second messenger" for metabolic control. They provided data indicating that in liver cells the anabolic effects of insulin may be mimicked by hypotonic swelling and the catabolic effects of glucagon by hyperosmotic shrinkage (Häussinger et al., 1993).

Study of the mechanisms involved in cell volume control has thus broad implications, and the investigation of the structure, function, and regulation of the membrane transport pathways involved in cell volume regulation can illuminate general features of the structure and function of cells.

References

Aabin, B., and Hoffmann, E. K. (1986). Inhibition of the volume- and Ca^{2+}-activated Cl^- transport pathway in Ehrlich ascites tumor cells. *Acta Physiol. Scand.* **128,** 42A.

Aabin, B., and Kristensen, B. I. (1987). Identification of calmodulin in Ehrlich ascites tumor cells and its Ca^{2+}-dependent binding to the cell membrane. *Acta Physiol. Scand.* **129,** 18A.

Ackerman, M. J., Wickman, K. D., and Clapham, D. E. (1994). Hypotonicity activates a native chloride current in Xenopus oocytes. *J. Gen. Physiol.* **103,** 153–179.

Adorante, J. S., and Miller, S. S. (1990). Potassium-dependent volume regulation in retinal pigment epithelium is mediated by Na,K,Cl cotransport. *J. Gen. Phys.* **96,** 1153–1175.

Agius, L., Peak, M., Beresford, G., Al-Habori, M., and Thomas, T. (1994). The role of ion content and cell volume in insulin action. *Biochem. Soc. Trans.* **22,** 518–522.

Al-Awqati, Q. (1989). Regulation of membrane transport by endocytotic removal and exocytotic insertion of transporters. *In* "Methods in Enzymology" (S. Fleischer and B. Fleischer, eds.), Vol. 172, pp. 49–54. Academic Press, San Diego, CA.

Al-Habori, M. (1994). Cell volume and ion transport regulation. *Int. J. Biochem.* **26,** 319–334.

Alper, S. L., Beam, K. G., and Greengard, P. (1980a). Hormonal control of Na,K-cotransport in turnkey erythrocytes: Multiple site phosphorylation of goblin, a high molecular weight protein of the plasma membrane. *J. Biol. Chem.* **225,** 4864–4871.

Alper, S. L., Palfrey, H. C., Dericmer, S. A., and Greengard, P. (1980b). Hormonal control of protein phosphorylation in turkey erythrocytes: Phosphorylation by cAMP-dependent and Ca²⁺-dependent protein kinases of distinct sites in globin, a high molecular weight protein of the plasma membrane. *J. Biol. Chem.* **255,** 11029–11039.

Alvarez, J., Montero, M., and Garcia-Sancho, J. (1992). High affinity inhibition of Ca²⁺-dependent K⁺ channels by cytochrome P-450 inhibitors. *J. Biol. Chem.* **267,** 11789–11793.

Anderson, M. P., and Welsh, M. J. (1990). Fatty acids inhibit apical membrane chloride channels in airway epithelia. *Proc. Natl. Acad. Sci. U.S.A.* **87,** 7334–7338.

Aronson, P. S. (1985). Kinetic properties of the plasma membrane Na⁺-H⁺ exchanger. *Annu. Rev. Physiol.* **47,** 545–560.

Axelrod, J. (1990). Receptor-mediated activation of phospholipase A₂ and arachidonic acid release in signal transduction. *Biochem. Soc. Trans.* **18,** 503–507.

Baker, A. J., Zornow, M. H., Sheller, M. S., Yaksh, T. L., Skilling, S. R., Smullin, D. H., Larson, A. A., and Kuczenski, R. (1991). Changes in extracellular concentrations of glutamate, aspartate, glycine, dopamine, serotonin, and dopamine metabolites after transient global ischemia in the rabbit brain. *J. Neurochem.* **57,** 1370–1379.

Bakker-Grunwald, T., Ogden, P., and Lamb, J. F. (1982). Effects of ouabain and osmolarity on bumetanide-sensitive potassium transport in Simian virus-transformed 3T3 cells. *Biochim. Biophys. Acta* **687,** 333–336.

Ballatori, N., and Boyer, J. L. (1992). Taurine transport in skate hepatocytes. II. Volume activation, energy and sulfhydryl dependence. *Am. J. Physiol.* **262,** G451–G460.

Bandarali, U., and Roy, G. (1992). Anion channels for amino acids in MDCK cells. *Am. J. Physiol.* **263,** C1200–C1207.

Baquet, A., Meijer, A. J., and Hue, L. (1991). Hepatocyte swelling increases inositol 1,4,5-triphosphate, calcium and cyclic AMP concentration but antagonizes phosphorylase activation by Ca²⁺-dependent hormones. *FEBS Lett.* **278,** 103–106.

Bear, C. E. (1990). A nonselective cation channel in rat liver cells is activated by membrane stretch. *Am. J. Physiol.* **258,** C421–C428.

Beck, J. S., Breton, S., Laprade, R., and Giebisch, G. (1991). Volume regulation and intracellular calcium in the rabbit proximal convoluted tubule. *Am. J. Physiol.* **260,** F861–F867.

Bennett, V. (1985). The membrane skeleton of human erythrocytes and its implications for more complex cells. *Annu. Rev. Biochem.* **54,** 273–304.

Bergh, C., Kelley, S. J., and Dunham, P. B. (1990). K-Cl cotransport in LK sheep erythrocytes: Kinetics of stimulation by cell swelling. *J. Membr. Biol.* **117,** 177–188.

Berridge, M. J. (1987). Inositol trisphosphate and diacylglycerol: Two interacting second messengers. *Annu. Rev. Biochem.* **56,** 159–193.

Berridge, M. J., and Irvine, R. F. (1989). Inositol phosphates and cell signalling. *Nature (London)* **341,** 197–205.

Bevan, C., Theiss, C., and Kinne, R. K. H. (1990). Role of Ca²⁺ in sorbitol release from rat inner medullary collecting duct (IMCD) cells under hypoosmotic stress. *Biochem. Biophys. Res. Commun.* **170,** 563–575.

Bianchini, L., and Grinstein, S. (1993). Regulation of volume-modulating ion transport systems by growth promoters. *In* "Interaction of Cell Volume and Cell Function" (F. Lang and D. Häussinger, eds.), pp. 249–277. Springer-Verlag, Berlin.

Bianchini, L., Woodside, M., Sardet, C., Pousségur, J., Takai, A., and Grinstein, S. (1991). Okadaic acid, a phosphatase inhibitor, induces activation and phosphorylation of the Na⁺/H⁺ antiport. *J. Biol. Chem.* **266,** 15406–15413.

Bickler, P. E. (1992). Cerebral anoxia tolerance in turtles: Regulation of intracellular calcium and pH. *Am. J. Physiol.* **263,** R1298–R1302.

Binder, H. J., and Murer, H. (1986). Potassium/proton exchange in brush-border membrane of rat ileum. *J. Membr. Biol.* **91,** 77–84.

Bize, I., and Dunham, P. B. (1994). Staurosporine, a protein kinase inhibitor, activates K-Cl cotransport in LK sheep erythrocytes. *Am. J. Physiol.* **266,** C759–C770.

Blatz, A. L., and Magleby, K. L. (1983). Single voltage-dependent chloride-selective channels of large conductance in cultured rat muscle. *Biophys. J.* **43,** 237–241.

Bonano, J. A. (1991). K^+/H^+ exchange, a fundamental cell acidifier in corneal epithelium. *Am. J. Physiol.* **260,** C618–C625.

Bookchin, R. M., Ortiz, O. C., and Lew, V. L. (1987). Activation of calcium-dependent potassium channels in deoxygenated sickled red cells. *Prog. Clin. Biol. Res.* **240,** 193–200.

Boron, W. F., Hogan, E. M., and Davis, B. A. (1994). Involvement of a G protein in the shrinkage-induced activation of Na-H exchange in barnacle muscle fibers. *In* "Cellular and Molecular Physiology of Cell Volume Regulation" (K. Strange, ed.), pp. 299–310. CRC Press, Boca Raton, Florida.

Bosma, M. M. (1989). Anion channels with multiple conductance levels in a mouse B lymphocyte cell line. *J. Physiol. (London)* **410,** 67–90.

Brehm, P., Kullberg, R., and Moody-Corbett, F. (1984). Properties of non-junctional acetylcholine receptor channels in innervated muscle of *Xenopus* larvae. *J. Physiol. (London)* **350,** 631–648.

Breitwieser, G. E., Altamirano, A. A., and Russell, J. M. (1990). Osmotic stimulation of Na^+-K^+-Cl^- cotransport in squid giant axon is $[Cl^-]_i$ dependent. *Am. J. Physiol.* **258,** C749–C753.

Breton, S., Beck, J. S., Cardinal, J., Giebisch, G., and Laprade, R. (1992). Involvement and source of calcium in volume regulatory decrease of collapsed proximal convoluted tubule. *Am. J. Physiol.* F656–F664.

Brugnara, C., Bunn, H. F., and Tosteson, D. C. (1986). Regulation of erythrocyte cation and water content in sickle cell anemia. *Science* **232,** 388–390.

Brugnara, C., Van Ha, T., and Tosteson, D. C. (1989). Role of chloride in potassium transport a K-Cl cotransport system in human red cells. *Am. J. Physiol.* **256,** C994–C1003.

Butt, A. G., Clapp, W. L., and Frizzel, R. A. (1990). Potassium conductances in tracheal epithelium activated by secretion and cell swelling. *Am. J. Physiol.* **258,** C630–C638.

Cabantchick, Z. I., and Greger, R. (1992). Chemical probes for anion transporters of mammalian cell membranes. *Am. J. Physiol.* **262,** C803–C827.

Cahalan, M. D., and Lewis, R. S. (1988). Role of potassium and chloride channels in volume regulation by Thymphocytes. *In* "Cell Physiology of Blood" (R. B. Gunn and J. C. Parker, eds.), pp. 281–301. Rockefeller Univ. Press, New York.

Cahalan, M. D., Ehring, G. R., Osipchuk, Y. V., and Ross, P. E. (1994). Volume-sensitive Cl^- channels in lymphocytes and multidrug-resistant cell lines. *J. Gen. Physiol.* **104,** 43a–44a.

Cala, P. M. (1980). Volume regulation by *Amphiuma* red blood cells. The membrane potential and its implications regarding the nature of the ion-flux pathways. *J. Gen. Physiol.* **76,** 683–708.

Cala, P. M. (1983a). Volume regulation by red blood cells: Mechanisms of ion transport. *Mol. Physiol.* **4,** 33–52.

Cala, P. M. (1983b). Cell volume regulation by *Amphiuma* red blood cells. The role of Ca^{+2} as a modulator of alkali metal/H^+ exchange. *J. Gen. Physiol.* **82,** 761–784.

Cala, P. M. (1986). Volume-sensitive alkali metal-H transport in *Amphiuma* red blood cells. *Curr. Top. Membr. Transp.* **26,** 79–99.

Cala, P. M., and Grinstein, S. (1988). Coupling between Na^+/H^+ and Cl^-/HCO^-_3 exchange in pH and volume regulation. In "Na^+/H^+ Exchange" (S. Grinstein, ed.), pp. 201–208. CRC Press, Boca Raton, FL.

Cala, P. M., and Maldonado, H. M. (1994). Regulatory Na/H exchange by *Amphiuma* red blood cells. *J. Gen. Physiol.* **105,** 1035–1054.

Cala, P. M., Mandel, L. J., and Murphy, E. (1986). Volume regulation by *Amphiuma* red blood cells: Cytosolic free Ca and alkali metal-H exchange. *Am. J. Physiol.* **250**, C423–C429.

Canessa, M., Fabry, M. E., Blumenfeld, N., and Nagel, R. L. (1987). Volume-stimulated, Cl⁻ dependent K⁺ efflux is highly expressed in young human red cells containing normal hemoglobin or HbS. *J. Membr. Biol.* **97**, 97–105.

Cantiello, H. F., Stow, J. L., Prat, A. G., and Ausiello, D. A. (1991). Actin filaments regulate Na channel activity. *Am. J. Physiol.* **261**, C882–C888.

Careaga-Houck, M., and Sprecher, H. (1990). Effects of a fish oil diet on the metabolism of endogenous (n-6) and (n-3) fatty acids in rat neutrophils. *Biochim. Biophys. Acta* **1947**, 29–34.

Chamberlin, M. E., and Strange, K. (1989). Anisosmotic cell volume regulation: A comparative view. *Am. J. Physiol.* **257**, C159–C173.

Chan, C.-C., Nicholson, D. W., Metters, K. M., Pon, D. J., and Rodger, I. W. (1994). Leukotriene D₄-induced increases in cytosolic calcium in THP-1 cells: Dependence on extracellular calcium and inhibition with selective leukotriene D₄ receptor antagonists. *J. Pharmacol. Exp. Ther.* **269**, 891–904.

Chan, H. C., Goldstein, J., and Nelson, D. J. (1992). Alternate pathways for chloride conductance activation in normal and cystic fibrosis airway epithelial cells. *Am. J. Physiol.* **269**, C1273–C1283.

Chen, P. Y., and Verkman, A. S. (1988). Sodium-dependent chloride transport in basolateral membrane vesicles isolated from rabbit proximal tubule. *Biochemistry* **27**, 655–660.

Cheung, R. K., Grinstein, S., Dosch, H.-M., and Gelfand, E. W. (1982). Volume regulation by human lymphocytes: Characterization of the ionic basis for regulatory volume decrease. *J. Cell. Physiol.* **112**, 189–196.

Chipperfield, A. R. (1986). The (Na⁺-K⁺-Cl⁻) co-transport system. *Clin. Sci.* **71**, 465–476.

Choi, D. W. (1988). Glutamate neurotoxicity and diseases of the central nervous system. *Neuron* **1**, 623–634.

Choi, D. W., and Rothman, S. M. (1990). The role of glutamate neurotoxicity in hypoxic ischemic neuronal death. *Annu. Rev. Neurosci.* **13**, 171–182.

Christensen, H. N., Hess, N., and Riggs, T. R. (1954). Concentration of taurine, β-alanine and triiodothyronine by ascites tumor carcinoma cells. *Cancer Res.* **13**, 124–127.

Christensen, O. (1987). Mediation of cell volume regulation by Ca²⁺ influx through stretch-activated channels. *Nature (London)* **330**, 66–68.

Christensen, O., and Hoffmann, E. K. (1992). Cell swelling activates K⁺- and Cl⁻-channels as well as nonselective stretch-activated cation channels in Ehrlich ascites tumor cells. *J. Membr. Biol.* **129**, 13–36.

Christensen, O., Simon, M., and Randlev, T. (1989). Anion channels in a leaky epithelium. A patch clamp study of choroid plexus. *Pfluegers Arch.* **415**, 37–46.

Christensen, S., Hoffmann, E. K., Saermark, T., and Simonsen, L. O. (1988). Inositol trisphosphate may be a second messenger in regulatory volume decrease in Ehrlich mouse ascites-tumour cells. *J. Physiol. (London)* **403**, 109P.

Christophersen, P. (1991). The Ca²⁺-activated K⁺ channel from human erythrocyte membranes: Single channel rectification and selectivity. *J. Membr. Biol.* **119**, 75–83.

Civan, M. M., Coca-Prados, M., and Peterson-Yantorno, K. (1994). Pathways signaling the regulatory volume decrease of cultured nonpigmented ciliary epithelial cells. *Invest. Ophthalmol. Visual Sci.* **35**, 2876–2886.

Cohen, D. M., and Gullans, S. R. (1994). Transcription factors and stress protein in renal epithelial cell adaptation to hyperosmotic stress. *In* "Cellular and Molecular Physiology of Cell Volume Regulation" (K. Strange, ed.), pp. 363–372. CRC Press, Boca Raton, Florida.

Colclasure, G. C., and Parker, J. C. (1992). Cytoplasmic protein concentration is the primary volume signal for swelling induced [K-Cl] cotransport in dog red cells. *J. Gen. Physiol.* **100**, 1–10.

Collins, K. D., and Washabaugh, M. W. (1985). The Hofmeister effect and the behaviour of water at interfaces. *Q. Rev. Biophys.* **18**, 323–422.

Cooper, K. E., Tang, J. M., Rae, J. L., and Eisenberg, R. S. (1986). A cation channel in frog lens epithelia responsive to pressure and calcium. *J. Membr. Biol.* **93**, 259–269.

Corasanti, J. G., Gleeson, D., and Boyer, J. L. (1990). Effects of osmotic stress on isolated rat hepatocytes. I. Ionic mechanisms of cell volume regulation. *Am. J. Physiol.* **258**, G290–298.

Corcia, C., and Armstrong, W. M. (1983). KCl cotransport: A mechanism for basolateral chloride exit in Necturus gallbladder. *J. Membr. Biol.* **76**, 173–182.

Cornet, M., Delpire, E., and Gilles, R. (1987). Study of microfilaments network during volume regulation process of cultured PC12 cells. *Pfluegers Arch.* **410**, 223–225.

Cornet, M., Delpire, E., and Gilles, R. (1988). Relations between cell volume control, microfilaments and microtubules networks in T2 and PC12 cultured cells. *J. Physiol.* (*London*) **83**, 43–49.

Cornet, M., Lambert, I. H., and Hoffmann, E. K. (1993a). Relation between cytoskeleton, hypo-osmotic treatment and volume regulation in Ehrlich ascites tumor cells. *J. Membr. Biol.* **131**, 55–66.

Cornet, M., Ubl, J., and Kolb, H. A. (1993b). Cytoskeleton and ion movements during volume regulation in cultured PC12 cells. *J. Membr. Biol.* **133**, 161–170.

Coulombe, A., and Coraboeuf, E. (1992). Large conductance chloride channels of newborn rat cardiac myocytes are activated by hypotonic media. *Phluegers Arch.* **422**, 143–150.

Cowley, B. D., Ferraris, J. D., Carper, D., and Burg, M. B. (1990). In vivo osmotic regulation of aldose reductase mRNA, protein and sorbitol content in rat renal medulla. *Am. J. Physiol.* **258**, F154–F161.

Cristol, J. P., Provencal, B., and Sirois, P. (1989). Leukotriene receptors. *J. Recep. Res.* **9**, 341–367.

Crooke, S. T., Mattern, M., Sarau, H. M., Winkler, J. D., Balcarek, J., Wong, A., and Bennett, C. D. (1989). The signal transduction system of the leukotriene D_4 receptor. *Trends Physiol. Sci.* **10**, 103–107.

Dahl, S. V., Hallbrucker, C., Lang, F., and Häussinger, D. (1991). Role of eicosanoids, inositol phosphates and extracellular Ca^{2+} in cell-volume regulation of rat liver. *Eur. J. Biochem.* **198**, 73–83.

Dascula, A., Nevo, Z., and Korenstein, R. (1992). Hyperosmotic activation of the Na^+-H^+ exchanger in a rat bone cell line: temperature dependence and activation pathways. *J. Physiol.* **456**, 503–518.

Davis, C. W., and Finn, A. L. (1985). Cell volume regulation in frog urinary bladder. *Fed. Proc. Fed. Am. Soc. Exp. Biol.* **44**, 2520–2525.

Davis, C. W., and Finn, A. L. (1987). Interactions of sodium transport, cell volume, and calcium in frog urinary bladder. *J. Gen. Physiol.* **89**, 687–702.

Davis, B. A., Hogan, E. M., and Boron, W. F. (1992). Role of G proteins in stimulation of Na-H exchange by cell shrinkage. *Am. J. Physiol.* **262**, C533–C536.

Delpire, E., and Lauf, P. K. (1991). Magnesium and ATP dependence of K-Cl cotransport in low K^+ sheep red blood cells. *J. Physiol.* (*London*) **441**, 219–231.

Delpire, E., Rauchman, M. I., Beier, D. R., Hebert, S. C., and Gullans, S. R. (1994). Molecular cloning and chromosome localization of a putative basolateral Na^+-K^+-$2Cl^-$ cotransporter from mouse inner medullary collecting duct (mIMCD-3) cells. *J. Biol. Chem.* **269**, 25677–25683.

Deutsch, C., and Chen, L. Q. (1993). Heterologous expression of specific K^+ channels in T lymphocytes: Functional consequences for volume regulation. *Proc. Natl. Acad. Sci. U.S.A.* **90**, 10036–10040.

Dho, S., Grinstein, S., Cichon, G., Covey, E. J., Su, W.-G., and Pace-Asciak, C. R. (1990). Hepoxilin A_3 induces changes in cytosolic calcium, intracellular pH and membranopotential in human neutrophils. *Biochem. J.* **266**, 63–68.

Diener, M., and Scharrer, E. (1993). The leukotriene D_4 receptor blocker. SK&F 104353, inhibits volume regulation in isolated crypts from the rat distal colon. *Eur. J. Pharmacol.* **238,** 217–222.

Diener, M., Nobles, M., and Rummel, W. (1992). Activation of basolateral Cl channels in the rat colonic epithelium during regulatory volume decrease. *Phluegers Archiv. European J. Physiol.* **421(6),** 530–538.

DiStefano, A., Wittner, M., Schlatter, E., Lang, H. J., Englert, H., and Greger, R. (1985). Diphenylamine-2-carboxylate, a blocker of the Cl⁻-conductive pathway in Cl⁻-transporting epithelia. *Pfluegers Arch.* **405,** Suppl. 1, S95–S100.

Doroshenko, P. (1991). Second messengers mediating activation of chloride current by intracellular GTPγS in bovine chromaffin cells. *J. Physiol. (London)* **436,** 725–738.

Doroshenko, P., and Neher, E. (1992). Volume-sensitive chloride conductance in bovine chromaffin cell membrane. *J. Physiol. (London)* **449,** 197–218.

Drewnowska, K., and Baumgarten, C. M. (1991). Regulation of cellular volume in rabbit ventricular myocytes: Bumetanaide, chlorothiazide, and ouabain. *Am. J. Physiol.* **260,** C122–C131.

Duhm, J. (1987). Furosemide-sensitive K^+ (Rb^+) transport in human erythrocytes: modes of operation, dependence of extracellular and intracellular Na^+, kinetics, pH dependence and the effect of cell volume and N-ethylmaleimide. *J. Membr. Biol.* **98,** 15–32.

Duhm, J., and Göbel, B. O. (1984). Na^+-K^+ transport and volume of rat erythrocytes under dietary K^+ deficiency. *Am. J. Physiol.* **246,** C20–C29.

Duncan, R., and Misler, S. (1989). Voltage-activated and stretch activated Ba conducting channels in an osteoblast-like cell line (umR 106). *FEBS Lett.* **251,** 17–21.

Dunham, P. B. (1995). Effects of urea on K-Cl cotransport in sheep red blood cells: evidence for two signals of swelling. *Am. J. Physiol.* **268,** C1026–C1032.

Dunham, P. B., and Ellory, J. C. (1981). Passive potassium transport in low potassium sheep red cells: Dependence upon cell volume and chloride. *J. Physiol. (London)* **318,** 511–530.

Dunham, P. B., Jessen, F., and Hoffmann, E. K. (1990). Inhibition of Na-K-Cl cotransport in Ehrlich ascites cells by antiserum against purified proteins of the cotransporter. *Proc. Natl. Acad. Sci. U.S.A.* **877,** 6828–6832.

Dunham, P. B., Klimczak, J., and Logue, P. J. (1993). Swelling activation of K-Cl cotransport in LK sheep erythrocytes: A three-state process. *J. Gen. Physiol.* **101,** 733–765.

Dürr, J. E., and Larsen, E. H. (1986). Indacrinone (MK-196)—a specific inhibitor of the voltage-dependent Cl⁻ permeability in toad skin. *Acta Physiol. Scand.* **127,** 145–153.

Eaton, W. A., and Hofrichter, J. (1987). Hemoglobin S gelation and sickle cell disease. *Blood* **70,** 1245–1266.

Eveloff, J. L., and Warnock, D. G. (1987). Activation of ion transport systems during cell volume regulation. *Am. J. Physiol.* **252,** F1–F10.

Falke, L. C., and Misler, S. (1989). Activity of ion channels during volume regulation by clonal N1E115 neuroblastoma cells. *Proc. Natl. Acad. Sci. U.S.A.* **86,** 3919–3923.

Feinstein, M. B., and Sha'afi R. I. (1983). Role of calcium in arachidonic acid metabolism and in the actions of arachidonic acid-derived metabolites. *In* "Calcium and Cell Function" (W. Y. Cheung, ed.), Vol. 4, pp. 337–376. Academic Press, New York.

Feit, P. W., Hoffmann, E. K., Schiødt, M., Kristensen, P., Jessen, F., and Dunham, P. B. (1988). Purification of proteins of the Na/Cl cotransporter from membranes of Ehrlich ascites cells using a bumetanice-Sepharose affinity column. *J. Membr. Biol.* **103,** 135–147.

Filipović, D., and Sackin, H. (1991a). A calcium-permeable stretch-activated cation channel in renal proximal tubule. *Am. J. Physiol.* **260,** F119–F129.

Filipović, D., and Sackin, H. (1991b). Strech and volume activated channels in isolated proximal tubule cells. *Am. J. Physiol.* **262,** F857–F870.

Fliegel, L., and Frøhlich, O. (1993). The Na^+/H^+ exchanger: An update on structure, regulation, and cardiac physiology. *Biochem. J.* **296,** 273–285.

Force, T., and Bonventre, J. V. (1994). Cellular signal transduction mechanisms. *In* "Cellular and Molecular Physiology of Cell Volume Regulation" (K. Strange, ed.), pp. 147–180. CRC Press, Boca Raton, Florida.

Ford-Hutchinson, A. W. (1991). Inhibition of leukotriene biosynthesis. *Ann. N.Y. Acad. Sci.* **629,** 133–142.

Ford-Hutchinson, A. W. (1994). Leukotriene C4 syntase and 5-lipoxygenase activating protein. Regulators of the biosythesis of sulfido-leukotrienes. *Ann. N.Y. Acad. Sci.* **744,** 78–83.

Foskett, J. K. (1990). Ca^{2+} modulation of Cl^- content controls cell volume in single solivary acinar cells during fluid secretion. *Am. J. Physiol.* **359,** C994–C1004.

Foskett, J. K. (1994). The role of calcium in the control of volume regulatory transport pathways. *In* "Cellular and Molecular Physiology of Cell Volume Regulation" (K. Strange, ed.), pp. 259–277. CRC Press, Boca Raton, FL.

Foskett, J. K., and Melvin, J. E. (1989). Activation of salivary secretion: Coupling of cell volume and $[Ca^{2+}]^i$ in single cells. *Science* **244,** 1582–1585.

Foskett, J. K., and Spring, K. R. (1985). Involvement of calcium and cytoskeleton in gallbladder epithelial cell volume regulation. *Am. J. Physiol.* **248,** C27–C36.

Foskett, J. K., Wong, M. M. M., Sue-A-Quan, G., and Robertson, M. A. (1994). Isomoticmodulation of cell volume and intracellular ion activities during stimulation of single exocrine cells. *J. Exp. Zool.* **268,** 104–110.

Freeman, C. J., Bookchin, R. M., Ortiz, O. E., and Lew, V. L. (1987). K-permeabilized human red cells lose an alkaline, hypertonic fluid containing excess K^+ over diffusible anions. *J. Membr. Biol.* **96,** 235–241.

Frelin, C., Chassande, O., and Lazdunski, M. (1986). Biochemical characterization of the $Na^+/K^+/Cl^-$ cotransport in chick cardiac cells. *Biochem. Biophys. Res. Commun.* **134,** 326–331.

Friedman, J. E., and Haddad, G. G. (1993). Major differences in Ca^{2+}_0 response to anoxia between neonatal and adult rat Ca1 neurons: Role of Ca^{2+}_0 and Na^+_0. *J. Neurosci.* **13,** 63–72.

Friedrich, F., Paulmilch, M., Kolb, H. A., and Lang, F. (1988). Inward rectifier K channel in renal epithelial cells (MDCK) activated by serotonin. *J. Membr. Biol.* **106,** 149–155.

Frizzell, R. A., Rechkemmer, G., and Shoemaker, R. L. (1986). Altered regulation of airway epithelial cell chloride channels in cystic fibrosis. *Science* **233,** 558–560.

Fugelli, K., and Rohrs, H. (1980). The effect of Na^+ and osmolality on the influx and steady state distribution of taurine and gamma-aminobutyric acid in flounder (*Platichthys flesus*) erythrocytes. *Comp. Biochem. Physiol. A* **67A,** 545–551.

Fugelli, K., and Thoroed, S. M. (1986). Taurine transport associated with cell volume regulation in flounder erythrocytes under anisosmotic conditions. *J. Physiol. (London)* **374,** 245–261.

Fugelli, K., and Zachariassen, K. E. (1976). The distribution of taurine, gamma-aminobutyric acid and inorganic ions between plasma and erythrocytes in flounder (*Platichtys flesus*) at different plasma osmolarities. *Comp. Biochem. Physiol.* **55A,** 173–177.

Furlong, T. J., and Spring, K. R. (1988). Volume-regulatory decrease by *Necturus,* gall bladder epithelium; basolateral exit of KCl by conductive pathways. *J. Gen. Physiol.* **92,** 27a.

Galkin, A. A., and Khodorov, B. I. (1988). The involvement of furosemid-sensitive ion counter transport system in the autoregulation of macrophage cell volume: Role of the cytoskeleton. *Biol. Membr.* **5,** 302–307.

Gamba, G., Miyanoshita, A., Lombardi, M., Lytton, J., Lee, W.-S., Hediger, M. A., and Hebert, S. C. (1994). Molecular cloning, primary structure and characterization of two members of the mammalian electroneutral sodium-(potassium)-chloride cotransporter family expressed in kidney. *J. Biol. Chem.* **269,** 17713–17722.

Garcia, J. J., Sánchez-Olea, R. S., and Pasantes-Morales, H. (1991). Taurine release associated to volume regulation in rabbit lymphocytes. *J. Cell. Biochem.* **45,** 207–212.

Garcia-Pérez, A., and Burg, M. B. (1991). Renal medullary organic osmolytes. *Physiol. Rev.* **71,** 1081.

Garcia-Romeu, F., Cossins, A. R., and Motais, R. (1991). Cell volume regulation by trout erythrocytes: Characteristics of the transport systems activated by hypotonic swelling. *J. Physiol. (London)* **440**, 547–567.

Gargus, J. J., and Slayman, C. W. (1980). Mechanism and role of furosemide-sensitive K⁺ transport in L cells: A genetic approach. *J. Membr. Biol.* **52**, 245–256.

Garlid, K. (1978). Unmasking the mitochondrial K/H exchanger. Swelling-induced K⁺ loss. *Biophys. Res. Comm.* **83**, 1450–1455.

Gazzola, G. C., Dall'Asta, V., Nucci, F. A., Rossi, P. A., Bussolati, O., Hoffmann, E. K., and Guidotti, G. G. (1991). Role of amino acid transport system A in the control of cell volume in cultured human fibroblasts. *Cell. Physiol. Biochem.* **1**, 131–142.

Geck, P. (1990). Volume regulation in Ehrlich cells. *In* "Cell Volume Regulation" (K. W. Beyenbach, ed.), pp. 26–59. Karger, Basel.

Geck, P., and Heinz, E. (1986). The Na-K-2Cl cotransport system. *J. Membr. Biol.* **91**, 97–105.

Geck, P., and Pfeiffer, B. (1985). Na⁺ + K⁺ + 2Cl⁻ cotransport in animal cells—Its role in volume regulation. *Ann. N.Y. Acad. Sci.* **456**, 166–182.

Gilles, R. (1983). Volume maintenance and regulation in animal cells: Some features and trends. *Mol. Physiol.* **4**, 3–16.

Gilles, R. (1987). Volume regulation in cells of eurohaline invertebrates. *Curr. Top. Membr. Transp.* **30**, 205–247.

Gilles, R., Hoffmann, E. K., and Bolis, L. eds. (1991). "Advances in Comparative and Environmental Physiology," Vol. 9. Springer Verlag, Berlin.

Glenney, J. R., Bretscher, A., and Weber, K. (1980). Calcium control of the intestinal microvillus cytoskeleton: Its implication for the regulation of microfilament organization. *Proc. Natl. Acad. Sci. U.S.A.* **77**, 6458–6462.

Goldstein, L., and Brill, S. R. (1991). Volume-activated taurine efflux from skate erythrocytes: Possible band 3 involvement. *Am. J. Physiol.* **260**, R1014–R1020.

Goldstein, L., and Kleinzeller, A. (1987). Cell volume regulation in lower vertebrates. *Curr. Top. Membr. Transp.* **30**, 181–204.

Goldstein, L., Brill, S. R., and Freund, E. V. (1990). Activation of taurine efflux in hypotonically stressed elasmobranch cells: Inhibition by disulfonates. *J. Exp. Zool.* **254**, 114–118.

Grassl, S. M., Holohan, P. D., and Ross, C. R. (1987). Cl⁻-HCO₃⁻ exchange in rat renal basolateral membrane vesicles. *J. Biol. Chem.* **262**, 2682–2687.

Greger, R., and Schlatter, E. (1983). Properties of the basolateral membrane of the cortical thick ascending limb of Henle's loop of rabbit kidney. A model for secondary active chloride transport. *Pfluegers Arch.* **396**, 325–334.

Greger, R., Gögelein, H., and Schlatter, E. (1988). Stimulation of NaCl secretion in the rectal gland of the dogfish *Squalus acanthias. Comp. Biochem. Physiol. A* **90A**, 733–739.

Grinstein, S., and Foskett, K. J. (1990). Ionic mechanisms of cell volume regulation in leucocytes. *Annu. Rev. Physiol.* **52**, 399–414.

Grinstein, S., and Rothstein, A. (1986). Mechanisms of regulation of the Na⁺/H⁺ exchanger. *J. Membr. Biol.* **90**, 1–12.

Grinstein, S., and Smith, J. D. (1989). Calcium induces charybdotoxin-sensitive membrane potential changes in rat lymphocytes. *Am. J. Physiol.* **22**, C197–C206.

Grinstein, S., and Smith, J. D. (1990). Calcium-independent cell volume regulation in human lymphocytes. Inhibition by charybdotoxin. *J. Gen. Physiol.* **95**, 97–120.

Grinstein, S., DuPré, A., and Rothstein, A. (1982a). Volume regulation by human lymphocytes. Role of calcium. *J. Gen. Physiol.* **79**, 849–868.

Grinstein, S., Clarke, C. A., and Rothstein, A. (1982b). Increased anion permeability during volume regulation in human lymphocytes. *Philos. Trans. R. Soc. London, Ser. B* **299**, 509–518.

Grinstein, S., Clarke, C. A., DuPré, A., and Rothstein, A. (1982c). Volume-induced increase of anion permeability in human lymphocytes. *J. Gen. Physiol.* **80**, 801–823.

Grinstein, S., Cohen, S., Sarkadi, B., and Rothstein, A. (1983). Induction of [86Rb] fluxes by Ca^{2+} and volume changes in thymocytes and their isolated membranes. *J. Cell. Physiol.* **116**, 352–362.

Grinstein, S., Rothstein, A., Sarkadi, B., and Gelfand, E. W. (1984). Responses of lymphocytes to anisotonic media: Volume-regulating behavior. *Am. J. Physiol.* **246**, C204–C215.

Grinstein, S., Goetz, J. D., Cohen, S., Rothstein, A., and Gelfand, E. W. (1985). Regulation of Na^+/H^+ exchange in lymphocytes. *Ann N.Y. Acad. Sci.* **456**, 207–219.

Grinstein, S., Mack, E., and Mills, G. B. (1986). Osmotic activation of the Na^+/H^+ antiport in protein kinase C-depleted lymphocytes. *Biochem. Biophys. Res. Commun.* **134**, 8–13.

Grinstein, S., Furuya, W., and Bianchini, L. (1992a). Protein kinases, phosphatases, and the control of cell volume. *News Physiol. Sci.* **7**, 232–237.

Grinstein, S., Woodside, M., Sardet, C., Pouysségur, J., and Rotin, D. (1992b). Activation of the Na^{pl}/H^+ antiporter during cell volume regulation. Evidence for a phosphorylation-independent mechanism. *J. Biol. Chem.* **267**, 23823–23828.

Grinstein, S., Woodside, M., Waddel, T. K., Downey, G. P., Orlowski, J., Pousségur, J., Wong, D. C. P., and Foskett, J. K. (1993). Focal localisation of the NHE-1 isoform of the Na^+/H^+ antiport: Assessment of effects on intracellular pH. *EMBO J.* **12**, 5209–5218.

Grinstein, S., Woodside, M., Goss, G. G., and Kapus, A. (1994). Osmotic activation of the Na^+/H^+ antiporter during volume regulation. *Biochem. Soc. Trans.* **22**, 512–516.

Gründer, S., Thiemann, A., Pusch, M., and Jentsch, T. J. (1992). Regions involved in the opening of the ClC-2 chloride channel by voltage and cell volume. *Nature (London)* **360**, 759–762.

Grygorczyk, R., Schwarz, W., and Passow, H. (1984). Ca^{2+}-activated K^+ channels in human red cells. Comparison of single-channel currents with ion fluxes. *Biophys. J.* **45**, 693–698.

Guggino, W. B. (1986). Functional heterogeneity in the early distal tubule of the Amphiuma kidney: Evidence for two modes of Cl^- and K^+ transport across the basolateral cell membrane. *Am. J. Physiol.* **250**, F430–F440.

Guharay, F., and Sachs, F. (1984). Stretch-activated single ion channel currents in tissue-cultured embryonic chick skeletal muscle. *J. Physiol. (London)* **352**, 685–701.

Gunn, R. B. (1992). Anion exchange mechanism of band 3 and related proteins. *Membr. Transp. Biol.* **5**, 233–261.

Haas, M. (1989). Properties and diversity of (Na-K-Cl) cotransporters. *Annu. Rev. Physiol.* **51**, 443–457.

Haas, M. (1994). The Na-K-Cl cotransporters. *Am. J. Physiol.* **267**, C869–C885.

Haas, M., and McManus, T. J. (1983). Bumetanide inhibits (Na + K + 2Cl) co-transport at a chloride site *Am. J. Physiol.* **245**, C235–C240.

Haas, M., and McManus, T. J. (1985). Effect of norepinephrine on swelling-induced potassium transport in duck red cells. Evidence against a volume regulatory decrease under physiological conditions. *J. Gen. Physiol.* **85**, 649–667.

Hall, A. C., and Ellory, J. C. (1986). Evidence for the presence of volume-sensitive KCl cotransport in young human red cells. *Biochim. Biophys. Acta* **858**, 317–320.

Hamill, O. P. (1983). Potassium and chloride channels in red blood cells. In "Single-Channel Recording" (B. Sakmann and E. Neher, eds.), pp. 451–471. Plenum, New York.

Han, E. S., Altenberg, G. A., and Reuss, L. (1994). Substrate-transport does not prevent P-glycoprotein-associated Cl^- currents activated by cell swelling. *Biophys. J.* **66**, A99.

Harbak, H., and Simonsen, L. O. (1993). Cell shrinkage couples agonist-induced receptor stimulation to activation of the Na^+-K^+-$2Cl^-$ co-transport system in Ehrlich mouse ascites tumour cells. *J. Physiol. (London)* **467**, 334P.

Harbak, H., and Simonsen, L. O. (1995). The K^+ channels activated during regulatory volume decrease (RVD) are distinct from those activated by Ca^{2+}-mobilizing agonists in Ehrlich mouse ascites tumour cells. *J. Physiol. (London)* (in press).

Häussinger, D., and Lang, F. (1991). Regulation of cell volume in the hepatic function: A mechanism for metabolic control. *Biochim. Biophys. Acta* **1071**, 331–350.

Häussinger, D., Gerok, W., and Lang, F. (1993). Cell volume and hepertic metabolism. *Adv. Comp. Environ. Physiol.* **14,** 33–67.

Häussinger, D., Newsome, W., von Dahl, S., Stoll, B., Noe, B., Schreiber, R., Wettstein, M., and Lang, F. (1994). The role of cell volume changes in metabolic regulation. *Biochem. Soc. Trans.* **22,** 497–502.

Hazama, A., and Okada, Y. (1988). Ca^{2+} sensitivity of volume-regulatory K^+ and Cl^- channels in cultured human epithelial cells. *J. Physiol. (London)* **402,** 687–702.

Hazama, A., and Okada, Y. (1990). Biphasic rises in cytosolic free Ca^{2+} in association with activation of K^+ and Cl^- conductance during the regulatory volume decrease in cultured human epithelial cells. *Pflüegers Arch.* **416,** 710–714.

Hegde, R. A., and Palfrey, H. C. (1992). Ionic effects on bumetamide binding to the activated Na/K/2Cl cotransporter: Selectivity and kinetic propertioes of ion binding sites. *J. Membr. Biol.* **126,** 27–37.

Hendil, K. B., and Hoffmann, E. K. (1974). Cell volume regulation in Ehrlich ascites tumor cells. *J. Cell. Physiol.* **84,** 115–125.

Henson, J. H., and Schatten, G. (1983). Calcium regulation of the actin-mediated cytoskeletal transformation of sea urchin coelomocytes. *Cell Motil.* **3,** 525–534.

Herbert, J. M., Augereau, J. M., Gleye, J., and Maffrand, J. P. (1990). Chelerythrine is a potent and specific inhibitor of protein kinase C. *Biochem. Biophys. Res. Commun.* **172,** 993–999.

Hladky, S. B., and Rink, T. J. (1977) pH equilibrium across the red cell membrane. *In* "Membrane Transport in Red Cells" (J. C. Ellory and V. L. Lev, eds.), pp. 115–135. Academic Press, London.

Hoffmann, E. K. (1977). Control of cell volume. *In* "Transport of Ions and Water Animals" (B. Gupta, J. Oschmann, and B. Wall, eds.), pp. 285–333. Academic Press, New York.

Hoffmann, E. K. (1978). Regulation of cell volume by selective changes in the leak permeabilities of Ehrlich ascites tumor cells. *Alfred Benzon Symp.* **11,** 397–417.

Hoffmann, E. K. (1983). Volume regulation by animal cells. *Semin. Ser.—Soc. Exp. Biol.* **17,** 55–80.

Hoffmann, E. K. (1985). Role of separate K^+ and Cl^- channels and of Na^+/Cl^-cotransport in volume regulation in Ehrlich cells. *Fed. Proc. Fed. Am. Soc. Exp. Biol.* **44,** 2513–2519.

Hoffmann, E. K. (1993). Control of volume regulatory ion transport processes in mammalian cell: Signalling by secondary messengers. *Alfred Benzon Symp.* **34,** 273–294.

Hoffmann, E. K., and Hendil, K. B. (1976). The role of amino acids and taurine in isosmotic intracellular regulation in Ehrlich ascites mouse tumour cells. *J. Comp. Physiol.* **108,** 279–286.

Hoffmann, E. K., and Kolb, A. (1991). The mechanisms of activation of regulatory volume responses after cell swelling. *Adv. Comp. Environ. Physiol.* **9,** 140–185.

Hoffmann, E. K., and Lambert, I. H. (1983). Amino acid transport and cell volume regulation in Ehrlich ascites tumour cells. *J. Physiol. (London)* **338,** 613–625.

Hoffmann, E. K., and Simonsen, L. O. (1989). Membrane mechanisms in volume and pH regulation in vertebrate cells. *Physiol. Rev.* **69,** 315–382.

Hoffmann, E. K., and Ussing, H. H. (1992). Membrane mechanisms in volume regulation in vertebrate cells and epithelia. *In* "Membrane Transport in Biology" (G. H. Giebisch, J. A. Schafer, H. H. Ussing, and P. Kristensen, eds.), Vol. 5, pp. 317–399. Springer-Verlag, Heidelberg.

Hoffmann, E. K., Sjöholm, C., and Simonsen, L. O. (1983). Na^+, Cl^- co-transport in Ehrlich ascites tumor cells activated during volume regulation (regulatory volume increase). *J. Membr. Biol.* **76,** 269–280.

Hoffmann, E. K., Simonsen, L. O., and Lambert, I. H. (1984). Volume-induced increase of K^+ and Cl^- permeabilities in Ehrlich ascites tumor cells. Role of internal Ca^{2+}. *J. Membr. Biol.* **78,** 211–222.

Hoffmann, E. K., Lambert, I. H., and Simonsen, L. O. (1986). Separate, Ca^{2+}-activated K^+ and Cl^- transport pathways in Ehrlich ascites tumor cells. *J. Membr. Biol.* **91,** 227–244.

Hoffmann, E. K., Lambert, I. H., and Simonsen, L. O. (1988). Mechanisms in volume regulation in Ehrlich ascites tumor cells. *Renal Physiol. Biochem.* **11,** 221–247.

Hoffmann, E. K., Simonsen, L. O., and Lambert, I. H. (1993). Cell volume regulation: Intracellular transmission. *In* "Interaction Cell Volume and Cell Function" (F. Lang and D. Häussinger, eds.), pp. 187–48. Springer-Verlag, Berlin.

Hoffmann, E. K., Jessen, F., and Dunham, P. B. (1994). The Na-K-2Cl cotransporter is in a permanently activated state in cytoplasts from Ehrlich ascites tumor cells. *J. Membr. Biol.* **138**, 229–239.

Hosoya, N., Mitsui, M., Yazama, F., Ishihara, H., Ozaki, H., Karaki, H., Hartshorne, D. J., and Mohri, H. (1993). Changes in the cytoskeletal structure of cultured smooth muscle cells induced by calyculin-A. *J. Cell Sci.* **105**, 883–890.

Hudson, R. L., and Schultz, S. G. (1988). Sodium-coupled glycine uptake by Ehrlich ascites tumor cells results in an increase in cell volume and plasma membrane channel activities. *Proc. Natl. Acad. Sci. U.S.A.* **85**, 279–283.

Hunter, M. (1990). Stretch-activated channels in the basolateral membrane of single proximal cells of frog kidney. *Pfluegers Arch.* **416**, 448–453.

Huxtable, R. J. (1992). The physiological actions of taurine. *Physiol. Rev.* **72**, 101–163.

Irvine, R. F. (1982). How is the level of free arachidonic acid controlled in mammals? *Biochem. J.* **204**, 3–16.

Jackson, P. S., and Strange, K. (1993). Volume-sensitive anion channels mediate swelling-activated inositol and taurine efflux. *Am. J. Physiol.* **265**, C1489–C1500.

Jackson, P. S., and Strange, K. (1994). Properties of VSOAC assessed by noise analysis and single channel measurements. *Annu. Meet. Soc. Gen. Physiol. 48th*, p. 49a.

Jackson, P. S., Morrison, R., and Strange, K. (1994). The volume-sensitive organic osmolyte-anion channel VSOAC is regulated by nonhydrolytic ATP binding. *Am. J. Physiol.* **267**, C1203–C1209.

Jacobs, M. H., and Stewart, D. R. (1942). The role of carbonic anhydrase in certain ionic exchanges involving the erythrocyte. *J. Gen. Physiol.* **25**, 539–552.

Jacobsen, J. G., and Smith, L. H. (1968). Biochemistry and physiology of taurine and taurine derivatives. *Physiol. Rev.* **48**, 424–511.

Jakobsen, L. D., Jensen, B. S., and Hoffmann, E. K. (1994). Regulation of the $Na^+/K^+/2Cl^-$ cotransporter in Ehrlich ascites tumor cells. *Acta Physiol. Scand.* **151**, 27A.

Jennings, M. L., and Al-Rohil, N. (1990). Kinetics of activation and inactivation of swelling-stimulated K^+/Cl^- transport. The volume-sensitive parameter is the rate constant for inactivation. *J. Gen. Physiol.* **95**, 1021–1040.

Jennings, M. L., and Schulz, R. K. (1991). Okadaic acid inhibition of KCl cotransport. Evidence that protein deophosphorylation is necessary for activation of transport by either cell swelling or N-ethylmaleimide. *J. Gen. Physiol.* **97**, 799–818.

Jensen, B. S., Jessen, F., and Hoffmann, E. K. (1993). Na^+, K^+, Cl^- cotransport and its regulation in Ehrlich ascites tumor cells. Ca^{2+}/calmedulin and protein kinase C dependent pathways. *J. Membr. Biol.* **131**, 161–178.

Jessen, F., and Hoffmann, E. K. (1992). Activation of the Na-K-2Cl cotransport system by reorganization of the actin filaments in Ehrlich ascites tumor cells. *Biochim. Biophys. Acta* **1110**, 199–201.

Joiner, C. H. (1993). Cation transport and volume regulation in sickle red blood cells. *Am. J. Physiol.* **264**, C251–C270.

Jørgensen, N. K., Lambert, I. H., and Hoffmann, E. K. (1994). On the role of Ca^{2+} and LTD_4 in regulatory volume decrease in Ehrlich ascites tumour cells. *Acta Physiol. Scand.* **151**, 47A.

Jørgensen, N. K., Lambert, I. H., and Hoffmann, E. K. (1995). Role of LTD_4 in the regulatory volume decrease response in Ehrlich Ascites Tumor Cells. *J. Membr. Biol.* (submitted).

Joyner, S. E., and Kirk, K. (1994). Two pathways for choline transport in eel erythrocytes: A saturable carrier and a volume-activated channel. *Am. J. Physiol.* **267**, R773–R779.

Kaji, D. (1986). Volume-sensitive K transport in human erythrocytes. *J. Gen. Physiol.* **88**, 719–738.

Kaji, D. (1993). Effect of membrane potential on K : Cl cotransport in human erythrocytes. *Am. J. Physiol.* **68**, C376–C382.

Kaji, D. M., and Tsukitani, Y. (1991). Role of protein phosphatase in activation of KCl cotransport in human erythrocytes. *Am. J. Physiol.* **260**, C176–C180.

Kapus, A., Grinstein, S., Wasan, S., Kandasamy, R., and Orlowski, J. (1994). Functional characterization of three isoforms of the Na^+/H^+ exchanger stably expressed in chinese hamster ovary cells. *J. Biol. Chem.* **269**, 1–9.

Kennedy, B. G. (1994). Volume regulation in cultured cells derived from human retinal pigment epithelium. *Am. J. Physiol.* **266**, C676–C683.

Kim, D. (1992). A mechanosensitive K$^+$ channel in heart cells. Activation by arachidonic acid. *J. Gen. Physiol.* **100**, 1021–1040.

Kimelberg, H. K. (1991). Swelling and volume control in brain astroglial cells. *Adv. Comp. Environ. Physiol.* **9**, 81–110.

Kimelberg, H. K., and Frangakis, M. V. (1986). Volume regulation in primary astrocyte cultures. *Adv. Biosci.* **61**, 177–186.

Kimelberg, H. K., and O'Connor, E. R. (1988). Swelling-induced depolarization of astrocyte potentials. *Glia* **1**, 219–224.

Kimelberg, H. K., Goderie, S. K., Higman, S., Pang, S., and Wanievski, R. A. (1990). Swelling-induced release of glutamate, aspartate, and taurine from astrocyte cultures. *J. Neurosci.* **10**, 1583–1591.

King, P. A., and Goldstein, L. (1983). Organic osmolytes and cell volume regulation in fish. *Mol. Physiol.* **4**, 53–66.

Kirk, K., and Kirk, J. (1993). Volume regulatory taurine release from a human lung cancer cell line: Evidence for amino acid transport via a volume-activated chloride channel. *FEBS Lett.* **336**, 153–158.

Kirk, K., and Kirk, J. (1994) Inhibition of volume-regulatory amino acid transport by Cl channel blockers in a human epithelial cell line. *J. Physiol. (London)* **475P**.

Kirk, K., Ellory, J. C., and Young, J. D. (1992). Transport of organic substrates via a volume-activated channel. *J. Biol. Chem.* **267**, 23475–23478.

Kolb, H. A., and Ubl, J. (1987). Activation of anion channels by zymosan particles in membranes of peritoneal macrophages. *Biochim. Biophys. Acta* **899**, 239–246.

Kramhøft, B., Hoffmann, E. K., and Feit, P. W. (1984). The effect of 'loop-diuretics' on a Na, Cl cotransport system activated during regulatory volume increase in Ehrlich ascites tumor cells. *Acta Physiol. Scand.* **121**, P2.

Kramhøft, B., Lambert, I. H., Hoffmann, E. K., and Jørgensen, F. (1986). Activation of Cl-dependent K transport in Ehrlich ascites tumor cells. *Am. J. Physiol.* **251**, C369–C379.

Kramhøft, B., Hoffmann, E. K., and Simonsen, L. O. (1994). pHi regulation in Ehrlich mouse ascites tumor cells: Role of sodium-dependent and sodium-independent chloride-bicarbonate exchange. *J. Membr. Biol.* **138**, 121–132.

Krarup, T., Jensen, B. S., and Hoffmann, E. K. (1994). Occlusion of Rb in the Na/K/2Cl cotransport system of Ehrlich ascites tumor cells. *Acta Physiol. Scand.* **151**, 28A.

Kregenow, F. M. (1971). The response of duck erythrocytes to nonhemolytic hypotonic media: Evidence of a volume controlling mechanism. *J. Gen. Physiol.* **58**, 372–395.

Kregenow, F. M. (1973). The response of duck erythrocytes to norepinephrine and an elevated extracellular potassium: Volume regulation in anisotonic media. *J. Gen. Physiol.* **61**, 509–527.

Kregenow, F. M. (1981). Osmoregulatory salt transporting mechanisms: Control of cell volume in anisotonic media. *Annu. Rev. Physiol.* **43**, 493–505.

Kregenow, F. M., Robbie, D. E., and Orloff, J. (1976). Effect of norepinephrine and hypotertonicity on K influx and cyclic AMP in duck erythrocytes. *Am. J. Physiol.* **231**, 306–312.

Kromphardt, H. (1963). Die aufnahme von taurine in Ehrlich-ascites tumorzellen. *Bichem. Z.* **343**, 283–293.

Kromphardt, H. (1965). Zur pH-Abhängigkeit des Transports neutraler Aminosäuren in Ehrlich-Ascites-Tumorzellen. *Biochem. Z.* **343**, 283–293.

Kubalski, A., Martinac, B., Ling, K.-Y., Adler, J., and Kung, C. (1993). Activities of a mechanosensitive ion channel in an E. coli mutant lacking the major lipoprotein. *J. Membr. Biol.* **131**(3), 151–160.

Kubo, M., and Okada, Y. (1992). Volume-regulatory Cl-channel current in cultured human epithelial cells. *J. Physiol. (London)* **456**, 351–371.

Kunze, H., and Vogt, W. (1971). Significance of phospholipase A for prostaglandin formation. *Ann. N.Y. Acad. Sci.* **180**, 123–125.

Kunzelmann, K., Pavenstadt, H., and Greger, R. (1989). Properties and regulation of chloride channels in cystic fibrosis and normal airway cells. *Pflüegers Arch.* **415,** 172–182.

Kwon, H. M. (1994). Molecular regulation of mammalian osmolyte transporters. *In* "Cellular and Molecular Physiology of Cell Volume Regulation" (K. Strange, ed.), pp. 383–394. CRC Press, Boca Raton, FL.

Lambert, I. H. (1984). Na+-dependent taurine uptake in Ehrlich ascites tumor cells. *Mol. Physiol.* **6,** 233–246.

Lambert, I. H. (1985). Taurine transport in Ehrlich ascites tumour cells. Specificity and chloride dependence. *Mol. Physiol.* **7,** 323–332.

Lambert, I. H. (1987). Activation of the volume- or ionophore A23187 plus Ca^{2+}-induced Cl^- transport pathway in Ehrlich mouse ascites tumour cells is mediated via leukotriene-D4. *J. Physiol. (London)* **394,** 85P.

Lambert, I. H. (1989). Leukotriene-D4 induced cell shrinkage in Ehrlich ascites tumor cells. *J. Membr. Biol.* **108,** 165–176.

Lambert, I. H. (1994). Eicosanoids and cell volume regulation. *In* "Cellular and Molecular Physiology of Cell Volume Regulation" (K. Strange, ed.), pp. 273–292. CRC Press, Boca Raton, FL.

Lambert, I. H., and Hoffmann, E. K. (1991). Regulation of taurine transport in Ehrlich ascites tumor cells. *J. Membr. Biol.* **131,** 67–79.

Lambert, I. H., and Hoffmann, E. K. (1993). Regulation of taurine transport in Ehrlich ascites tumor cells. *J. Membr. Biol.* **131,** 67–79.

Lambert, I. H., and Hoffmann, E. K. (1994). Swelling activates separate taurine and chloride channel in Ehrlich mouse ascites tumor cells. *J. Membr. Biol.* **142,** 289–298.

Lambert, I. H., Simonsen, L. O., and Hoffmann, E. K. (1984). Volume regulation in Ehrlich ascites tumor cells: pH sensitivity of the regulatory decrease, and role of the Ca^{2+}-dependent K^+ channel. *Acta Physiol. Scand.* **120,** 46A.

Lambert, I. H., Hoffmann, E. K., and Christensen, P. (1987). Role of prostaglandins and leukotrienes in volume regulation by Ehrlich ascites tumor cells. *J. Membr. Biol.* **98,** 247–256.

Lambert, I. H., Hoffmann, E. K., and Jørgensen, F. (1989). Membrane potential, anion and cation conductances in Ehrlich asites tumor cells. *J. Membr. Biol.* **111,** 113–132.

Lang, F., Paulmichl, M., Völkl, H., Gstrein, E., and Friedrich, F. (1987). Electrophysiology of cell volume regulation. *In* "Molecular Nephrology: Biochemical Aspects of Kidney Function" (Z. Kovacevic and W. G. Guder, eds.), pp. 133–139. Springer-Verlag, Berlin and New York.

Lang, F., Völkl, H., and Häussinger, D. (1990). General principles in cell volume regulation. *In* "Cell Volume Regulation" (K. W. Beyenbach, ed.), pp. 1–26. Karger, Basel.

Lang, F., Paulmichl, M., Pfeilschifter, J., Friedrich, F., Wöll, E., Waldegger, S., Ritter, M., and Tschernko, E. (1991). Cellular mechanisms of bradykinin-induced hyperpolarization in renal epitheloid MDCK-cells. *Biochim. Biophys. Acta* **1073,** 600–608.

Lansman, J. B., Hallam, T. J., and Rink, T. J. (1987). Single stretch-activated ion channels in vascular endothelial cells as mechanotransducers? *Nature (London)* **325,** 811–813.

Larsen, A. K., Jensen, B. S., and Hoffmann, E. K. (1994). Activation of protein kinase C during cell volume regulation in Ehrlich mouse ascites tumor cells. *Biochim. Biophys. Acta* **1222,** 477–482.

Larson, M., and Spring, K. R. (1987). Volume regulation in epithelia. *Curr. Top. Membr. Transp.* **30,** 105–123.

Lassing, I., and Lindberg, U. (1985). Specific interaction between phosphatidylinositol-4,5-bisphosphate and profilaction. *Nature (London)* **318,** 472–474.

Lau, K. R., Hudson, R. L., and Schultz, S. G. (1984). Cell swelling increases a barium-inhibitable potassium conductance in the basolateral membrane of *Necturus* small intestine. *Proc. Natl. Acad. Sci. U.S.A.* **81,** 3591–3594.

Lauf, P. K. (1985a). On the relationship between volume- and thiol-stimulated K^+Cl^- fluxes in red cell membranes. *Mol. Physiol.* **8,** 215–234.

Lauf, P. K. (1985b). $K^+ : Cl^-$ cotransport: Sulphydryls, divalent cations, and the mechanism of volume activation in a red cell. *J. Membr. Biol.* **88,** 1–13.

Lauf, P. K., and Bauer, J. (1987). Direct evidence for chloride dependent volume reduction in macrocytic sheep reticulocytes. *Biochem. Biophys. Res. Commun.* **144**, 849–855.

Lauf, P. K., McManus, T. J., Haas, M., Forbush, B., III, Duhm, J., Flatman, P. W., Saier, M. H., Jr., and Russell, J. M. (1987). Physiology and biophysics of chloride and cation cotransport across cell membranes. *Fed. Proc., Fed. Am. Soc. Exp. Biol.* **46**, 2377–2394.

Lauf, P. K., Bauer, J., Adragna, N. C., Fujise, H., Zade-Oppen, A. M. M., Ryu, K. H., and Delpire, E. (1992). Erythrocyte K-Cl cotransport: Properties and regulation. *Am. J. Physiol.* **263**, C917–C932.

Lauritzen, L., Hoffmann, E. K., Hansen, H. S., and Jensen, B. (1993). Dietary (n-3) and (n-6) fatty acids are equipotent in stimulating volume regulation in Ehrlich ascites tumor cells. *Am. J. Physiol.* **264**, C109–C117.

Law, R. O. (1991). Amino acids as volume-regulatory osmolytes in mammalian cells. *Comp. Biochem. Physiol. A* **99A**, 263–277.

Law, R. O., and Burg, M. B. (1991). The role of organic osmolytes in the regulation of mammalian cell volume. *Adv. Compar. Environ. Physiol.* **9**, 189–226.

Leite, M. V., and Goldstein, L. (1987). Ca^{2+} ionophore and phorbol ester stimulate taurine efflux from skate erythrocytes. *J. Exp. Zool.* **242**, 95–97.

Levinson, C. (1990). Regulatory volume increase in Ehrlich ascites tumor cells. *Biochim. Biophys. Acta* **1021**, 1–8.

Levinson, C. (1991). Regulatory volume increase in Ehrlich ascites tumor cells is mediated by the 1Na : 1K : 2Cl cotransport system. *FASEB J.* **4**, A564.

Levitan, I. B. (1994). Modulation of ion channels by protein phosphorylation and dephosphorylation. *Annu. Rev. Physiol.* **56**, 193–212.

Lew, V. L., and Bookchin, R. M. (1986). Volume, pH, and ion-content regulation in human red cells: Analysis of transient behavior with an integrated model. *J. Membr. Biol.* **92**, 57–74.

Lewis, R. S., Ross, P. E., and Cahalan, M. D. (1993). Chloride channels activated by osmotic stress in T lymphocytes. *J. Gen. Physiol.* **101**, 801–826.

Lewis, S. A., and de Moura, J. L. C. (1982). Incorporation of cytoplasmic vesicles into apical membrane of mammalian urinary bladder epithelium. *Nature (London)* **297**, 685–688.

Lewis, S. A., and Donaldson, P. (1990). Ion channels and cell volume regulation: Chaos in an organized system. *News Physiol. Sci.* **5**, 112–119.

Li, M., McCann, J. D., Anderson, M. P., Clancy, J. P., Lieddtke, C. M., Nairn, A. C., Greengard, P., and Welsh, M. J. (1989). Regulation of chloride channel by protein kinase C in normal and cystic fibrosis airway epithelia. *Science* **244**, 1353–1356.

Lise, B., and de Rouffinac, C. (1985). Urinary concentrating ability: Insights from comparative anatomy. *Am. J. Physiol.* **249**, R643–R666.

Livne, A., and Hoffmann, E. K. (1990). Cytoplasmic acidification and activation of Na^+/H^+ exchange during regulatory volume decrease in Ehrlich ascites tumour cells. *J. Membr. Biol.* **114**, 153–157.

Livne, A., Grinstein, S., and Rothstein, A. (1987). Volume-regulating behavior of human platelets. *J. Cell. Physiol.* **131**, 354–363.

Lohr, J. W., and Grantham, J. J. (1986). Isovolumetric regulation of isolated S2 proximal tubules in anisotonic media. *J. Clin. Invest.* **78**, 1165–1172.

Lukacs, G. L., and Moczydlowski, E. (1990). A cloride channel from lobster walking leg nerves: characterization of single-channel properties in planar bilayers. *J. Gen. Physiol.* **96**, 707–733.

Luna, E. J., and Hitt, A. L. (1992). Cytoskeleton-plasma membrane interactions. *Science* **258**, 955–964.

Lytle, C., and Forbush, B., III (1992a). Is $[Cl]_i$ the switch controlling Na-K-2Cl cotransport in shark rectal gland? *Biophys. J.* **61**, A384.

Lytle, C., and Forbush, B., III (1992b). The Na-K-Cl cotransport protein in shark rectal gland. II. Regulation by direct phosphorylation. *J. Biol. Chem.* **267**, 25438–25443.

Lytle, C., and McManus, T. J. (1986). A minimal kinetic model of (Na + K + 2Cl) cotransport with ordered binding and glide symmetry. *J. Gen. Physiol.* **88**, 36a.

Lytle, C., Xu, J.-C., Biemesderfer, D., Haas, M., and Forbush, B., III (1992). The Na-K-Cl cotransport protein of shark rectal gland. I. Development of monoclonal antibodies, immunoaffinity purification, and partial biochemical characterization. *J. Biol. Chem.* **267,** 25428–25437.

Macey, R. I. (1984). Transport of water and urea in red blood cells. *Am. J. Physiol.* **246,** C195–C203.

Mac Lean-Fletcher, S. D., and Pollard, T. D. (1980). Mechanism of action of cytochalasin B on actin. *Cell* **20,** 329–341.

MacLeod, R. J. (1994). How an epithelial cell swells in a determinant of the signaling pathways that activate RVD. *In* "Cellular and Molecular Physiology of Cell Volume Regulation" (K. Strange, ed.), pp. 191–200. CRC Press, Boca Raton, FL.

MacLeod, R. J., and Hamilton, J. R. (1990). Regulatory volume increase in isolated mammalian jejunal villus is due to bumetanide sensitive NaKCl cotransport. *Am. J. Physiol.* **258,** G665–G674.

MacLeod, R. J., and Hamilton, J. R. (1991). Separate K and Cl transport pathways are activated for regulatory volume decrease in jejunal villus cells. *Am. J. Physiol.* **260,** G405–G415.

MacLeod, R. J., Lembessis, P., and Hamilton, J. R. (1992a). Effect of protein kinase C inhibitors on Cl$^-$ conductance required for volume regulation after L-alanine cotransport. *Am. J. Physiol.* **262,** C950–C955.

MacLeod, R. J., Lembessis, P., and Hamilton, J. R. (1992b). Differences in Ca^{2+}-mediation of hypotonic and Na$^+$-nutrient regulatory volume decrease in suspensions of jejunal enterocytes. *J. Membr. Biol.* **130,** 23–31.

MacRobbie, E. A. C., and Ussing, H. H. (1961). Osmotic behavior of the epithelial cells of frog skin. *Acta Physiol. Scand.* **53,** 348–365.

Mahnensmith, R. L., and Aronson, P. S. (1985). The plasma membrane sodium-hydrogen exchanger and its role in physiological and pathophysiological processes. *Circ. Res.* **56,** 773–788.

Maldonado, H. C., and Cala, P. M. (1994). Labeling of the amphiuma erythrocyte K$^+$/H$^+$ exchanger with H$_2$DIDS. *Am. J. Physiol.* **267,** C1002–C1012.

Margalit, A., and Livne, A. A. (1991). Lipoxygenase product controls the regulatory volume decrease of muman plateles. *Platelets* **2,** 207–214.

Margalit, A., and Livne, A. A. (1992). Human platelets exposed to mechanical stresses express a potent lipoxygenase product. *Thromb. Haemostasis* **68,** 589–594.

Margalit. A., Sofer, Y., Grossman, S., Reynaud, D., Pace-Asciak, C. R., and Livne, A. A. (1993a). Hepoxilin A$_3$ is the endogenous lipid mediator opposing hypotonic swelling of intact human platelets. *Proc. Natl. Acad. Sci. U.S.A.* **90,** 2589–2592.

Margalit, A., Livne, A. A., Funder, J., and Granot, Y. (1993b). Initiation of RVD response in human platelets: Mechanical-biochemical transduction involves petussis-toxin-sensitive G protein and phospholipase A$_2$. *J. Membr. Biol.* **136,** 303–311.

Marnela, K.-M., Timonen, M., and Lähdesmäki, P. (1984). Mass spectrometric analyses of brain synaptic peptides containing taurine. *J. Neurochem.* **43,** 1650–1653.

Mason, M. J., Smith, J. D., De Jesus Garcia-Soto, J., and Grinstein, S. (1989). Internal pH-sensitive site couples Cl$^-$-HCO$^-_3$ exchange to Na$^+$-H$^+$ antiport in lymphocytes. *Am. J. Physiol.* **256,** C428–C433.

Mastrocola, T., Lambert, I. H., Kramhöft, B., Rugolo, M., and Hoffmann, E. K. (1993). Volume regulation in human fibroblasts: Role of Ca^{2+} and 5-lipoxygenase products. *J. Membr. Biol.* **136,** 55–62.

McCann, J. D., Li, M., and Welsh, M. J. (1989a). Identification and regulation of whole-cell chloride currents in airway epithelium. *J. Gen. Physiol.* **94,** 1015–1036.

McCarty, N. A., and O'Neil, R. G. (1991a). Calcium-dependent control of volume regulation in renal proximal tubule cells. I. Swelling-activated Ca^{2+} entry and release. *J. Membr. Biol.* **123,** 149–160.

McCarty, N. A., and O'Neil, R. G. (1991b). Calcium-dependent control of volume regulation in renal proximal tubule cells. II. Roles of dihydropyridine-sensitive and -insensitive Ca^{2+} entry pathways. *J. Membr. Biol.* **123,** 161–170.

McCarty, N. A., and O'Neil, R. G. (1992). Calcium signaling in cell volume regulation. *Physiol. Rev.* **72**(4), 1037–1061.

McManus, M. L., and Churchwell, K. B. (1994). Clinical significance of cellular osmoregulation. *In* "Cellular and Molecular Physiology of Cell Volume Regulation" (K. Strange, ed.), pp. 63–77. CRC Press, Boca Raton, FL.

Melmed, R. N., Karanian, P. J., and Berlin, R. D. (1981). Control of cell volume in the J774 macrophage by microtubule disassembly and cyclic AMP. *J. Cell Biol.* **90,** 761.

Mercer, R. W., and Hoffman, J. F. (1985). Bumetanide-sensitive Na/K cotransport in ferret red blood cells. *Biophys. J.* **47,** 157a.

Meyer, M., Maly, K., Uberall, F., Hoflacher, J., and Grunicke, H. (1991). Stimulation of K⁺ transport system by Haras. *J. Biol. Chem.* **266,** 8230–8235.

Miller, C. E., Moezydlowski, ■., Latorre, R., and Phillips, M. (1985). Charybdotoxin, a protein inhibitor of single Ca^{2+}-activated K⁺ channels from mammalian skeletal muscle. *Nature (London)* **313,** 316–318.

Mills, J. W. (1987). The cell cytoskeleton: Possible role in volume control. *Curr. Top. Membr. Transp.* **30,** 75–101.

Mills, J. W., and Lubin, M. (1986). Effect of adenosine 3′,5′-cyclic monophosphate on volume and cytoskeleton of MDCK cells. *Am. J. Physiol.* **250,** C319–C324.

Mills, J. W., Schwiebert, E. M., and Stanton, B. A. (1994). The cytoskeleton and cell volume regulation. *In* "Cellular and Molecular Physiology of Cell Volume Regulation" (K. Strange, ed.), pp. 241–258. CRC Press, Boca Raton, FL.

Minton, A. P. (1983). The effect of volume occupancy upon the thermodynamic activity of proteins: Some biochemical consequences. *Mol. Cell. Biochem.* **55,** 119–140.

Minton, A. P. (1994). Influence of macromolecular crowding on intracellular association reactions. Possible role in volume regulation. *In* "Cellular and Molecular Physiology of Cell Volume Regulation" (K. Strange, ed.), pp. 181–190. CRC Press, Boca Raton, FL.

Minton, A. P., Colclasure, G. C., and Parker, J. C. (1992). Model for the role of macromolecular crowding in regulation of cellular volume. *Proc. Natl. Acad. Sci. U.S.A.* **89,** 10504–10506.

Mong, S., Wu, H.-L., Hogaboom, G. K., Clark, M. A., Stadel, J. M., and Crooke, S. T. (1984). Regulation of ligand binding to leukotriene D₄ receptors: Effects of cations and guanine nucleotides. *Eur. J. Pharmacol.* **106,** 241–253.

Montrose-Rafizadeh, C., and Guggino, W. B. (1991). Role of intracellular calcium in volume regulation by rabbit medullary thick ascending limb cells. *Am. J. Physiol.* **260,** F402–F409.

Morris, C. E. (1990). Mechanosensitive ion channels. *J. Membr. Biol.* **113,** 93–107.

Musch, M. W., and Field, M. (1989). K-independent Na-Cl cotransport in bovine tracheal epithelial cells. *Am. J. Physiol.* **256,** C658–C665.

Nakahari, T., Murakami, M., Yoshida, H., Miyamoto, M., Sohma, Y., and Imai, Y. (1990). Decrease in rat submandibular acinar cell volume during ACh stimulation. *Am. J. Physiol.* **258,** G878–G886.

Needleman, G. J., Raz, A., Minkes, M. S., Ferrendelli, J. A., and Sprecher, H. (1979). Triene prostaglandins: Prostacyclin and thromboxane biosynthesis and unique biological properties. *Proc. Natl. Acad. Sci. U.S.A.* **76,** 944–948.

Neufeld, E. J., and Majerus, P. W. (1983). Arachidonate release and phosphatidic acid turnover in stimulated human platelets. *J. Biol. Chem.* **258,** 2461–2467.

O'Donnell, M. E. (1989). Regulation of Na-K-Cl cotransport in endothelial cells by atrial natriuretic factor. *Am. J. Physiol.* **257,** C36–C44.

O'Donnell, M. E. (1993). Role of Na-K-Cl cotransport in vascular endothelial cell volume. *Am. J. Physiol.* **264,** C1316–C1326.

O'Grady, S. M., Palfrey, H. C., and Field, M. (1987). Characteristics and functions of Na-K-Cl cotransport in epithelial tissues. *Am. J. Physiol.* **253,** C177–C192.

Okada, Y., and Hazama, A. (1989). Volume-regulatory ion channels in epithelial cells. *NIPS* **4,** 238–242.

Okada, Y., Yada, T., Ohno-Shosaku, T., and Oiki, S. (1987). Evidence for the involvement of calmodulin in the operation of Ca-activated K channel in mouse fibroblasts. *J. Membr. Biol.* **96,** 121–128.

Okada, Y., Hazama, A., and Yuan, W. (1990). Stretch-induced activation of Ca^{2+}-permeable ion channels is involved in the volume regulation of hypotonically swollen epithelial cells. *Neurosci. Res.* **12,** S5–S13.

Olivieri, O., Vitoux, D., Galacteros, F., Bachir, D., Bloquit, Y., Beuzard, Y., and Brugnara, C. (1992). Hemoglobin variants and activity of the (K^+Cl^-) cotransport system in human erythrocytes. *Blood* **79,** 793–797.

O'Neill, W. C. (1989). Cl-dependent K transport in a pure population of volume regulating human erythrocytes. *Am. J. Physiol.* **256,** C858–C864.

O'Neill, W. C., and Klein, J. D. (1992). Regulation of vascular endothelial cell volume by Na-K-2Cl cotransport. *Am. J. Physiol.* **262,** C436–C444.

Orlowski, J., Kandasamy, R. A., and Shull, G. E. (1992). Molecular cloning of putative members of the Na/H exchanger gene family. cDNA cloning, deduced amino acid sequence, and mRNA tissue expression of the rat Na/H exchanger NHE-1 and two structurally related proteins. *J. Biol. Chem.* **267,** 9331–9339.

Pace-Asciak, C. R., and Asorta, S. (1989). Biosynthesis, catabolism, and biological properties of HPETEs, hydroperoxide derivatives of arachidonic acid. *Free Radical Biol. Med.* **7,** 409–433.

Palfrey, H. C., and O'Donnell, M. E. (1992). Characteristics and regulation of the Na/K/2Cl cotransporter. *Cell. Physiol. Biochem.* **2,** 293–307.

Palfrey, H. C., and Pewitt, E. B. (1993). The ATP and Mg^{2+} dependence of Na-K-2Cl cotransport reflects a requirement for protein phosphorylation: Studies using calyculin A. *Pfluegers Arch.* **425,** 321–328.

Parent, L., Cardinal, J., and Sauvé, R. (1988). Single-channel analysis of a K channel at basolateral membrane of rabbit proximal convoluted tubule. *Am. J. Physiol.* **254,** F1–F9.

Paris, S., and Pouysségur, J. (1986). Growth factors activate the bumetanide-sensitive $Na^+/K^+/Cl^-$ cotransport in hamster fibroblasts. *J. Biol. Chem.* **261,** 6177–6183.

Parker, J. C. (1993a). Urea alters set point volume for K-Cl cotransport. Na-H proton exchange, and Ca-Na exchange in dog red blood cells. *Am. J. Physiol.* **265,** C447–C452.

Parker, J. C. (1993b). In defense of cell volume? *Am. J. Physiol.* **265,** C1191–C1200.

Parker, J. C. (1994). Coordinated regulation of volume-activated transport pathways. *In* "Cellular and Molecular Physiology of Cell Volume Regulation" (K. Strange, ed.), pp. 311–321. CRC Press, Boca Raton, FL.

Parker, J. C., and Dunham, P. B. (1989). Passive cation transport. *In* "Red Blood Cell Membranes" (P. Agre and J. C. Parker, eds.), pp. 507–521. Dekker, New York.

Parker, J. C., McManus, T. J., Starke, L. C., and Gitelman, H. J. (1990). Coordinated regulation of Na/H exchange and [K-Cl] cotransport in dog red cells. *J. Gen. Physiol.* **96,** 1141–1152.

Parker, J. C., Colclasure, G. C., and McManus, T. J. (1991). Coordinated regulation of shrinkage-induced Na/H proton exchange and swelling-induced [K-Cl] cotransport in dog red cells. Further evidence from activation kinetics and phosphatase inhibition. *J. Gen. Physiol.* **98,** 869–880.

Parker, J. C., Dunham, P. B., and Minton, A. P. (1995). Effects of ionic strength on the regulation of Na/H exchange and K-Cl cotransport in dog red blood cells. *J. Gen. Physiol.* (in press).

Pasantes-Morales, H., Moran, J., and Schousboe, A. (1990). Volume-sensitive release of taurine from cultured astrocytes: Properties and mechanism. *Glia* **3,** 427–432.

Pasantes-Morales, H., Murray, R. A., Sánchez-Olea, R., and Moran, J. (1994). Regulatory volume decrease in cultured astrocytes II. Permeability pathway to amino acids and polyols. *Am. J. Physiol.* **266,** C172–C178.

Payne, J. A., and Forbush, B., III (1994). Alternatively spliced isoforms of the putative renal Na-K-Cl cotransporter are differentially distributed within the rabbit kidney. *Proc. Natl. Acad. Sci. U.S.A.* **91,** 4544–4548.

Pedersen, S. F., Kramhøft, B., Jørgensen, N. K., and Hoffmann, E. K. (1994). The Na^+/H^+ exchange system in Ehrlich ascites tumor cells. Effects of cell volume, phosphorylation, and calcium. *Acta Physiol. Scand.* **151,** 26A.

Petersen, O. H. (1988). The control of ion channels and pumps in exocrine acinar cells. *Comp. Biochem. Physiol. A* **90A,** 717–721.

Petersen, O. H., and Gallacher, D. V. (1988). Electrophysiology of pancreatic and salivary acinar cells. *Annu. Rev. Physiol.* **50,** 65–80.

Pewitt, E. B., Hegde, R. S., Haas, M., and Palfrey, H. C. (1990). The regulation of Na/K/2Cl cotransport and bumetanide binding in avian erythrocytes by protein phosphorylation and dephosphorylation. *J. Biol. Chem.* **265,** 20747–20756.

Pierce, S. K., and Politis, A. D. (1990). Ca^{2+}-activated cell volume recovery mechanisms. *Annu. Rev. Physiol.* **52,** 27–42.

Pierce, S. K., and Rowland-Faux, L. M. (1992). Ionomycin produces an improved volume recovery by an increased efflux of taurine from hypoosmotically stressed molluscan red blood cells. *Cell Calcium* **13,** 321–327.

Pierce, S. K., Politis, A. D., Cronkite, D. H., Rowland, L. M., and Smith, L. H., Jr. (1989). Evidence of calmodulin involvement in cell volume recovery following hypo-osmotic stress. *Cell Calcium* **10,** 159–169.

Prakash, V., Loucheux, C., Scheufele, S., Gorbunoff, M. J., and Timasheff, S. N. (1981). Interactions of proteins with solvent components in 8 M urea. *Arch. Biochem. Biophys.* **210,** 455–464.

Pritchard, K., and Moody, C. J. (1986). Caldesmon: A calmodulin binding actin regulatory protein. *Cell Calcium* **75**(5-6), 309–327.

Rasola, A., Galieta, L. J., Gruenert, D. C., and Romeo, G. (1994). Volume-sensitive chloride currents in four epithelial cell lines are not directly correlated to the expression of the MDR-1 gene. *J. Biol. Chem.* **269,** 1432–1436.

Reuss, L. (1983). Basolateral KCl cotransport in a NaCl-absorbing epithelium. *Nature (London)* **305,** 723–726.

Reuss, L., and Cotton, C. U. (1994). Volume regulation in epithelia: Transcellular transport and cross-talk. *In* "Cellular and Molecular Physiology of Cell Volume Regulation" (K. Strange, ed.), pp. 31–47. CRC Press, Boca Raton, Florida.

Richter, E. A., Cleland, P. J. F., Rattigan, S., and Clark, M. G. (1987). Contraction-associated translocation of protein kinase C. *FEBS Lett.* **217,** 232–236.

Rink, T. J., Sanchez, A., Grinstein, S., and Rothstein, A. (1983). Volume restoration in osmotically swollen lymphocytes does not involve changes in free Ca^{2+} concentration. *Biochim. Biophys. Acta* **762,** 593–596.

Robertson, M. A., and Foskett, J. K. (1994). Na^+ transport pathways in secretory acinar cells: Membrane cross talk mediated by $[Cl^-]_i$. *Am. J. Physiol.* **267,** C146–C156.

Robertson, R. P. (1986). Characterization and regulation of prostaglandin and leukotriene receptors: An overview. *Prostaglandins* **31,** 395–411.

Rochette, C., Nicholson, D. W., and Metters, K. M. (1993). Identification and target-size analysis of the leukotriene D_4 receptor in the human THP-1 cell line. *Biochim. Biophys. Acta* **1177,** 283–290.

Rosales, O. R., and Sumpio, B. E. (1992). Changes in cyclic strain increase inositol trisphosphate and diacylglycevol in encothelial cells. *Am. J. Physiol.* **262,** C956–C962.

Rossi, J. P. F. C., and Schatzmann, H. J. (1982). Is the red cell calcium pump electrogenic? *J. Physiol. (London)* **327,** 1–15.

Rothstein, A., and Mack, E. (1990). Volume-activated K^+ and Cl^- pathways of dissociated epithelial cells (MDCK): role of Ca^{2+}. *Am. J. Physiol.* **258,** C827–C834.

Rothstein, A., and Mack, E. (1992). Volume-activated calcium uptake: Its role in cell volume regulation of Madin-Darby canine kidney cells. *Am. J. Physiol.* **262,** C339–C347.

Rotin, D., Mason, M. J., and Grinstein, S. (1991). Channels, antiports, and regulation of cell volume in lymphoid cells. *Adv. Comp. Environ. Physiol.* **9,** 118–139.

Roy, G., and Malo, C. (1992). Activation of amino acid diffusion by a volume increase in cultured kidney (MDCK) cells. *J. Membr. Biol.* **130,** 83–90.

Rugolo, M., Mastocola, T., Flamigni, A., and Lenaz, G. (1989). Chloride transport in human fibroblasts is activated by hypotonic shock. *Biochem. Biophys. Res. Commun.* **160,** 1330–1338.

Russell, J. M. (1983). Cation-coupled chloride influx in squid axon. Role of potassium and stoichiometry of the transport process. *J. Gen. Physiol.* **81,** 909–925.

Sachs, F. (1987). Baroreceptor mechanisms at the cellular level. *Fed. Proc., Fed. Am. Soc. Exp. Biol.* **46,** 12–16.

Sachs, F. (1988). Mechanical transduction in biological systems. *CRC Crit. Rev. Biomed. Eng.* **16,** 141–169.

Sachs, F. (1991). Mechanical transduction by membrane ion channels: A mini review. *Mol. Cell. Biochem.* **104,** 57–60.

Sackin, H. (1987). Stretch-activated potassium channels in renal proximal tubule. *Am. J. Physiol.* **253,** F1253–F1262.

Sackin, H. (1989). A stretch-activated K^+ channel sensitive to cell volume. *Proc. Natl. Acad. Sci. U.S.A.* **86,** 1731–1735.

Sackin, H. (1994). Stretch-activated ion channels. *In* "Cellular and Molecular Physiology of Cell Volume Regulation" (K. Strange, ed.), pp. 215–240. CRC Press, Boca Raton, FL.

Saltin, B., Sjögaard, G., Strange, S., and Juel, C. (1987). Redistribution of K^+ in the human body during muscular exercise; its role to maintain whole body homeostasis. *In* "Man in Stressful Environments: Thermal and Work Physiology" (K. Shiraki and M. K. Yousef, eds.), pp. 247–267. Thomas, Springfield, IL.

Samuelsson, B. (1983). Leukotrienes: Mediators of immediate hypersensitivity reactions and inflammation. *Science* **220,** 568–575.

Samuelsson, B., Haeggström, J. Z., and Wetterholm, A. (1991). Leukotriene biosynthesis. *Ann. N.Y. Acad. Sci.* **629,** 89–99.

Sánchez-Olea, R., Pasantes-Morales, H., Lazaro, A., and Cereijido, M. (1991). Osmolarity-sensitive release of free amino acids from cultured kidney cell (MDCK). *J. Membr. Biol.* **121,** 1–9.

Sardet, C., Franchi, A., and Pousségur, J. (1989). Molecular cloning, primary structure and expression of the human growth factor-activatable Na^+/H^+ antiporter. *Cell* (*Cambridge, Mass.*) **56,** 271–280.

Sardet, C., Counillin, L., Franchi, A., and Pousségur, J. (1990). Growth factors induce phosphorylation of the Na^+/H^+ antiporter a, glycoprotein of 110 kD. *Science* **247,** 723–726.

Sargeant, S., Sohn, D. H., and Kim, H. D. (1989). Volume-activated Na/H exchange activity in fetal and adult pig red cells: Inhibition by cyclic AMP. *J. Membr. Biol.* **109,** 209–220.

Sarkadi, B., and Parker, J. C. (1991). Activation of ion transport pathways by changes in cell volume. *Biochim. Biophys. Acta* **1071,** 407–427.

Sarkadi, B., Mack, E., and Rothstein, A. (1984a). Ionic events during the volume response of human peripheral blood lymphocytes to hypotonic media. I. Distinctions between volume-activated Cl^- and K^+ conductance pathways. *J. Gen. Physiol.* **83,** 497–512.

Sarkadi, B., Mack, E., and Rothstein, A. (1984b). Ionic events during the volume response of human peripheral blood lymphocytes to hypotonic media. II. Volume- and time-dependent activation and inactivation of ion transport pathways. *J. Gen. Physiol.* **83,** 513–527.

Sarkadi, B., Cheung, R., Mack, E., Grinstein, S., Gelfand, E. W., and Rothstein, A. (1985). Cation and anion transport pathways in volume regulatory response of human lymphocytes to hyposmotic media. *Am. J. Physiol.* **248,** C480–C487.

Sasaki, S., Ishibashi, I., Yoshiyama, N., and Shiigai, T. (1988). KCl co-transport across the basolateral membrane of rabbit renal proximal straight tubules. *J. Clin. Invest.* **81,** 194–199.

Sauvé, R., Simoneau, C., Monette, R., and Roy, G. (1986). Single-channel analysis of the potassium permeability in HeLa cancer cells: Evidence for a calcium activated potassium channel of small unitary conductance. *J. Membr. Biol.* **92,** 269–282.

Sauvé, R., Parent, L., Simoneau, C., and Roy, G. (1988). External ATP triggers a biphasic activation process of a calcium-dependent K^+ channel in cultured bovine aortic endothelial cells. *Pfluegers Arch.* **412,** 469–481.

Scherer, R. W., and Breitwieser, G. E. (1990). Arachidonic acid metabolites alter G protein-mediated signal transduction in heart. Effects on muscarinic K^+ channels. *J. Gen. Physiol.* **96,** 735–755.

Schlichter, L. C., and Sakellaropoulos, G. (1994). Intracellular Ca^{2+} signaling induced by osmotic shock in human T lymphocytes. *Exp. Cell Res.* **215**, 211–222.

Schousboe, A., Sánchez-Olea, R., Moran, J., and Pasantes-Morales, H. (1991). Hyposmolarity-induced taurine release in cerebellar granule cells is associated with diffusion and not with high-affinity transport. *J. Neurosci. Res.* **30**, 661–665.

Schultz, S. G. (1989a). Intracellular sodium activities and basolateral membrane potassium conductances of sodium-absorbing epithelial cells. *Curr. Top. Membr. Transp.* **34**, 21–44.

Schultz, S. G. (1989b). Volume preservation. Then and now. *News Physiol. Sci.* **4**, 169–172.

Schultz, S. G., Hudson, R. L., and Lapointe, J.-Y. (1985). Electrophysiological studies of sodium cotransport in epithelia: Towards a cellular model. *Ann. N.Y. Acad. Sci.* **456**, 127–135.

Schwarze, W., and Kolb, H. A. (1984). Voltage-dependent kinetics of an ion channel of large conductance in macrophages and myotube membranes. *Pfluegers Arch.* **402**, 281–291.

Shennan, D. B., McNeillie, S. A., and Curran, D. E. (1994). The effect of a hyposmotic shock on amino acid efflux from lactating rat mammary tissue: Stimulation of taurine and glycine efflux via a pathway distinct from anion exchange and volume-activated anion channels. *Exp. Physiol.* **79**, 797–808.

Shorofsky, S. R., Field, M., and Fozzard, H. A. (1984). Mechanism of Cl secretion in canine trachea: Changes in intracellular chloride activity with secretion. *J. Membr. Biol.* **81**, 1 8.

Stephens, A. W. (1985). Cellular volume control. In "The Kidney: Physiology and Pathophysiology" (D. W. Seldin and G. Giebisch, eds.), pp. 91–115. Raven Press, New York.

Sigal, E. (1991). The molecular biology of mammalian arachidonic acid metabolism. *Am. J. Physiol.* **260**, L13–L28.

Simonsen, L. O., Brown, A. M., Christensen, S., Harbak, H., Svane, P. C., and Hoffmann, E. K. (1990). Thrombin and bradykinin mimic the volume response induced by cell swelling in Ehrlich mouse ascites tumour cells. *Renal Physiol. Biochem.* **13**, 176.

Sjölander, A., and Grönroos, E. (1994). Leukotriene D$_4$-induced signal transduction. *Ann. N.Y. Acad. Sci.* **744**, 155–160.

Sjölander, A., Grönroos, E., Hammarström, S., and Andersson, T. (1990). Leukotriene D$_4$ and F$_4$ induce transmembrane signaling in human epithelial cells. *J. Biol. Chem.* **265**, 20976–20981.

Smardo, F. L., Burg, M. B., and García-Pérez, A. (1992). Kidney aldose reductase gene transcription is osmotically regulated. *Am. J. Physiol.* **262**, C776–C782.

Smith, W. L. (1989). The eicosanoids and their biochemical mechanisms of action. *Biochem. J.* **259**, 315–324.

Sokabe, M., Sachs, F., and Jing, Z. (1991). Quantitative video microscopy of patch clamped membranes: Stress, strain, capacitance, and stretch activation. *Biophys. J.* **59**, 722–728.

Solc, C. K., and Wine, J. J. (1991). Swelling-induced and depolarization-induced Cl channels in normal and cystic fibrosis epithelial cells. *Am. J. Physiol.* **261**, C658–C674.

Solis, J. M., Herranz, A. S., Herreras, O., Menedez, N., and del Rio, M. R. (1990). Weak organic acids induce taurine release through an osmotic-sensitive process in *in vivo* rat hippocampus. *J. Neurosci. Res.* **26**, 159–167.

Sorota, S. (1992). Swelling-induced chloride-sensitive current in canine atrial cells revealed by whole-cell patch-clamp. *Circ. Res.* **70**, 679–687.

Spring, K. R., and Hoffmann, E. K. (1992). Cell volume control. In "The Kidney: Physiology and Pathophysiology" (D. W. Seldin and G. Giebisch, eds.), pp. 147–169. Raven Press, New York.

Starke, L. C., and Jennings, M. L. (1993). K-Cl cotransport in rabbit red cells: Further evidence for regulation by protein phosphatase type 1. *Am. J. Physiol.* **264**, C118–C124.

Starke, L. C., and McManus, T. J. (1990). Intracellular free magnesium determines the volume regulatory set point in duck red cells. *FASEB J.* **4**, A818.

Steidl, M., Ritter, M., and Lang, F. (1991). Regulation of potassium conductance by prostaglandin in cultured renal epitheloid (Madin-Darby canine kidney) cells. *Pfluegers Arch* **418**, 431–436.

Stein, W. D. (1986). Intrinsic, apparent, and effective affinities of co- and countertransport systems. *Am. J. Physiol.* **250**, C523–C533.

Stockbridge, L. L., and French, A. S. (1988). Stretch-activated cation channels in human fibroblasts. *Biophys. J.* **54**, 187–190.

Stossel, T. P. (1989). From signal to pseudopod. How cells control cytoplasmic actin assembly. *J. Biol. Chem.* **264**, 18261–18264.

Strange, K., ed. (1994a). "Cellular and Molecular Physiology of Cell Volume Regulation." CRC Press, Boca Raton, FL.

Strange, K., Morrison, R., Shrode, L., and Putnam, R. (1993). Mechanism and regulation of swelling-activated inositol efflux in brain glial cells. *Am. J. Physiol.* **265**, C244–C256.

Sun, A., Grossman, E. B., Lombard, M., and Hebert, S. C. (1991). Vasopressin alters the mechanism of apical Cl entry from Na : Cl to Na : K : 2Cl cotransport in mouse medullary thick ascending limb. *J. Membr. Biol.* **120**, 83–94.

Suzuki, M., Kawahara, K., Ogawa, A., Morita, T., Kawaguchi, Y., Kurihara, S., and Sakai, O. (1990). $[Ca^{2+}]_i$ rises via G protein during regulatory volume decrease in rabbit proximal tubule cells. *Am. J. Physiol.* **258**, F690–F696.

Szirmai, M., Sarkadi, B., Szasz, I., and Gardos, G. (1988). Volume regulatory mechanisms of human platelets. *Haematologia* **21**, 33–40.

Tanglietti, V., and Teselli, M. (1988). A study of stretch activated channels in the membrane of frog oocyte: Interaction with Ca^{2+} ions. *J. Physiol. (London)* **407**, 311–328.

Taylor, C. W., and Richardson, A. (1991). Structure and function of inositol trisphosphate receptors. *Pharmacol. Ther.* **51**, 97–137.

Thomas-Young, R. J., Smith, T. C., and Levinson, C. (1993). Regulatory volume decrease in Ehrlich ascites tumor cells is not mediated by a rise in intracellular calcium. *Biochim. Biophys. Acta* **1146**, 81–86.

Thornhill, W. G., and Laris, P. C. (1984). KCl loss and cell shrinkage in the Ehrlich ascites cell induced by hypotonic media, 2-deoxyglucose, and propranolol. *Biochim. Biophys. Acta* **773**, 207–218.

Thoroed, S. M., and Fugelli, K. (1993). Characterization of the Na^+-dependent taurine influx in flounder erythrocytes. *J. Comp. Physiol. B* **163**, 307–316.

Thoroed, S. M., and Fugelli, K. (1994a). The Na^+-independent taurine influx in flounder erythrocytes and its association with the volume regulatory taurine efflux. *J. Exp. Biol.* **186**, 245–268.

Thoroed, S. M., and Fugelli, K. (1994b). Free amino compounds and cell volume regulation in erythrocytes from different marine fish species under hypoosmotic conditions: The role of a taurine channel. *J. Comp. Physiol. B* **164**, 1–10.

Thoroed, S. M., Lambert, I. H., Hansen, H. S., and Hoffmann, E. K. (1994). Cell swelling stimulates the Ca^{2+}-sensitive cytosolic phospholipase A_2 in Ehrlich ascites tumour cells. *Acta Physiol. Scand.* **151**, 25A.

Thurston, J. H., Hauhart, R. E., and Naccarato, E. F. (1981). Taurine: Possible role in osmotic regulation of mammalian heart. *Science* **214**, 1373–1374.

Torchia, J., Lytle, D. J., Pon, D. J., Forbush, B., III, and Sen, A. K. (1992). The Na-K-Cl cotransporter of avian salt gland. Phosphorylation in response to cAMP-dependent and calcium-dependent secretagogues. *J. Biol. Chem.* **267**, 25444–25450.

Tse, C.-H., Levine, S. A., Yan, C. C. H., Brant, S. R., Nath, S., Pousségur, J., and Donowitz, M. (1994). Molecular properties, kinetics and regulation of mammalian Na^+/H^+ exchangers. *Cell. Physiol. Biochem.* **4**, 282–300.

Tse, M., Levine, S., Yun, C., Brant, S., Counillon, L. T., Pousségur, J., and Donowitz, M. (1993). Structure/function studies of the epithelial isoforms of the mammalian Na^+/H^+ exchanger gene family. *J. Membr. Biol.* **135**, 93–108.

Tseng, G. N. (1992). Cell swelling increases membrane conductance of canine cardiac cells: Evidence for a volume-sensitive Cl channel. *Am. J. Physiol.* **262**, C1056–C1068.

Ubl, J., Murer, H., and Kolb, H.-A. (1988). Ion channels activated by osmotic and mechanical stress in membranes of opossum kidney cells. *J. Membr. Biol.* **104**, 223–232.

Ubl, J., Murer, H., and Kolb, H.-A. (1989). Simultaneous recording of cell volume, membrane current and membrane potential: effect of hypotonic shock. *Pfluegers Arch.* **415**, 381–383.

Uchida, S., Yamauchi, A., Preston, A. S., Kwon, H. M., and Handler, J. S. (1993). Medium tonicity regulates expression of the Na^+- and Cl^--dependent betaine transport in Madin-Darby canine kidney cells by increasing transcription of the transporter gene. *J. Clin. Invest.* **91**, 1604.

Ussing, H. H. (1982a). Volume regulation of frog skin epithelium. *Acta Physiol. Scand.* **114**, 363–369.

Ussing, H. H. (1982b). Pathways for transport in epithelia. *In* "Functional Regulation at the Cellular and Molecular Levels" (R. A. Corradino, ed.), pp. 285–297. Elsevier/North-Holland, Amsterdam.

Ussing, H. H. (1985). Volume regulation and basolateral co-transport of sodium, potassium, and chloride ions in frog skin epithelium. *Pfluegers Arch.* **405**, Suppl. 1, S2–S7.

Ussing, H. H. (1986). Epithelial cell volume regulation illustrated by experiments in frog skin. *Renal Physiol.* **9**, 38–46.

Valverde, M. A., Diaz, M., Sepulveda, F. V., Gill, D. R., Hyde, S. C., and Higgins, C. F. (1992). Volume-regulated chloride channel associated with the human multidrug resistant P-glycoprotein. *Nature (London)* **355**, 830–833.

Van den Bosch, H. (1980). Intracellular phospholipases A. *Biochim. Biophys. Acta* **604**, 191–246.

Vander Meulen, J. A., Klip, A., and Grinstein, S. (1987). Possible mechanism for cerebral edema in diabetic ketoacidosis. *Lancet* **1**, 306.

Van Dorp, D. A. (1967). Aspects of the biosynthesis of prostaglandins. *Prog. Biochem. Pharmacol.* **3**, 71–82.

Van Rossum, G. D. V., and Russo, M. A. (1981). Ouabain-resistant mechanism of volume control and the ultra structural organization of liver slices recovering from swelling *in vitro*. *J. Membr. Biol.* **59**, 191–209.

Vegesna, R. V. K., Mong, S., and Crooke, S. T. (1988). Leukotriene-D_4-induced activation of protein kinase C in rat basophilic leukemia cells. *Eur. J. Pharmacol.* **147**, 387–396.

von Harsdorf, R., Lang, R., Fullerton, M., Smith, A. I., and Woodcock, E. A. (1988). Right atrial dilation increases inositol-(1,4,5) trisphosphate accumulation. *FEBS Lett.* **233**, 201.

Wakabayashi, S., Sardet, C., Fafournoux, P., Counillon, L., Meloche, S., Pagès, G., and Pousségur, J. (1992a). Structure formation of the growth factor-activatable Na^+/H^+ exchanger (NHE-1). *Rev. Physiol. Biochem. Pharmacol.* **119**, 157–186.

Wakabayashi, S., Fafournoux, P., Sardet, C., and Pousségur, J. (1992b). The Na^+/H^+ antiporter cytoplasmic domain mediates growth factor signals and controls "H^+-sensing." *Proc. Natl. Acad. Sci. U.S.A.* **89**, 2424–2428.

Walsh, W. (1989). Role of glial cells in the regulation of the brain ion microenvironment. *Prog. Neurobiol. (Oxford)* **33**, 309–333.

Watanabe, T., Shimizu, T., Miki, I., Sakanaka, C., Honda, Z.-I., Seyama, Y., Teramoto, T., Matsushima, T., Ui, M., and Kurokawa, K. (1990). Characterization of the guinea pig lung membrane leukotriene D_4 receptor solubilized in an active form. *J. Biol. Chem.* **265**, 21237–21241.

Watson, P. A. (1991). Function follows form: Generation of intracellular signals by cell deformation. *FASEB J.* **5**, 2013–2019.

Weber, K., and Osborne, M. (1982). The cytoskeleton. *Natl. Cancer Inst. Mongr.* **60**, 31–36.

Welling, P. A., and O'Neill, R. G. (1990). Cell swelling activates basolateral membrane Cl and K conductance in rabbit proximal tubule. *Am. J. Physiol.* **258**, F951–F962.

Welsh, M. J. (1986). An apical-membrane chloride channel in human tracheal epithelium. *Science* **232**, 1648–1650.

Welsh, M. J. (1987). Electrolyte transport by airway epithelia. *Physiol. Rev.* **67**, 1143–1184.

Welsh, M. J. (1990). Abnormal regulation of ion channel in cystic fibrosis epithelia. *FASEB J.* **4**, 2718–2725.

Whalley, D. W., Hool, L. C., Ten Eick, R. E., and Rasmussen, H. H. (1993). Effect of osmotic swelling and shrinkage on Na^+-K^+ pump activity in mammalian cardiac myocytes. *Am. J. Physiol.* **265**, C1201–C1210.

Winkler, J. D., Mong, S., and Crooke, S. T. (1987). Leukotriene D4-induced homologues desentization of calcium mobilization in rat basophilic leukemia cells. *J. Pharmacol. Exp. Ther.* **244**, 449–455.

Wong, M. M. Y., and Foskett, J. K. (1991). Oscillations of cytosolic sodium during calcium oscillations in exocrine ascinar cells. *Science* **254,** 1014–1016.

Wong, P. Y.-K., and Cheung, W. Y. (1979). Calmodulin stimulates human platelet phospholipase A2. *Biochem. Biophys. Res. Commun.* **90,** 473–480.

Wong, S. M. E., and Chase, H. S., Jr. (1986). Role of intracellular calcium in cellular volume regulation. *Am. J. Physiol.* **250,** C841–C852.

Wong, S. M. E., DeBell, M. C., and Chase, H. S., Jr. (1990). Cell swelling increases intracellular free [Ca] in cultured toad bladder cells. *Am. J. Physiol.* **258,** F292–F296.

Worrell, R. T., Butt, A. G., Cliff, W. H., and Frizzell, R. A. (1989). A volume-sensitive chloride conductance in human colonic cell line T84. *Am. J. Physiol.* **256,** C1111–C1119.

Xu, J.-C., Lytle, C., Zhu, T. T., Payne, J. A., Benz, E., Jr., and Forbush, B., III (1994). Molecular characterization and functional expression of the bumetanide-sensitive Na-K-Cl cotransporter. *Proc. Natl. Acad. Sci. U.S.A.* **91,** 2201–2205.

Yamaguchi, D. T., Green, J., Kleeman, C. R., and Muallem, S. (1989). Characterization of volume-sensitive, calcium-permeating pathway in the osteosarcoma cell line UMR-106-01. *J. Biol. Chem.* **264,** 4383–4390.

Yamamoto, S. (1989). Mammalian lipoxygenases: Molecular and catalytic properties. *Prostaglandins, Leukotrienes Essent. Fatty Acids* **35,** 219–229.

Yamauchi, A., Uchida, S., Kwon, H. M., Preston, A. S., Robey, R. B., García-Pérez, A., Burg, M. B., and Handler, J. S. (1992). Cloning of a Na^+ and Cl^--dependent betaine transporter that is regulated by hypertonicity. *J. Biol. Chem.* **267,** 649–652.

Yamauchi, A., Uchida, S., Preston, A. S., Kwon, H. M., and Handler, J. S. (1993). Hypertonicity stimulates transcription of the gene for the Na^+/myo-inositol cotransporter in MDCK cells. *Am. J. Physiol.* **264,** F20–F23.

Yancey, P. H. (1994). Compatible and conteracting solutes. *In* "Cellular and Molecular Physiology of Cell Volume Regulation" (K. Strange, ed.), pp. 81–109. CRC Press, Boca Raton, FL.

Yang, X.-C., and Sachs, F. (1989). Block of stretch activated ion channels in *Xenopus oocytes* by gadolinium and calcium ions. *Science* **243,** 1068–1071.

Yantorno, R. E., Carre, D. A., Coca-Prados, M., Krupin, T., and Civan, M. M. (1992). Whole cell patch clamping of ciliary epithelial cells during anisosmotic swelling. *Am. J. Physiol.* **262,** C501–C509.

Youmans, S. J., and Barry, C. R. (1991). Effects of valinomycin on vanadate-sensitive and vanadate-resistant H^+ transport in vescicles from turtle bladder epithelium: Evidence for a K^+/H^+ exchanger. *Biochem. Biophys. Res. Comm.* **176,** 1285–1290.

Zeuthen, T. (1991). Secondary active transport of water across ventricular cell membrane of choroid plexus epithelium of *Necturus maculosus*. *J. Physiol. (London)* **444,** 153–173.

Zimmerman, S. B., and Harrison, B. (1987). Macromolecular crowding increases binding of DNA polymerase to DNA. An adaptive effect. *Proc. Natl. Acad. Sci. U.S.A.* **84,** 1871–1875.

Zimmerman, S. B., and Minton, A. P. (1993). Macromolecular crowding: Biochemical, biophysical, and physiological consequences. *Annu. Rev. Biophys. Biomol. Struct.* **22,** 27–65.

Ziyadeh, F. N., Mills, J. W., and Kleinzeller, A. (1992). Hypotonicity and cell volume regulation in shark rectal gland: Role of organic osmolytes and F-actin. *Am. J. Physiol.* **262,** F468–F480.

Bacterial Stimulators of Macrophages

Sunna Hauschildt[1] and Bernhard Kleine
Institut für Immunbiologie der Universität, D-79104 Freiburg, Germany

Our current understanding of the interaction between bacteria and macrophages, cells of the immune system that play a major role in the defense against infection, is summarized. Cell-surface structures of Gram-negative and Gram-positive bacteria that account for these interactions are described in detail. Besides surface structures, soluble bacterial molecules, toxins that are derived from pathogenic bacteria, are also shown to modulate macrophage functions. In order to affect macrophage functions, bacterial surface structures have to be recognized by the macrophage and toxins have to be taken up. Subsequently, signal transduction mechanisms are initiated that enable the macrophage to respond to the invading bacteria. To destroy bacteria, macrophages employ many strategies, among which antigen processing and presentation to T cells, phagocytosis, chemotaxis, and different bactericidal mechanisms are considered to be the main weapons.

KEY WORDS: Macrophage surface receptors, Signal transduction, Antigen processing and presentation, Phagocytosis, Chemotaxis, Bactericidal mechanisms.

I. Introduction

The protection of an individual against destructive effects of bacteria is one of the primary duties of the eukaryotic immune system. Eukaryotic organisms have been confronted with hostile prokaryotes throughout their existence. As a result, defense mechanisms against these microbes can be found in single-cell organisms like slime molds or amoebas as well as in the most complex mammalian species.

Although each vertebrate individual as a whole has to be protected against the invasion of bacteria, it tolerates local colonization by many

[1] Present address: Institut für Zoologie der Universität Leipzig, D-04103 Leipzig, Germany.

bacteria. Under normal conditions, a steady-state situation is achieved in which the presence of commensal bacteria can even be beneficial to the eukaryotic host. Bacteroides present in the digestive tract may help to metabolize large proteins and carbohydrates so that amino acids and sugars can be taken up and used by the host (Cady *et al.*, 1989). The presence of bacteria and of bacterial products might even help to maintain some basic activities of the immune system (Pabst *et al.*, 1982). Animals raised under germ-free conditions on a low-molecular-weight diet, devoid of bacteria and their components, could be shown to lack immunoglobulin G (IgG) subclasses which are necessary to support Fc receptor-mediated uptake (Hooijkaas *et al.*, 1985). It has been suggested that one sleep phase, slow-wave sleep, depends on the presence of somnotropic derivatives of the bacterial peptidoglycan released by macrophages after phagocytosis and digestion of bacteria (Johannsen *et al.*, 1991). Newborns who have not had the opportunity to deal with bacteria, therefore, sleep more lightly. They start to develop deeper sleeping phases only after their encounter with the bacterial world.

The antibacterial defense of mammals is accomplished by several steps. The first defense line is physical blockage, which is achieved by nonpermeable skin, by low pH in the gut, or by flushing the eyes or the urethra. The epithelial surface in the lung removes bacteria. Such physical hindrances are supported by enzymes like lysozyme. This peptidoglycan-degrading enzyme can destroy the cell wall of any bacteria that is not protected by capsules or other surface sheaths. Below the surface of skin and mucosa, a second defense barrier is present that consists of cells which engulf bacteria and digest them intracellularly. Most of this bacterial uptake is performed by macrophages, but Langerhans cells are equally able to phagocytose bacteria, to digest them, and to process them for antigen presentation in lymphoid organs. There the original Langerhans cell becomes the interdigitating dendritic cell (Moll *et al.*, 1993). Lymphocytes, which recognize the bacterial antigens on the surface of dendritic cells, are activated and mature into effector cells which either produce antibacterial immunoglobulins or destroy cells expressing bacterial antigens. The immunoglobulins provide the next line of the antibacterial defense. Bound to their bacterial target structure, they have labeled the bacteria, now called opsonized, for complement attack as well as for enhanced and more efficient phagocytosis. Nonopsonized bacteria can resist intracellular digestion but once opsonized they are destroyed by the phagocytes. Immunoglobulins, which are also secreted into the fluids by the lacrymal or salivary glands, further block the adhesion of bacteria to mucosal or epithelial surfaces.

The center of antibacterial defenses in vertebrate organisms is the macrophage. It is attracted by bacterial chemoattractants and migrates into bacterial foci. It engulfs bacteria, kills them intracellularly, and digests them.

Through the release of soluble mediators it stimulates other immune cells to move to the site of bacterial contamination and to initiate immune responses that result in antibacterial immunoglobulin production or in the appearance of cytotoxic effector cells.

Macrophages arise in the bone marrow from hemopoietic stem cells by differentiation via the mycloid precursor cells. Colony-stimulating factors (CSF) such as macrophage (M-) CSF and granulocyte/macrophage (GM) CSF support the propagation of precursor cells and their maturation into monocytes. These cells leave the bone marrow and circulate in large numbers (10–20% of the white blood cells) in the blood. Attracted by chemokines and epithelial adhesion molecules, monocytes become activated: they adhere to the blood vessels and penetrate the epithelial walls (Wheeler *et al.*, 1993). They move along the chemokine gradient toward infectious foci or other inflammatory sites. They take up intact bacteria or remnants of cell tissues. In this way they clean up the infected or damaged sites. Finally, they become seeded within tissues. These resident cells are called macrophages.

Depending on the organ in which the macrophage becomes resident, functional differences have been found; for example, alveolar macrophages and peritoneal macrophages differ in their ability to phagocytose bacteria (Riveau *et al.*, 1986). By analyzing the Fcγ receptor type I (CD64) of monocytes, different subpopulations could be observed (Parant *et al.*, 1984). Monocytes and macrophages can, therefore, no longer be regarded as a monolithic block of identical cells because they show functional differences that depend on the history of each cell.

The interactions between bacteria and macrophages are determined by individual bacterial molecules and their counterparts on the surface or in the interior of macrophages. The bacterial molecules in question are membrane and cell wall components as well as molecules secreted by bacteria. The *soluble* bacterial molecules known to modulate macrophage functions are all toxins that are derived from pathogenic bacteria. Molecules secreted from normally harmless bacteria which can affect macrophage functions have not been described. In contrast, the *surface structures* of all bacteria are able to stimulate certain monocyte and macrophage functions. The major components of the Gram-negative and Gram-positive bacterial cell walls, lipopolysaccharide and peptidoglycan, respectively, also represent the major macrophage stimuli. Cell wall proteins, porins, lipoproteins, and other outer membrane proteins (OMP) from Gram-negative bacteria, lipoteichoic acids from Gram-positive bacteria, glycolipids and a lipoarabinomannan from *Mycobacteria* have also been implicated in macrophage activation. For obvious reasons, the surface components of infectious bacteria and their effects on macrophage functions have found much more interest in the literature than the stimulation of macrophages by surface compounds of commensal bacteria.

In this chapter, we describe those components of bacteria known to affect monocytes and macrophages. Macrophages will be regarded as targets and as effector cells in bacteria–macrophage interactions. Focusing on the bacterial interactions with macrophages, we disregard one aspect of the origin of bacterial virulence: the colonization of the host as a function of bacterial adhesion to surfaces and of the penetration of the physical and structural defense barriers. Macrophages do not participate in this line of antibacterial defense. Since nonpathogenic bacteria only rarely overcome these structural hindrances and do not encounter macrophage defense reactions, this chapter will deal almost exclusively with bacteria that are pathogenic for humans and for vertebrate animals.

II. Bacteria-Affecting Macrophages

A. General Considerations

The bacterial cell wall provides the structural elements that determine the fate of the individual bacteria in macrophage-bacteria interactions. One component of these structures is a plethora of polysaccharides composed of almost every possible link between two different hexoses or pentoses and as diverse as bacterial genera or strains might be. While only one type of conjugation between two amino acids is possible, eight different glycosidic links are possible between two hexoses. This diversity is enhanced exponentially by adding further sugar residues. In order to deal with this wealth of structural diversity, a system of repeating units has evolved, with smaller subunits being synthesized cytoplasmatically and transferred by undecaprenol carriers to the periplasmic space to be added to the growing polysaccharides. Such repeating units are found in capsular polysaccharides, teichoic and lipoteichoic acids, lipopolysaccharides (LPS), and further bacterial carbohydrates.

The second feature of the bacterial cell wall is the peptidoglycan (PG), which consists of N-acetylmuramic acid (MurNAc)-N-acetylglucosamine dimers in large polymeric chains with small oligopeptides chains attached. By covalent linkage between two adjacent oligopeptides, a peptidoglycan layer is created. While the PG of Gram-negative bacteria exists as a monolayered structure, the PG of Gram-positive bacteria is multilayered. The third element of bacterial surfaces are proteins comprising outer membrane proteins in Gram-negative bacteria, lipoproteins, proteinaceous fimbriae, or pili in both Gram-negative or positive types. A special bacterial surface element is displayed by the lipid layers of mycobacteria and related actino-mycetes which contain waxes and glycosides with characteristic very long chain and branched fatty acids.

While these structural features are present in all bacteria, capsules can be found in only a limited number of strains, especially in pathogenic strains. Capsules may be polymers of complex oligosaccharide units of amino acids, amino sugars, or acidic amino sugars. For the majority of the encapsulated bacteria, the capsule displays protection against phagocytosis, and makes the bacterium more virulent. There are, however, other struc-tural elements which, like capsules, protect against or confer resistance to phagocytosis.

B. Gram-Positive Bacteria

1. Cell-Membrane Components of Gram-Positive Bacteria

The most pronounced component of the cell wall of Gram-positive bacteria is a multilayered peptidoglycan which confers solidity and firmness on the bacterium. Lipoteichoic acids (LTA) extend from the plasma membrane through this peptidoglycan into the extracellular space, while teichoic acids (TA) are directly attached to the PG. LTA and PG are potent stimulators of macrophages. Capsular polysaccharides and some membraneous poly-peptides have also been described as being involved in the interactions of Gram-positive bacteria with the host's immune system, especially mono-cytes and macrophages.

a. Peptidoglycans PGs are built of β1,4-linked D-2-deoxy-2-(N-acetyl)amino-glucopyranose (glucosamine : GlcNAc) polymers. Every sec-ond GlcNAC residue is substituted at the C3 position by a lactic acid in ether linkage. The [3-lactyl]GlcNAc is called N-acetylmuramic acid. The lactic acid residue of the muramic acid bears a pentapeptide which is characteristic of the bacterial strain. The structure of the *S. aureus* pentapep-tide is L-alanyl-D-γ-glutamyl-L-lysyl-D-alanyl-D-alanine. The ε-amino group of the lysyl residue is further substituted by a pentapeptide of glycine residues. Cross-linking of adjacent polymers is achieved by a transpeptidase reaction which replaces the terminal D-alanine of one polymer with the amino-terminal glycine of the next polymer (see Fig. 1A). The extent to which MurNAc is substituted by these peptides, and the PG cross-linked, is especially high in pathogenic strains (Davis, 1990). There are considerable variations in the PG structure.

PG and partial structures of PG are potent stimulators of the immune system. Similar to other essential structural elements of bacterial cell walls,

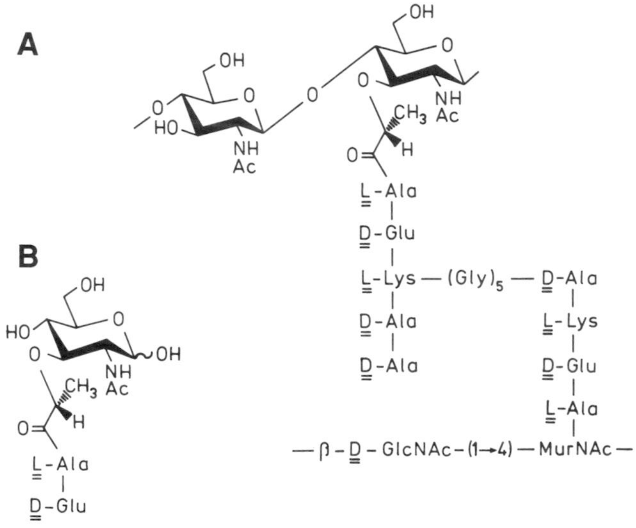

FIG. 1   (A) Section of a model of *Staphylococcus aureus* peptidoglycan. (B) Muramyl dipeptide, the active partial structure of peptidoglycan.

they can reproduce most of the major signs and symptoms associated with bacterial infections, such as fever, inflammation, leukocytosis, acute phase responses, and lymphocyte activation. Among the cells mediating the host responses, macrophages play the major role.

The smallest partial structure of PG which still exhibits full stimulatory capacity for macrophages is muramyl dipeptide: *N*-acetyl-muramoyl-L-alanyl-D-α-glutamic acid (Fig. 1B). It induces increased phagocytosis (McCoy *et al.,* 1985), adherence to plastic or glass, cytostasis or cytotoxicity against tumor cell lines (Tenu *et al.,* 1987; Daemen *et al.,* 1989), secretion of cytokines such as interleukin-1 (IL-1) (Beck *et al.,* 1990), IL-6, IL-8, tumor necrosis factor-α (TNF-α) (Kurzman *et al.,* 1993), prostaglandin synthesis and release, and release of lysosomal enzymes. Human monocytes can secrete IL-1 or prostaglandin (PGE$_2$) only after stimulation with muryamyl dipeptide (MDP) in the presence of interferon-γ (IFN-γ) (Fidler *et al.,* 1990).

MDP-triggered biochemical changes include phospholipase A$_2$ (PLA$_2$) stimulation, protein kinase C (PKC) translocation, enhanced expression of class II major histocompatibility antigens (Nesmeyanov *et al.,* 1990), or nitrite secretion. Some murine strains show different responsiveness to PG and to MDP, while most other rodents are responsive to both stimuli (Nagao *et al.,* 1992).

One of the major drawbacks of MDP when applied to humans has been its pyrogenicity, a feature which is shared by lipopolysaccharides and which is mediated by "endogenous pyrogen" (EP) (Dinarello *et al.*, 1974): IL-1, IL-6, or TNF-α. For this reason synthetic MDP analogs lacking this property have been synthesized. When the terminal carboxy group of D-glutamate was esterified by *n*-butanol, the resulting analog was apyrogenic (Chedid *et al.*, 1982). This murabutide can no longer induce EP. The methyl ester of MDP (murametide) is similarily apyrogenic. In contrast to murabutide, murametide is an inducer of EP (Parant *et al.*, 1984). Using murametide, it was also possible to show that IL-1-induced pyrogenic responses are not necessarily accompanied by somnogenic responses (Cady *et al.*, 1989; Riveau *et al.*, 1986).

One MDP analog, MTP-PE, which consists of MurNAc-L-alanyl-D-[γ-phosphatidylethanolaminyl]glutamyl-L-alaninamide (Fig. 2), has been shown to be active in antitumor responses (Asano and Kleinerman, 1993). Encapsulated in liposomes, this L-MTP-PE is currently used in clinical studies with patients suffering from melanomas or osteosarcoma (Asano *et al.*, 1994; Kleinerman *et al.*, 1993). The antitumor effect is mediated by macrophages shown to produce IL-1α or β, TNF-α, IL-6, or IL-8 upon treatment with L-MTP-PE (Asano *et al.*, 1994).

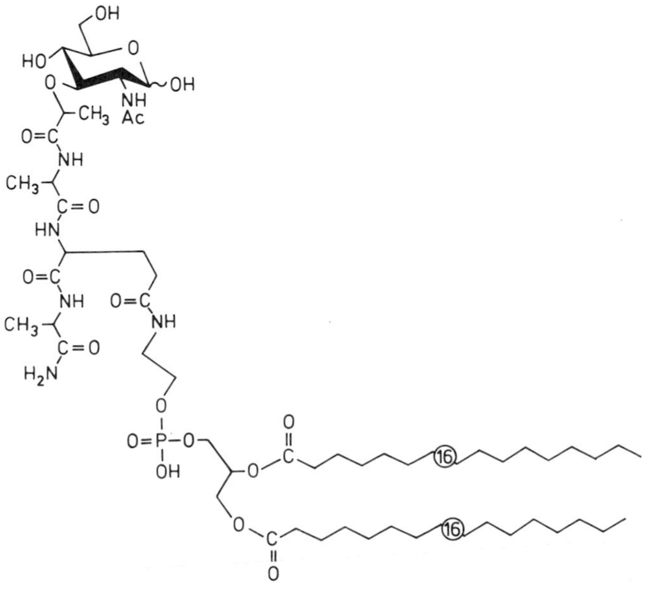

FIG. 2   MTP-PE, a synthetic analog of MDP with antitumor activity.

b. Teichoic Acids and Lipoteichoic Acids   Teichoic acids display another general structural element of the Gram-positive bacterial cell wall. They may be either bound to the peptidoglycan (teichoic acids *in sensu stricto*) or inserted into the cytoplasmic membrane (lipoteichoic acids). TA and LTA differ in several aspects: While in most of the structures analyzed LTAs are composed of a polyglycerol phosphate backbone, TAs display several backbone elements: polyglycerol phosphate, and/or polyribitol phosphate backbones. In *Staphylococci,* mixed polymers of polyglycerolribitol phosphate could be identified, also. The glycerol is often substituted at the C2 position by alanyl-, monosaccharide, or di-N-acetylglucosamine (di-GlcNAc) residues (Endl *et al.,* 1983). Unusual polymers such as poly(galactosylglycerol phosphate) (Potekhina *et al.,* 1991) or poly(erythritol phosphate) (Potekhina *et al.,* 1993) have been described in some species. In *Staphylococcus auricularis,* TA is composed of a GlcNAc-phosphate polymer (Endl *et al.,* 1983). Polymeric ribitol phosphate TAs were observed, for example, in *Listeria monocytogenes.* The substituents of the ribitol or glycerol phosphate determine the *O*-specific serotypes of *Listeria,* as can be seen in Table I (Uchikawa *et al.,* 1986). In *Norcadia,* glycerol phosphate-GalNAc-glycerol phosphate polymers were found (Tul'skaya *et al.,* 1993).

The teichoic acids of pneumococci are especially complex: Figure 3 shows the structure of the pneumococcal lipoteichoic acid. However, in these bacteria the repeating unit is shared by the teichoic and the lipoteichoic acids (Fischer *et al.,* 1993). This repeating unit, also called C-polysaccharide, is recognized by the C-reactive protein in human serum.

The integration of lipoteichoic acids into the cell membrane is achieved in different ways. The phosphomonoester end of the polymer is linked to a lipid which may either be a glycolipid, and/or a phosphatidylglycolipid (Fischer *et al.,* 1990). The complex lipid anchor formed by pneumococci is shown in Fig. 3 (Behr *et al.,* 1992). TAs are directly attached to the 6-position of muramic acid residues of the peptidoglycan by special bridge regions which differ from LTA glycolipids (Coley *et al.,* 1978; Kaya *et al.,* 1985).

TABLE I

Serotype Determining Structures of the *Listeria monocytogenes* Teichoic Acid[a]

Serotype	Structure of the repeating unit
3a	-1-[GlcNAc-α1,4]ribitol-5-phosphate
4b	-4-[Gal-α1,6][Glc-β1,3]GlcNAc-β1,2-ribitol-5-phosphate
4f	-4-[Gal-α1,6][GlcNAc-α1,3][b]GlcNAc-β1,2-ribitol-5-phosphate
6	-4-GclNAc-β1,4-ribitol-5-phosphate
7	-1-ribitol-5-phosphate

[a] According to Uchikawa *et al.,* 1986.
[b] Substitution at C3 to 70%.

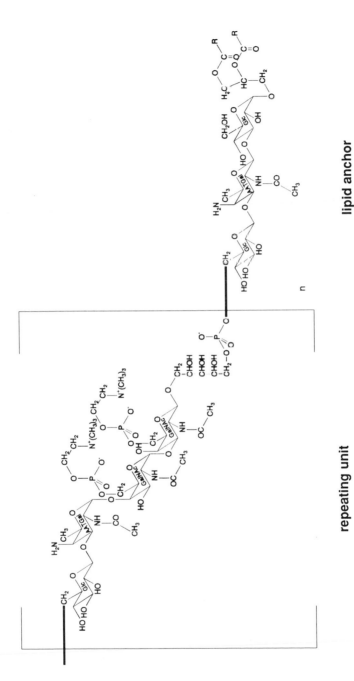

repeating unit　　　　　　　　　**lipid anchor**

FIG. 3　Model of the *Pneumococcus* lipoteichoic acid.

Intact LTAs have been found to constitute potent macrophage activators: They induce cytokine release (IL-1β, IL-6, IL-8, and TNF-α) from human monocytes (Bhakdi *et al.*, 1991; Standiford *et al.*, 1994; Mattsson *et al.*, 1993) and from mouse macrophages (Kuwano *et al.*, 1993). They induce tumor cytotoxicity (Kuwano *et al.*, 1993), release of lysosomal enzymes (Harrop *et al.*, 1980), nitrite production (Kuwano *et al.*, 1993), and stimulate the respiratory burst (Tarsi-Tsuk and Levy, 1990; Levy *et al.*, 1990). They also stimulate luminol-mediated chemiluminescence in human monocytes (Ohshima *et al.*, 1988).

These effects are, however, not observed with all of the LTA isolates, but vary according to their origin: LTA from *Bacillus subtilis, Streptococcus pyogenes, S. sanguis, S. pneumoniae, Staphylococcus aureus, Enterococcus faecalis, E. hirae,* and from *Listeria monocytogenes* induced TNF-α production. Large amounts of nitrite, however, were secreted upon stimulation with LTA from *B. subtilis, L. monocytogenes,* and the different streptococcal LTA, while the *E. faecalis* and *E. hirae* LTA provoked less pronounced nitrite responses. The *S. aureus* LTA failed to induce any nitrite secretion (Keller *et al.*, 1992).

It has been observed that a lipid anchor is required to induce these reactions in macrophages. Teichoic acids which do not bear the lipid anchor and LTA chemically modified by removing the glyceride from LTA do not stimulate any of these reactions (Keller *et al.*, 1992). These findings demonstrate that LTAs behave like lipopolysaccharides from Gram-negative bacteria where most of the biological activity resides in the lipid A portion (see Section II,C).

c. Surface Proteins of Gram-Positive Bacteria   The surface proteins of Gram-positive bacteria have been treated less extensively than outer membrane proteins of Gram-negative species. The proteins analyzed in most detail are the M-proteins of streptococci (Fischetti, 1989) because antibodies against these proteins confered protection against further infection by streptococci in man (McCarty, 1990), whereas antibodies against other components such as capsular polysaccharides, peptidoglycan, or teichoic acids, were not protective (McCarty, 1990).

To a large extent, M-proteins are composed of repetitive elements forming a very large helical rod. Figure 4 shows that the M-proteins consist of an N-terminal nonhelical region of a few amino acids followed by the so-called A repeat. This component contains repetitive elements of 14 amino acids, each of them forming four helical turns. The B repeat consists of 5 repetitive elements of 21 or 25 amino acids forming 6 or 7 turns of the helix. Two of these elements are identical, and the other two differ in one or two amino acids. The C repeat contains two larger elements forming 11 or 10 helical turns and one smaller component of 9 helical turns. A D

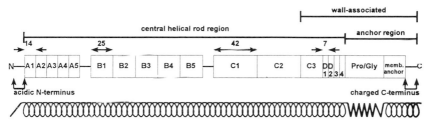

FIG. 4   Structural features of M proteins of streptococci (according to Fischetti, 1989).

region of 26 amino acids is the innermost helical structure of the M-protein. No proline destroys the helical structure in the A, B, C, or D region. These regions are thought to be exposed above the carbohydrate (teichoic acid) layer. The anchor region by which M-proteins are integrated in the cell membrane consists of a proline/glycine-rich part reaching through the peptidoglycan, and a membrane domain which is then followed by a small cytoplasmic tail of six charged amino acids. The helical rod region of the M-protein forms a prototype coiled-coil structure. For thermodynamic reasons, a structure of M-protein dimers is more likely to occur than a single coiled-coil helix (Fischetti, 1989).

The different M-proteins display antigenic variations (55 immunological types of streptococcal M-antigens; McCarty, 1990). Different M6 isolates showed variations in the number of the A repeats. Compared with the M6 structure, M5 structures contained a smaller N-terminal region, their A repeats forming only two helical turns. There are only four B elements, and the D region is larger. The anchor regions with the proline/glycine-rich stretch and the membrane domain were equal in size. An M24 isolate contained a very large N-terminal element, and no A repeats (Fischetti, 1989). Such antigenic variations might allow bacteria to grow in hosts which have already developed antibodies against M proteins of other group A streptococci.

One possible pathological effect of M-proteins is suggested by homologies to human proteins such as myosin, tropomyosin, keratin, or laminin. Antibodies raised against M-proteins cross-react with these proteins (Fischetti, 1989; Cunningham et al., 1992). Anti-M5 antibodies react strongly with skeletal muscle myosin and ventricular myosin (Manjula and Fischetti, 1986). Such cross-reactive antibodies might be involved in myocarditis or other inflammatory lesions. During immune reactions against streptococcal M-proteins, self-tolerance for coiled-coil type proteins might be broken, self-proteins recognized as antigens, and the cells expressing these antigens destroyed.

The presence of the M-proteins renders the bacteria resistant to phagocytosis. When bacteria expressing the M-protein (M^+) and those lacking an

M-protein (M^-) were mixed and incubated with human blood, the M^+ grew, whereas the M^- bacteria were phagocytosed and killed. The phagocytes did not "care" about M^+ bacteria. Fischetti (1989) suggested that this behavior of phagocytes is mediated by acidic amino acids in the N-terminal region of the different M-proteins. Electrostatic repulsion between phagocytic membranes and the most exposed M-protein structures may be a possible explanation for this type of resistance to phagocytosis. Alternatively, many surface receptors of phagocytes might be inactive in an acidic environment. Unlike many other plasmid-encoded bacterial compounds determining virulence, the M-protein is chromosomally located. There is one *emm* gene per bacterium.

2. Soluble Components of Gram-Positive Bacteria

Exotoxins of Gram-positive strains are among the most potent toxins known today. One microgram of *Clostridium botulinum* neurotoxin is lethal for man (Swartz, 1990b). This toxin is coded for by the bacterial chromosome. There are, however, different toxins coded for by plasmids. Potentially toxic bacteria that do not harbor the plasmids encoding the toxins are relatively harmless for the mammalian host.

There are several types of toxins: The prominent toxin of *Staphylococci* is a pore-forming hemolysin, the pore resembling that formed by the membrane attack complex (Novick, 1990). Other erythrocyte-lysing toxins are present in *Staphylococci* (β-toxin, a sphingomyelinase; Novick, 1990), *C. perfringens* (θ-toxin; Swartz, 1990b), or in *C. tetani* (tetanolysin; Swartz, 1990b).

Leucocidins, consisting of so-called S and F components, are present in *Staphylococci* (Novick, 1990), *C. perfringens* (v toxin; Swartz, 1990b), as well as in the Gram-negative *Pseudomonase aeruginosa* (Lory, 1990). Upon binding of the S component to the ganglioside GM_1 on a cell membrane, the cellular (endogenous) phospholipase A_2 is activated. Products of PLA_2 activity then bind the leucocidins' F component, induce the formation of a potassium channel, and finally lead to cytolysis of the cell (Novick, 1990).

An equally widespread toxin has ADP-ribosylating activity, the prototype of which is the *Bordetella pertussis* toxin (PT). The binding (B) subunits of PT bind to sialic acid containing N-linked sugars of surface glycoproteins. This attachment mediates the uptake of the enzymatic A (or S1) subunit into the cell, which then ADP-ribosylates and thus inactivates the guanosine triphosphate (GTP)-binding protein (G_i), an inhibitor of adenyl cyclase (AC). AC thus becomes activated, leading to an enhanced production of cAMP (Robbins and Pittman, 1990). Enhanced cAMP levels may influence various functions, depending on the cells affected. Hormonal actions may be affected: Hypoglycemia may result from a stimulation of insulin production by epinephrine on Langerhans islet cells in infected animals while under healthy conditions epinephrine inhibits insulin release. Immune reactions that may be affected include capillary permeability, lymphocytosis, macrophage migration, and phagocytosis (Robbins and Pittman, 1990). ADP-ribosylating toxins have also been found in the following bacteria:

C. botulinum:	the C3 enzyme ADP-ribosylates the small GTP-binding proteins rho and rac (Swartz, 1990b)
C. perfringens:	the toxin ADP-ribosylates globular actin (Swartz, 1990b)
Corynebacterium diphtheriae:	the diphtheria toxin (DT) ADP-ribosylates specifically the EF-2 elongation factor of eukaryotic cells (Collier, 1990)
(Gram-negative bacteria:)	
P. aeruginosa:	the ADP-ribosyltransferase exoenzyme S is a virulence factor (Lory, 1990)
V. cholerae:	the enterotoxic cholera toxin (CT) ADP-ribosylates and thus activates the stimulatory G_s-protein, which leads to enhanced cAMP levels (Falkow and Mekalanos, 1990)
E. coli:	the heat-labile toxin (LT) binds the CT to the GM_1 ganglioside and induces ADP-ribosylation of a G-protein, thus enhancing cAMP (Falkow and Mekalanos, 1990)

There are further exotoxins which do not fall into the three types mentioned above. While *B. pertussis* stimulates host adenyl cyclase via PT, it also produces its own adenylate cyclase which can enter phagocytic cells and is thought to reduce phagocytic activity (Robbins and Pittman, 1990). The exotoxin of *Bacillus anthracis* is composed of three polypeptides: the edema factor with calmodulin-dependent adenylate cyclase activity, the protective antigen, and the lethal factor (Swartz, 1990a). The *C. tetani* neurotoxin binds to GD_{1b} and GT_{1b} gangliosides containing sialic acid, thereby opening a pore through which the enzymatic chain is translocated. The latter is then intraaxonally transported to the neuronal dendrites where it blocks the inhibitory neurotransmitters glycine and γ-aminobutyric acid (Swartz, 1990b). The α-toxin of *C. perfringens* is a calcium-dependent phospholipase

C which activates platelet-activating factor and destroys leukocytes by cleaving lecithin in their cell membranes (Swartz, 1990b).

Recently, the severe actions of several pyrogenic exotoxins from group A streptococci and from *S. aureus* have been linked to direct effects on T lymphocytes and monocytes and macrophages (Table II). These toxins stimulate subpopulations of T cells expressing specific variable genes of some V gene families. Owing to a bivalent action on these T-cell receptors and on MHC class II antigens on macrophages, T cells become strongly activated for some time. Afterward the T cells are eliminated and repertoire holes are observed. Owing to their polyclonal activation, such toxins have been called superantigens (Lee and Schlievert, 1991; Sjögren, 1991; Chatila *et al.*, 1991).

C. Gram-Negative Bacteria

1. Cell-Membrane Components of Gram-Negative Bacteria

Compared with Gram-positive bacteria, the cell wall of Gram-negative strains is more complex (Fig. 5). They contain a characteristic additional lipid bilayer called the outer membrane (OM). Between the (inner) cytoplasmic membrane and the outer membrane, a single-layered "sacculus" of peptidoglycan is found. It contains diaminopimelate instead of lysine and often lacks the peptide bridges found in Gram-positive bacteria. Direct links between tetrapeptides have been observed instead (Davis, 1990). The OM contains several unique compounds such as lipopolysaccharides, lipoproteins (LP), and specialized outer membrane pore-forming proteins.

In *Enterobacteriaceae,* the bacteria most intensively studied, wild-type lipopolysaccharides form the outer leaflet of OM while the inner leaflet is composed of phospholipids. Rough mutants of Enterobacteriaceae, which lack most of the polysaccharide of wild-type LPSs, contain phospholipids in both leaflets of the OM (Nikaido, 1979). In other Gram-negative bacteria that do not possess large polysaccharides, for example, *Neisseria,* phospholipids and LPSs might be present in both leaflets of the outer membrane.

Firm connections between the outer membrane and the PG are established by an outer membrane protein (OMP-A) (Benz and Bauer, 1988) and by a lipoprotein (Braun and Rehn, 1969; Braun and Hantke, 1974). This LP is the most abundant protein in *E. coli* with about 600,000 copies per bacterium (van Alphen *et al.,* 1978). About one third of the LP molecules are covalently bound to the peptidoglycan while its N-terminal lipopeptide is integrated into the inner leaflet of the OM (Braun and Hantke, 1975). The lipid moiety of LP is strongly involved in host–bacteria interactions and its synthetically prepared analogs have been analyzed (see later discus-

TABLE II

Some Microbial Toxins Which Act as Superantigens[a]

Microorganism	Protein	Abbreviation	M_r	Active concentrations	V_β specificity
Streptococcus aureus	Enterotoxin B	SEB	28,300	1–10 ng/ml	3, 12. 14, 15, 17, 20
	Enterotoxin C$_2$	SEC$_2$	26,000	1–10 ng/ml	12, 13.1, 17
	Enterotoxin D	SED	27,300	0.1–1 ng/ml	
	Enterotoxin E	SEE	29,600	0.1–1 ng/ml	8
	Toxic shock syndrome toxin-1	TSST-1	22,000	1–10 ng/ml	2
Streptococcus pyogenes	Erythrogenic toxin A	ETA	29,200	0.1–1 ng/ml	8
	Erythrogenic toxin B	ETB	27,000	1000 ng/ml	8
	M protein	pepM	33,000	1000 ng/ml	8
P. aeruginosa	Exotoxin A		66,000	10–100 ng/ml	
Mycoplasma arthritidis	soluble mitogen	MAM	~15,000	1–10 ng/ml	

[a] From Fleischer (1991).
[b]

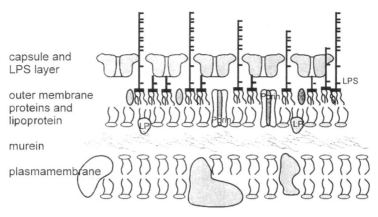

FIG. 5   Model of the cell wall of Gram-negative bacteria.

sion) in large detail for structure and function relationships. While only the M-protein of Gram-positive bacteria has been implicated in host–bacteria interactions, several outer membrane proteins of Gram-negative bacteria have been found to modulate immune cell functions (Table III). In some bacteria, capsules have been found (compare Table IV). They always seem to serve the same purpose: to confer resistance to phagocytosis or to permit intraphagosomal survival.

TABLE III

Outer Membrane Proteins of Gram-Negative Bacteria[a]

Protein	Characteristics	Function
OMP-A	Structural protein	Retains the outer membrane to the peptidoglycan via ionic interactions
OMP-B	Regulatory protein for the expression of OMP-C and OMP-F	Contains two proteins: the EnvZ enzyme with kinase and phosphatase activity and the OMP-R DNA-binding protein
OMP-C	Porin	
OMP-F	Porin	
OMP-%	DNA-binding protein	(See OMP-B)
LamB	Porin, phage λ-receptor	Maltose uptake
PhoE	(Phosphate uptake ?)	Induced by phosphate starvation

[a] According to Benz and Bauer (1988).

a. Lipopolysaccharides Lipopolysaccharides (Rietschel and Brade, 1992) are the endotoxins of Gram-negative bacteria. They are constituents of the bacterial cell wall, in which they are localized in the outer leaflet of the outer membrane. LPS provide essential physiological functions for the bacteria (e.g., rigidity of the cell wall, protection against antibiotics, and host defense strategies). They are antigens and, thus, possess binding sites for antibodies and for other serum factors. LPSs are important for bacterial virulence and are potent immunostimulators (e.g., activation of B lymphocytes and granulocytes).

The common architecture of LPS is sketched in Fig. 6. They consist of three regions, which are differentiated by their structure, genetics, biosynthesis, and biological properties. The molecule is anchored in the outer membrane by a lipid moiety termed lipid A, to which a more or less complex saccharide portion is linked. This saccharide consists of an oligosaccharide unit consisting of up to fifteen monosaccharides (the core region), to which a polysaccharide made up of repeating units is linked (the O-antigen). This type of LPS is called S-form LPS, and it was identified in bacteria of remote origin, e.g., Enterobacteriaceae, Vibrionaceae, and Pseudomonadaceae.

The second type of LPS, termed R-form LPS, consists only of the core region and lipid A. It was first obtained from bacteria like *Salmonella minnesota* and *Escherichia coli* which possessed a defect in the gene cluster encoding for the synthesis of the O-antigen (*rfb* locus) (Galanos *et al.*, 1977). However, today several species are known expressing R-form LPS without any known defect in LPS biosynthesis (e.g., *Chlamydia, Bordetella pertussis, Haemophilus influenzae*).

The carbohydrate backbone of lipid A (Rietschel *et al.*, 1992; Takayama and Qureshi, 1992) consists of a β-(1-6)-linked amino-hexose disaccharide. This comprises in most cases D-glucosamine (GlcN, Fig. 7). In several LPSs (e.g., *Campylobacter jejuni*), 2,3-dideoxy-2,3-diamino-D-glucose (GlcN3N) was identified, which may be present as a constituent of pure or hybrid (with GlcN) disaccharides. The disaccharide backbone is in most cases phosphorylated, namely at positions C-1 of the reducing and C-4' of the nonreducing aminohexose, and these phosphate groups may be further substituted.

Characteristic acyl residues of lipid A are (R)-3-hydroxy fatty acids, which are partly esterified at their OH-groups by other fatty acids (3-

FIG. 6   Scheme of the general structure of bacterial LPS.

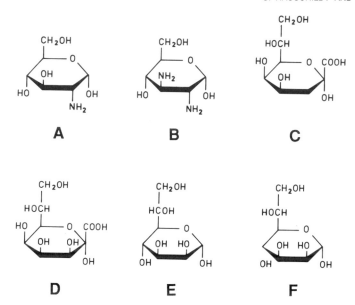

FIG. 7   Formulas of (A), 2-amino-2-deoxy-D-glucopyranose (D-glucosamine, GlcN); (B), 2,3-dideoxy-2,3-diamino-D-glucopyranose (GlcN3N); (C), 3-deoxy-D-*manno*-octulopyranosonic acid (Kdo); (D), D-*glycero*-D-*talo*-octulopyranosonic acid (Ko); (E), L-*glycero*-D-*manno*-heptopyranose; (F), D-*glycero*-D-*manno*-heptopyranose.

acyloxyacyl groups). In Fig. 8 the structures of lipid A from *E. coli* and *Chromobacterium violaceum* are shown, representing two types of fatty acid distribution over the GlcN residues. In *E. coli* lipid A, six fatty acids are nonsymmetrically distributed. The nonreducing GlcN II carries an ester-linked (at O-3′) (R)-3-hydroxy tetradecanoic acid that is esterified with tetradecanoic acid (3-tetradecanoyloxytetradecanoic acid) and an amide-linked 3-dodecanoyloxytetradecanoic acid. The reducing GlcN I carries unsubstituted tetradecanoic acid in ester (O-3) and amide linkage. In the lipid A of *C. violaceum,* the fatty acids are symmetrically distributed. Both GlcN residues carry an ester-linked (O-3, O-3′) 3-hydroxydecanoic acid and an amide-linked 3-dodecanoyloxydodecanoic acid. It should be noted that lipid A from various LPSs differs in the amount, quality, and distribution of fatty acids.

The characteristic constituents of the core region (Holst and Brade, 1992) are also shown in Fig. 7. All LPSs contain at least one residue of 3-deoxy-D-*manno*-octulopyranosonic acid (Kdo), which links the core region to the lipid A moiety. Mutants that do not contain Kdo are not viable. Only one case is so far known (LPS of *Acinetobacter calcoaceticus*) in which this Kdo residue is replaced by D-*glycero*-D-*talo*-octulopyranosonic acid (Ko). This

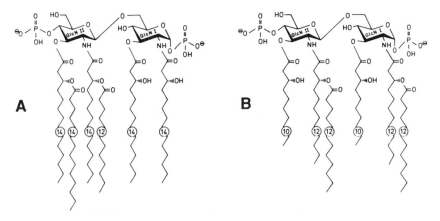

FIG. 8   Structures of the lipid A from (A) *E. coli* and (B) *C. violaceum.*

LPS also contains Kdo residues. Other typical components of the core region are heptopyranoses, either in the L, D or in the D, D configuration. However, there are core regions in which both heptoses are present and others which lack these compounds. The core region is negatively charged, owing to (in addition to Kdo) either the substitution with phosphate groups or, in the case of nonphosphorylated cores, with other acidic compounds, for example, uronic acids. The phosphate substitution of the core region is an important structural parameter; however, many of the structural features are to date not known.

The O-antigen (Galanos *et al.,* 1977; Lüderitz *et al.,* 1982; Jann and Jann, 1984) comprises in many cases a heteropolysaccharide of repeating units made up of between two and eight monosaccharides. Homopolymers have also been identified. The length of the O-antigen varies in different bacteria; it may contain up to 50 repeating units. A large variety of constituents have been identified: neutral sugars, amino sugars (also with substituted amino groups), deoxy sugars, sugar acids (also in the amidated form), and sugar phosphates. In addition, some positions may be *O*-acetylated or *O*-methylated. From this variety of constituents the synthesis of a broad structural diversity of O-antigens is furnished, which defines the variety of O-antigenic factors. A repeating unit may carry several of these O-antigenic factors, and an O-antigenic factor can be determined by a monosaccharide moiety or even by parts of it. The serological specificity of S-form bacteria is determined by the O-factors, of which many have already been structurally characterized.

b. Lipoproteins and Lipopeptides   The lipoprotein of *E. coli* is composed of 58 amino acids (Fig. 9). It is synthesized from a prolipoprotein equipped

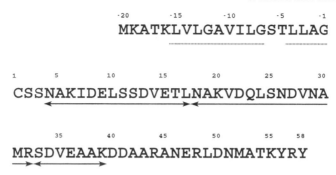

FIG. 9   Sequence of the *E. coli* (pro-)lipoprotein (Braun and Bosch, 1973).

with an additional 20-amino acid peptide which functions as a signal peptide and mediates the cross-membrane transport of the prolipoprotein into the periplasmic space. In the periplasmic space the signal peptide is replaced by a long-chain fatty acid bound to the new N-terminal cysteine residue. A glyceride is further linked to the cysteine by a thioether (Braun and Bosch, 1973). The protein moiety of lipoprotein is composed of helical units like the M-proteins shown in Fig. 4. One 13-amino acid and one 14-amino acid-long unit (4–17, 18–32), each forming four helical turns, may have evolved by gene duplication. One 7-amino acid stretch (33–39), forming two helical turns seems to have been multiplied four times. However, deletions obscure the original helical structure (Braun and Bosch, 1972). The lipoprotein lacks proline, which would destroy the helical secondary structure. Synthetic peptides which span the region from AA 2–20 were shown to form helices, while peptides of the C-terminal part were randomly folded (Loleit *et al.*, 1994).

   Most of the biological activity of lipoprotein resides within the N-terminal lipoamino acid (Bessler *et al.*, 1977). The natural product bears a variety of fatty acids at the C-2 position of the glycerol with up to 30% unsaturated and cylopropane fatty acids, similar to bacterial phospholipids, in which the C-2 of the glycerol is mostly substituted by unsaturated fatty acids. These unsaturated fatty acids are, however, not necessary for the interactions of lipoproteins with the immune system: Synthetic lipopeptide analogs carrying three saturated palmitoyl groups (called Pam$_3$Cys = S-2,3-bispalmi-toyloxypropyl-*N*-palmitoyl-cysteine; see Fig. 10) are as active as natural lipopeptides (Bessler *et al.*, 1985).

   Structural analogs of the natural lipopeptides have been prepared to analyze the structure and function relationship. Analogs have been tested for, for example, lymphocyte mitogenicity *in vitro* (Bessler *et al.*, 1985), adjuvant activity *in vivo* (Lex *et al.*, 1986), or for their capacity to activate

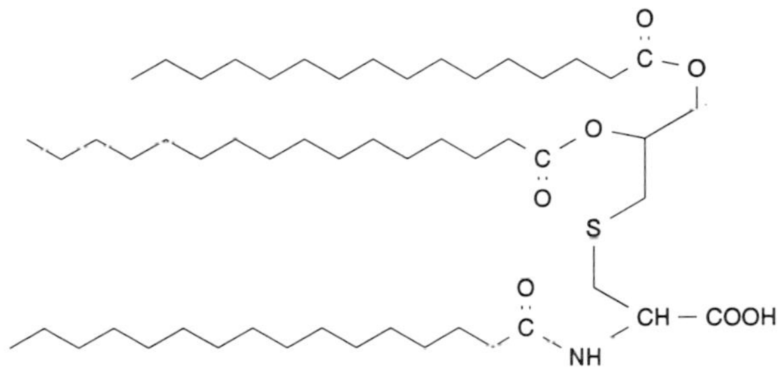

FIG. 10 S-2,3-bispalmitoyloxypropyl-N-palmitoyl-cysteine, the bioactive lipid of synthetic lipopeptides.

macrophages, leading to cytokine production, phagocytosis, major histocompatibility complex (MHC) expression (Hauschildt et al., 1990a) and cytotoxicity in vitro (Hoffmann et al., 1988, 1989). The lipoamino acid has two optically active carbon atoms: the C 2 of the glycerol and the α-C of the cysteine. Of the four Pam3Cys stereoisomers, the R,R configuration identical to the natural configuration was most potent, while the S,S configuration was the least active isomer. S,R and R,S isomers proved to be mediocre stimuli (Wiesmüller et al., 1990).

The substitution of sulfur by a methylene group led to the active carba-analog (Metzger et al., 1991). A lack of fatty acids reduced the biological activity of the lipopeptide analog dramatically. To test the importance of the ester link between the glycerol and the two FA substituents, a synthetic product was constructed with a 1,2-dicarboxy-6-amino-hexane backbone and two C16 alcohols in ester linkage while the amino group was substituted by an amide-linked palmitic acid. This molecule did not exhibit biological activity, and it was concluded that the ester links between glycerol and its FA substituents are indeed essential for the biological function of the lipopeptide (J. Metzger, W. G. Bessler, and G. Jung, unpublished results).

While the lipoamino acid could efficiently be replaced only by its carba-analog, it became clear that the residues attached to the carboxy terminal of Pam3Cys are less important. In the absence of any further AA, however, the lipoamino acid alone was only marginally active. Adding a single serine to Pam3Cys restored the biological activity (Bessler et al., 1985). Several dipeptides such as Ser-Ser, Asn-Ala, or Ala-Gly attached to Pam3Cys were found to be as active as the natural tetrapeptide Ser-Ser-Asn-Ala of the natural lipopeptide. The Pam3Cys-Lys-Lys-Lys-Lys as well as the Pam3Cys-Glu-Glu-Glu-Glu derivatives carrying four positive or four negative charges, respectively, are water-soluble analogs and potent stimulators in biological assays. Water solubility could be enhanced by the addition of polyoxyethylene (polyethyleneglycol), which did not influence the biological activity of the Pam$_3$Cys-Ser-(Glu) moiety (W. Bessler, unpublished results; Kleine et al., 1994).

c. Outer Membrane Proteins of Gram-Negative Bacteria

Unlike the cell-wall proteins of Gram-positive bacteria, among which only the M-proteins have been shown to be involved in host–microbe interactions, several outer membrane proteins of Gram-negative bacteria (Table III) have been found to modulate host responses. It has even been claimed that endotoxicity, including lethal toxicity, could be evoked by outer membrane proteins of Gram-negative strains (Goodman and Sultzer, 1979). However, such an "endotoxin protein" has been shown to be a mixture of several OMPs possibly contaminated with endotoxin preparations (Goldman et al., 1981). The most abundant OMP-A protein is required to bind the outer membrane to the peptidoglycan (Benz and Bauer, 1988). Other OMPs, OMP-C and OMP-F, form water-filled diffusion pores allowing low molecular solutes to enter the bacterium (Benz and Bauer, 1988). The OMP-C porin protein is a virulence factor of Shigella flexneri (Bernardini et al., 1993).

OMP-F and peptides thereof have been shown to be involved in host–bacteria interactions, to be mitogenic for mouse and human lymphocytes, and to stimulate macrophages to become cytotoxic for tumor target lines (Vordermeier and Bessler, 1987; Vordermeier et al., 1987, 1990). Such porins are found in all Gram-negative bacteria. The porin of P. aeruginosa is a prominent virulence factor of this strain. It forms very small pores so that even low-molecular-weight antibiotic substances are excluded, which renders these bacteria resistant to antibiotic treatments (Benz and Bauer, 1988). The porin of S. typhimurium reacts in vitro with the C1q complement component, either in its soluble form or in its membrane-attached form found on peritoneal exudate macrophages (Latsch et al., 1990).

A 15-kDa OMP has been found firmly attached to LPSs (Geyer et al., 1979). A 39-kDa OMP from P. mirabilis also forms stable complexes with LPS. These complexes have been found to reduce the LPS-induced superoxide anion and IL-1 production in macrophages (Weber et al., 1992). In contrast, the suppression of MHC class II expression by LPS in these cells was not affected, and the TNF-mediated macrophage cytotoxicity against tumor target lines was even enhanced (Weber et al., 1993). Anti-LPS antibody responses in mice are predominently T-cell independent and of the IgM type. When LPS, however, is injected in complex with the 39-kDa protein, anti-LPS antibodies of IgG isotypes, predominently IgG2a, were observed. Thus, the presence of the 39-kDa protein, which by itself is a strong antigen, renders the anti-LPS antibody production T-cell dependent (Nixdorff et al., 1992).

d. Capsules of Gram-Negative Bacteria

There are many more encapsulated strains among Gram-negative than among Gram-positive bacteria

(Jann and Jann, 1990). Isolates of *Neisseria,* Enterobacteriaceae, *Vibrio, Haemophilus influenzae, Pasteurella,* and *Bacteroides fragilis* possess capsules (Table IV). The majority of these capsules are composed of acidic polysaccharides. The capsules of *E. coli, Haemophilus influenzae,* and *Neisseria meningitidis* have been analyzed in detail. These bacteria share two features with group B streptococci and *S. pneumoniae:* the acidic capsular polysaccharides and the capacity to produce meningitis in humans. Among

TABLE IV

Encapsulated Bacteria

Species	Capsules	Function
Gram-positive bacteria		
Bacillus anthrax	D-Glutamate polymer	Antiphagocytotic
Pneumococcal c.	>80 Type-specific polysaccharides	Capsules determine pathogenicity
Staphylococcal c.	Glucosaminuronic acid polymer (only in few strains and isolates)	?
Streptococcal c.	Hyaluronate capsule	Mimicry of the extracellular matrix (?)
Gram-negative bacteria		
Bacteroides fragilis	Polysaccharide capsule of L-fucose, D-galactose, DL-quinovasamine, D-glucosamine	Phagocytosis resistance
Haemophilus influenza	Capsular polysaccharides, several types; e.g., ribose-ribitol phosphate or glucose-ribitol phosphate	Antiphagocytotic
Neisseria	Group-specific polysaccharides containing sialic acid	?
Pseudomonas aeruginosa	Polymer of mannuronic acid and guluronic acid: alginate	Protection from phagocytosis and from antibody and complement attack
Yersinia pestis	Fraction 1 antigen: lipid-protein-polysaccharide	Intracellular survival in macrophages

100 different capsular types, only K1, K5, and K12 are associated with the disease. K1 is a polymer of N-acetylneuraminic acid in α-2,8 linkage. K5 is composed of a glucuronic acid-(1,4)-α-Glc-NAc-β1,4 polymer, while K12 is built of a rhamnose-(1,2)-α-Rha-(1,5)-β-(7/8-OAc)KDO-α2,3 polymer. The capsule of *H. influenzae* B (HIB), like the *E. coli* K100 type, is composed of a ribose-ribitol phosphate polymer. The attachment of capsular polysaccharides (PS) to the bacterial membrane is achieved by lipid moieties linked to the reducing sugar moieties of the polysaccharide. One type of capsule binds to the core region of LPS. The polysaccharide moiety of the enterobacterial common antigen (ECA) has also been found (partially) attached to the LPS core (Kuhn *et al.*, 1988). Other capsular PSs are linked to phosphatidic acid by an acid-labile phosphodiester link. The extent to which the capsular PSs are bound to these lipids varies. Between 20 and 50% of the PSs are found linked to phosphatidic acid.

Owing to the mimicry of host structures, these capsules are nonimmunogenic: K1 is identical to the carbohydrate region of the embryonic form of the neuronal cellular adhesion molecule (N-CAM). K5 is identical to the first intermediates of heparin. The K1 structure is common to *E. coli*, group B streptococci, and *N. meningitidis* B, which is the pathogenic type of *N. meningitidis*. These capsules prevent the binding of O-antigen-specific antibodies. They also provide a binding site for factor H and C3b which results in the breakdown of the amplification loop of the alternative pathway of complement activation (Moxon and Kroll, 1990). The streptococcal M-protein (Section II,B,1), which binds factor H with high affinity, inactivates the alternative pathway in a similar way (Cross, 1990).

It has also been suggested that direct interactions between phagocytes and capsules are impaired by the hydrophilic nature of capsules. Negative charges on both the bacterial and the monocytic surface may lead to mutual repulsion and render the bacteria resistant to phagocytosis.

2. Soluble Components of Gram-Negative Bacteria

Two exotoxins of Gram-negative bacteria have already been mentioned (see Section II,B,2): cholera toxin and the heat-labile *E. coli* enterotoxin. These proteins share a common mode of action as well as a high structural similarity (about 70% of the nucleotide and amino acid sequences are conserved). CT consists of two types of subunits: an A subunit of 27215 Da with two disulfide-linked fragments (A_1 and A_2), and a B subunit of 11,677 Da. The holotoxin is composed of one A subunit and five B subunits. After the binding of a B subunit to a ganglioside, the A1 fragment is translocated into the cell, where it catalyzes ADP-ribosylation of the G_s-protein which regulates cAMP synthesis. The LT gene (*elt*) could be cloned and efficiently be expressed in enterotoxogenic *E. coli* strains as well as in other bacteria. However, the expression of the CT gene (*ctx*) was much reduced in *E. coli* compared with *Vibrio cholerae*.

When the toxin regulator gene (*toxR*) was cotransfected into *E. coli,* the expression of CT was enhanced, suggesting that the toxR protein binds to a cholera-specific promoter DNA. This promoter contains the heptamer sequence TTTTGAT, tandemly repeated three to eight times. Less toxogenic strains either lack toxR or have fewer heptamer repeats. Avirulent strains lack the *ctx* gene. It should be noted that the *ctx* gene is chromosomally located while the LT gene (*elt*) is plasmid-encoded. The production of LT in toxogenic strains is about 200-fold less than the production of CT in virulent strains (Mekalanos, 1985). CT blocks mucosal-associated bactericidal activity and impairs the function of macrophages and neutrophils, thus allowing the bacteria to grow and destroy the digestive tract.

Dysentery results from the action of another bacterial toxin, the shiga toxin of *Shigella dysenteriae* type 1 and of closely related shiga-like toxins (SLT). The chromosomally encoded shiga toxin is composed of a single enzymatic A unit with 293 amino acids (MG = 32,335 Da) and of four or five binding B units, each consisting of 69 amino acids (MG = 7691 Da). A bacterial protease cleaves the precursor form of an active A unit into two fragments of 27 kDa and 4 kDa. After the binding of the activated toxin to Gal-α1-4-Gal containing glycosphingolipids, the 27-kDa fragment enters the cytosol of target cells by endocytic uptake via clathrin-coated pits. Within the cell, the toxin inhibits protein synthesis. The specific *N*-glycosidase activity of the toxin splits one particular adenosine residue of the 28S RNA of the 60S ribosomal subunit. This mode of action is similar to the toxic action of ricin. It looks as if the adenosine residues cleaved off by both toxins are necessary for the binding of elongation factors (O'Brien *et al.,* 1992).

D. Special Bacterial Taxa

So far Gram-positive and Gram-negative bacteria have been compared. There are, however, other genera of bacteria such as mycobacteria and related norcardia and corynebacteria, that are distinct from either Gram-positive or Gram-negative bacteria. Their surface components consist of a variety of lipids, waxes, or glycolipids. This surface layer of lipids is attached to the peptidoglycan by a polymer, arabinogalactan, substituted by mycolic acid (MA) residues. These MA are characteristic for mycobacteria and related Actinomycetes. They consist of α-alkyl-β-hydroxy branched, long-chain fatty acids mainly in THREO configuration.

Another hydrophobic surface component, termed "cord factor," was identified as 6,6'-dimycolyltrehalose (Fig. 11). Usually, mixtures of MA

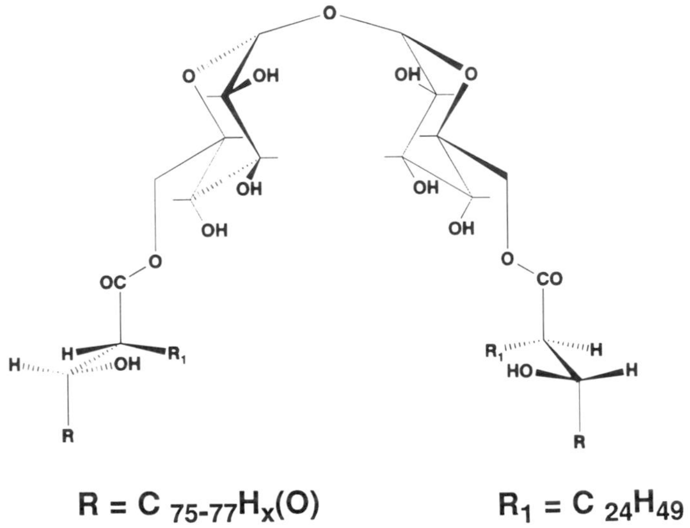

$$R = C_{75-77}H_x(O) \qquad\qquad R_1 = C_{24}H_{49}$$

FIG. 11   Cord factor, a major toxin of mycobacteria.

with different chain lengths are found, the chain length distribution of the MA being strain-specific. In *M. tuberculosis,* the majority of MAs possess 78 or 80 carbon atoms. In these bacteria the substituent in the α-position has been identified as a $C_{24}H_{49}$ alkyl chain (Wolinski, 1990). The cord factor is a major toxin of mycobacteria: When it is extracted by organic solvents, the bacteria remain viable but become avirulent. It affects leukocyte migration and doses of about 10 μg are lethal for mice. Mycobacteria passaged *in vitro* for a long time (years) are less virulent, and contain smaller amounts of cord factor than bacteria freshly isolated from animals or passaged for shorter periods. The toxicity of the cord factor in mice may be due to effects on the hepatic lipid metabolism, on microsomal enzymes, or on mitochondria (Wolinski, 1990).

Other bacterial surface lipids which have been implicated in the interactions of bacteria with macrophages include a sulfolipid (Fig. 12) as well as a phenolic glycolipid (Fig. 13). The sulfolipid was shown to inhibit the priming of cultured human monocytes (Pabst *et al.,* 1988), whereas the phenolic glycolipid acts as a OH' radical scavenger and blocks its antimicrobial activity (Neill and Klebanoff, 1988).

The outcome of an infection by *M. tuberculosis* or by *M. leprae* is partly dependent on the virulence of the bacteria, and on the number of bacteria inoculated. On the other hand, it also depends on the immune status of the host. In primary, exudative lesions, *M. tuberculosis* can infect macrophages and multiply within these cells. In later stages, when delayed-type

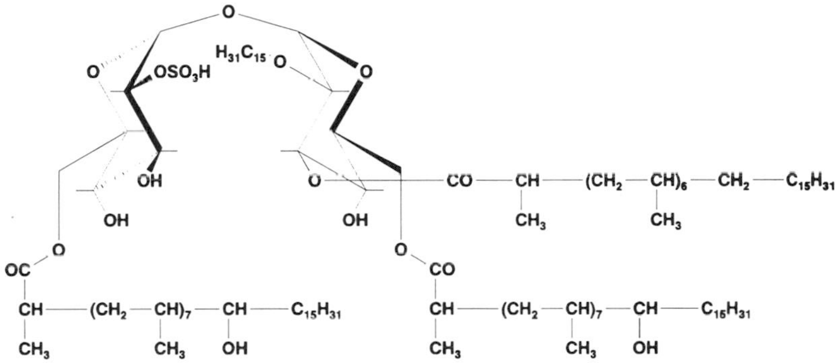

FIG. 12   The sulfolipid from *M. tuberculosis.*

hyperreactivity (DTH) immune responses have been mounted against bacterial proteins [e.g., tuberculin, purified protein derivative (PPD)], the intracellular bacteria are killed by activated macrophages. Upon contact with bacterial products, such activated macrophages become concentrically arranged in the form of epithelioid cells forming the tubercles characteristic of the disease. These cells may fuse into giant cells with many nuclei arranged at their periphery and viable mycobacteria visible in the cytoplasm. A mantle of lymphocytes and proliferating fibroblasts is formed around multiple layers of epitheloid cells.

Similar granulomatous lesions are also found in other mycobacterial infections, for example, leprosy (Wolinski, 1990). In lepromatous leprosy, however, bacteria multiply within "foamy" macrophages (so-called "lepra cells"). A cell-mediated immune response against bacterial protein (lepromin) is barely detectable. In the tuberculoid (granulomatous) phase of leprosy, few lepra cells and intracellular bacteria are found, and an antilepromin DTH reaction in the patient's skin can be evoked (Wolinski, 1990). The lepromatous phase is characterized by a lack of interferon-γ-producing T_H1 cells while T_H2 cells secreting IL-4 and IL-10 are present. Macrophages cannot be activated because IFN-γ is missing. By injecting purified protein derivative or IL-2, macrophages and T lymphocytes were recruited which mounted an antibacterial immune response and induced phagocytosis and destruction of bacteria (Kaplan, 1993). The different T-cell subpopulations involved and the nature of cytokines secreted have been recently reviewed (Salgame *et al.,* 1992; Kaplan, 1993).

While live mycobacteria are dangerous for the vertebrate host, killed bacteria have been found to constitute potent stimuli of the immune system. The complete Freund's adjuvant (CFA) contains killed mycobacteria and mineral acids. This adjuvant has been used in numerous cases to induce

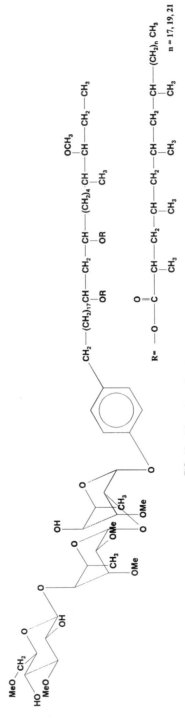

FIG. 13 The phenolic glycolipid of *M. tuberculosis*.

immune responses against antigens which, when given alone, would not induce any humoral or cellular responses. The use of CFA is, however, forbidden for human purposes because large granulomas are induced at the injection site. The search for the minimal structure of mycobacteria which could substitute for the intact bacteria in CFA led to the finding that a peptidoglycan substructure could be used. This minimal PG component was muramyl dipeptide which has already been described (see Section II,B,1).

III. Macrophages as Targets in Bacteria–Macrophage Interactions

A. General Considerations

Pathogenic bacteria have evolved many creative means to either avoid uptake by the macrophage or to invade the host cells. On the other hand, macrophages have developed means to take up bacteria and to destroy them. Uptake is facilitated by receptors present on the surface of the macrophage. To destroy pathogens that enter macrophages and replicate within them, activation of macrophages seems to be a prerequisite. According to Adams and Hamilton (1984), macrophage activation can be defined as acquisition of competence to complete complex functions. Signal transduction mechanisms underlying macrophage activation will be discussed, with lipopolysaccharide serving as an example for illustrating bacteria-induced signal transduction events.

B. Macrophage Surface Receptors

The plasma membrane of macrophages contains receptors that mediate attachment and entry of bacteria into the cell. Among these receptors are those that recognize (1) the Fc region of the IgG molecule, (2) the third component of complement (C3), and (3) mannose/fucose-N-acetylglucosamine-terminated glycoproteins. Coating bacteria with antibodies or C3 (opsonization) precedes their binding to the respective receptors. Macrophages express three classes of Fc receptors for IgG: FcγRI (CD64) with high affinity for monomeric IgG (Anderson, 1982), the rank order for human isotypes being $IgG_1 > IgG_3 > IgG_4 > IgG_2$; Fc$\gamma$RII (CD32) (Rosenfeld et al., 1985), and FcγRIII (CD16) (Clarkson and Ory, 1988), which have low affinity for monomeric IgG, but can effectively bind complexed IgG (Unkeless et al., 1988). In contrast to the human FcγRI and II, the murine FcγRI/II receptors recognize the isotypes IgG_{2a} and IgG_{2b}, respectively (Unkeless *et al.*, 1981). In addition, macrophages express the low affinity receptor for IgE, FcΣ RII (CD23), which has striking homology to several lectin-binding proteins (Ikuta *et al.*, 1987).

Receptors involved in complement-mediated phagocytosis are CR1 (CD35), CR3 (CD11b/CD18), and CR4 (CD11c/CD18). In contrast to Fc receptors that are constitutively present and functional on the membrane surface, surface expression of CR1 (Wright and Silverstein, 1982) and CR3, which are maintained in the cytoplasm, is upregulated upon stimulating the cells. CR1 is the ligand for the complement components C3b/C4b, while CR3 as well as CR4 recognize the complement fragment iC3b (Myones *et al.*, 1988). CR4 and CR3 are both members of the integrin supergene family of leukocyte cell adhesion molecules (Leu-CAM) (Ross *et al.*, 1989). These heterodimers share a β_2 chain, CD18, and are therefore called β_2 integrins. The α-chain contains binding sites for Ca^{2+} and Mg^{2+}. These cations seem to have a role in iC3b binding by CR3 and CR4; a high-affinity calcium binding site has been found on CR3 (Graham and Brown, 1991).

Opsonization of bacteria by complement components and/or specific antibodies has the advantage that a large variety of bacteria can be recognized by a limited number of Fc receptors on macrophages and on other phagocytic cells. In addition, phagocytosis can also occur via lectin-like proteins specific for oligosaccharides terminating in fucose, mannose, and galactose (Sharon, 1984; Ofek and Sharon, 1988; Gordon *et al.*, 1988). The mannosyl–fucosyl receptor involved in the binding and phagocytosis of certain bacteria (Speert *et al.*, 1988) is perhaps the best characterized of these receptors (Ezekowitz and Stahl, 1988). Its surface expression increases as human monocytes mature into macrophages. It is downregulated when macrophages are activated by interferon-γ (Gordon and Mokoena, 1989). Activation also leads to the downregulation of the transferrin receptor, by which iron is taken up by macrophages. Since iron is essential for bacterial survival, a decreased iron uptake diminishes bacterial replication, thus leading to an efficient kill of the bacteria once taken up by the macrophage.

Macrophages also possess receptors that recognize bacterial cell wall components. One of these components is lipopolysaccharide, the major integral constituent of the outer membrane of Gram-negative bacteria such as enterobacteriaceae, neisseriaceae, and chlamydiaceae (Lüderitz *et al.*, 1982). Only recently, CD14 has been identified as a receptor to which LPS, when complexed with the plasma protein, LPS-binding protein (LBP) binds (Tobias *et al.*, 1986, 1993). CD14 is a glycosylphosphatidylinositol (GPI)-anchored membrane protein, which also exists in a soluble form lacking the GPI anchor (Bazil *et al.*, 1986). Whether the binding of the LBP–LPS complex to CD14 alone is sufficient to activate macrophages remains to be established. It is possible that binding of LPS to CD14 is followed by a subsequent interaction of the LPS–CD14 complex with a different receptor that enables transmembrane signalling (Lee *et al.*, 1993). Alternatively, LPS may interact directly with other receptors without help from LBP or CD14.

C. Adhesion and Entry

One strategy bacteria use to overcome host defenses is to enter, multiply, and persist within host cells, thereby escaping humoral defense mechanisms. Macrophages often serve as host cells. On the one hand, they are well suited as an intracellular habitat because they are long-lived and highly phagocytic cells. Yet on the other hand, they are potent effectors of antimicrobial defenses so that intracellular bacteria must develop mechanisms for protecting themselves from the hostile environment. Bacterial internalization often begins with adhesion of the bacteria to the macrophage cell surface. Binding to integrin receptors, such as the complement receptors CR3 and CR4, appears to provide a general pathway for the entry of bacteria into the macrophage. Both *Legionella pneumophila* (Horwitz, 1992) and *Leishmania major* (Mosser and Edelson, 1985) can bind the complement fragment iC3b which is recognized by CR3. Complement receptors have also been shown to play a role in the uptake of bacteria such as *Mycobacterium tuberculosis* (Schlesinger *et al.*, 1990), *Mycobacterium leprae* (Schlesinger and Horwitz, 1990), *Bordetella pertussis* (Relman *et al.*, 1990), and certain *Salmonella* species (Ishibashi and Arai, 1990).

There is evidence that receptor–ligand interaction between macrophage and bacteria governs the postphagocytic fate of an internalized bacteria (Edelson, 1982; Joiner *et al.*, 1990; Falkow *et al.*, 1992; Conlan and North, 1992).

Entering host cells via complement receptors seems a safe route for bacteria, namely because they bypass the onset of an oxidative burst and the release of toxic oxygen metabolites (Wright and Silverstein, 1983; Yamamoto and Johnston, 1984). Consistent with this hypothesis, bacteria such as *Mycobacterium leprae* (Holzer *et al.*, 1986) and *Legionella pneumophila* (Jacobs *et al.*, 1984) fail to trigger an oxidative burst upon entry into the macrophage.

However, involvement of a complement receptor in the uptake of bacteria does not necessarily save bacteria from intracellular killing. As shown recently, binding of *Listeria monocytogenes* to CR3 during phagocytosis leads to bacterial killing, which is abolished when binding to CR3 is blocked (Drevets *et al.*, 1993). Since ligation of Fc receptors is often accompanied by an oxidative burst (Wright and Griffin, 1985), the type of opsonization may determine a bacteria's fate. For instance, intracellular killing of *Staphy-*

lococcus aureus by an oxygen-dependent mechanism is stimulated by engagement of Fc receptors on human monocytes (Zheng *et al.*, 1993).

The route of uptake may also govern the fate of bacteria within intracellular compartments. This is consistent with the finding that viable *Mycobacteria tuberculosis* ingested by macrophages enter phagosomes, which fail to fuse with lysosomes (Armstrong and D'Arcy Hart, 1971). When bacteria pretreated with antibodies are phagocytosed, fusion of phagosome-lysosome can be observed. However, the fusion does not lead to killing of the bacteria (Armstrong and D'Arcy Hart, 1975). *Mycobacterium leprae* (Frehel and Rostogi, 1987) and *Mycobacterium tuberculosis* (Goren *et al.*, 1976) also prevent phagosome-lysosome fusion.

Despite recent advances in understanding the binding of bacterial cell wall components to macrophage surfaces, the subsequent fate of these components in macrophages remains poorly characterized. The compounds best studied so far with respect to their uptake and distribution are lipopolysaccharides (Kang *et al.*, 1990) and lipopeptides. Lipopeptides, potent macrophages, and B-cell activators constitute the N-terminal part of lipoprotein and are considered to be responsible for the biological activity of lipoprotein, which like LPS, is a major cell wall component of Gram-negative bacteria (Bessler *et al.*, 1977). In order to follow the routes and to determine the distribution of lipopeptides within macrophages, an appropriate method to use is electron energy loss spectroscopy (EELS) (Joy and Maher, 1981; Wolf *et al.*, 1989). The electron microscope method is based on the fact that electrons interact with the target atoms upon passing through ultrathin cryosections of samples. Electrons lose characteristic, element-specific amounts of energy by exciting inner shell electrons of target atoms. The electron energy spectrum of the transmitted beam shows discontinuities (edges) at characteristic energies. By analyzing the amplitudes of the characteristic edges taken from different cellular compartments, qualitative information about the elemental distribution can be obtained. Fluorinating lipopeptides help to improve the detection of these substances within the cells, since cells contain hardly any fluorine and fluorine can easily be detected by its specific energy spectrum.

Using this method, lipopeptides were found to be rapidly taken up by macrophages (Wolf *et al.*, 1989; Uhl *et al.*, 1991). They are found attached to the plasma membrane, within the cytoplasm, in the nuclear membrane and nucleus 2 min after stimulation. Prolonging the incubation time up to 20 min is not associated with the appearance of lipopeptides in other cellular compartments, such as lysosomes, vesicles, and vacuoles. This observation and the finding that lipopeptides are found in the nucleus only 2 min after stimulation argues against a receptor-mediated endocytosis or pinocytosis. Whether the uptake into the nucleus is associated with gene regulation remains to be clarified. The fact that lipopeptides are found in different cell compartments might indicate that the interaction of lipopeptides with intracellular compartments is important for cell activation. In order to gain entry into the interior of cells, lipopeptides have to pass plasma membranes.

To gain more detailed information about the interaction of lipopeptide molecules with the plasma membrane of macrophages, the freeze fracture technique was used. Cells are frozen at −196°C in the presence of a cryoprotectant to prevent ice crystal formation. Then the frozen block is cracked with a knife blade. The fracture plane often passes through the hydrophobic middle of lipid bilayers, thereby exposing the interior of cell membranes (Fig. 14). Attachment of lipopeptides to the cell membranes of macrophages

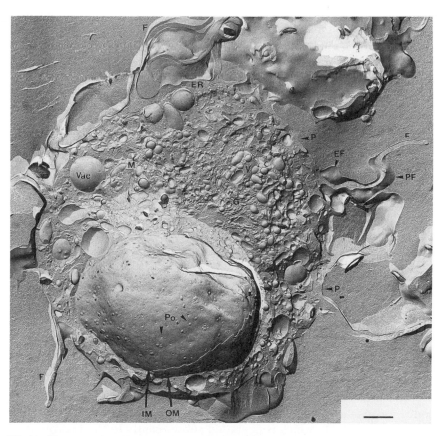

FIG. 14 Electron micrograph of a freeze-fractured bone marrow-derived macrophage. ER, endoplasmic reticulum; F, filopodium; G, Golgi cisternae; M, mitochondrion; P, plasma membrane; PF, plasmatic face of the plasma membrane; EF, extraplasmatic face of the plasma membrane; Po, pores of the nuclear membrane; OM, outer nuclear membrane; IM, inner nuclear membrane; V, vesicle; Vac, vacuole. Bar = 1 μm. (Reproduced from Uhl *et al.*, 1992.)

is followed by rapid changes of the plasma membrane structure (Uhl *et al.*, 1992). In both leaflets of the membrane (*E*- and *P*-face) the structural alteration consists of a transient decrease in particle density. This effect is not due to a loss of proteins, but is caused by lateral diffusion of single particles which subsequently aggregate. It remains to be clarified how far these changes and the presence of lipopeptides within macrophages are prerequisites for initiating the activation process by lipopeptides.

D. Signal Transduction Mechanisms

Once exposed to bacteria or bacterial components, macrophages increase their bactericidal or tumoricidal functions. To perform these functions, macrophages must recognize and respond to bacteria. Macrophage-specific responses are mediated through biochemical events triggered by the interaction between the bacteria and the macrophage. Intracellular pathogens can actively alter biochemical events and thus cause deactivation of mononuclear phagocytes (Reiner, 1994). This chapter describes some of the intracellular signal transduction mechanisms which are involved in bacteria-induced macrophage activation. LPS will serve as an example for illustrating bacteria-induced intracellular events.

One of the early events initiated by LPS is the phospholipase C (PLC)-catalyzed hydrolysis of phosphatidylinositol-4,5-bisphosphate (PIP_2), which generates inositol 1,4,5-trisphosphate ($InsP_3$) and 1,2-diacylglycerol (DAG) (Berridge and Irvine, 1984). $InsP_3$ and its phosphorylation product (inositol 1,3,4,5-tetrakisphosphate), have been linked in many cell systems to the elevation of intracellular levels of $[Ca^{2+}]_i$ from intracellular and extracellular stores, respectively. DAG is known to activate protein kinase C (Nishizuka, 1984), and the activation of this serine–threonine kinase is potentiated by increased cytosolic Ca^{2+} concentrations (Bell, 1986; Nishizuka, 1986). Several studies have suggested that treatment of macrophages with LPS results in the activation of PLC as measured by the formation of $InsP_3$. The response was modest and occurred 1–20 min after activation (Prpic *et al.*, 1987; Chang *et al.*, 1990). It is not clear how far this slow turnover of $InsP_3$, which according to Hurme *et al.* (1992) and Drysdale *et al.* (1987) is not accompanied by a rise in $[Ca^{2+}]_i$, is an obligatory cellular event in LPS-triggered activation (Chang *et al.*, 1990). The principal source of DAG production in macrophages is not PIP_2, but rather phosphatidylcholine (PC) (Sebaldt *et al.*, 1990), which is degraded subsequently to the hydrolysis of PIP_2. Guanine nucleotide regulatory proteins (G-proteins), often termed Gp, may be involved in coupling receptor activation to the hydrolysis of PIP_2 (Wallace and Fain, 1985; Uhing *et al.*, 1986; Litosch *et al.*, 1985; Straub and Gershengron, 1986; Cockcroft and Gomperts, 1985). These G-proteins bind guanosine diphosphate (GDP) in their resting state and become activated when GDP is exchanged for guanosine triphosphate as a result of stimulus–receptor interaction. G-proteins are heterotrimers with unique α-subunits, but shared β- and γ-subunits (Spiegel, 1987). The α-subunits range in size from 45 to 49 kDa and can be ADP-ribosylated by bacterial toxins, such as pertussis/cholera toxin (Gilman, 1987; Becker et al., 1986). G-proteins coupled to the PLC are mostly unresponsive to bacterial toxins; for example, the toxins have no effect on G-proteins of the Gq family, which activate the PLC β-1 selectively (Sternweis and Smrka, 1992). Currently there is no clear evidence that Gp proteins are involved in pathways induced by LPS. Although the involvement of adenylate cyclase-coupled G_i proteins in the regulation of cells by LPS has been suggested (Jakway and DeFranco, 1986), it is unclear by which mechanisms LPS might modulate the function of the protein.

Several of the LPS-induced effector functions have been reported to be sensitive to inhibitors of protein kinase C (Kovacs et al., 1988; Hurme and Serkkola, 1991; Novotney et al., 1991). PKC, a serine–threonine kinase, is involved in numerous cellular functions (Nishizuka, 1986). At present the mammalian PKC family consists of 12 different polypeptides, namely, α, βI, βII, γ, δ, ε, ζ, η, θ, ι, λ, and μ. They differ in localization, cofactor dependence, and substrate range, indicating that each might have a separate and unique function in the cell (Dekker and Parker, 1994). The N-terminal half of the 80-kDa protein is considered to be the regulatory domain, whereas the C-terminal half is considered to contain the catalytic domain. The regulatory domain has binding sites for Ca^{2+}, 1,2DAG, and phosphatidylserine, and binding sites for ATP have been found in the catalytic domain.

The regulatory domain can be cleaved from the catalytic domain by the Ca^{2+}-dependent neutral protease calpain (Kishimoto et al., 1989). The properties of PKC present in mouse macrophages have not yet been fully characterized. In the J744 macrophage cell line, LPS induces the cleavage of PKC, which results in the production of the regulatory and catalytic domain fragments (Novotney et al., 1991), whereas PKC of cells of the $P388D_1$ macrophage cell line do not undergo proteolytic cleavage upon stimulation by LPS.

As suggested by Chen et al. (1992), there may be two alternative pathways by which LPS activates PKS. Either activation occurs via DAG and $InsP_3$, or via calpain-catalyzed cleavage of PKC, yielding the active 40-kDa catalytic domain (PKM). The cleavage of PKC to PKM by calpain frees it from regulatory constraint and releases PKM into the cytosol, where it can phosphorylate a distinct set of proteins. A role for LPS in activating calpain has not yet been elucidated.

LPS induces the transcription, translation, and myristoylation of three proteins in macrophages with apparent molecular masses of 40, 42, and 68 kDa (Aderem *et al.*, 1986, 1988). All three proteins are substrates of PKC, and myristoylation of the 68-kDa protein enhances the rate of phosphorylation via PKC. Weinstein *et al.* (1991, 1992) showed that LPS induces rapid protein phosphorylation of a number of distinct proteins in macrophages. Two of the phosphoproteins have been identified as isoforms of the mitogen-activated protein kinases (MAPK1 and MAPK2). Because MAPK1 and MAPK2 are implicated in the regulation of a broad range of cellular processes (Cobb *et al.*, 1991; Thomas, 1992), the activation of these kinases in LPS-stimulated macrophages may be a critical step leading to antibacterial responses. The same may be true for tyrosine phosphorylation induced by the LPS-LBP complex binding to CD14 (Weinstein *et al.*, 1993; Han *et al.*, 1993). The tyrosine protein kinases (TPKs) can be divided into two groups: one group, the "receptor TPKs" that possess extracellular domains which generally bind polypeptide hormones, and another group, the "nonreceptor TPKs" that lack extracellular sequences (Bolen *et al.*, 1992). The nonreceptor TPKs, however, are often associated with cell-surface ligand proteins. There appear to be at least eight distinct families of nonreceptor TPKs. These include the Scr family, which seems to function as an early response surface receptor ligation-sensitive intracellular kinase and which in turn contains at least nine members. The members Scr, Fyn, Lyn, HcK, and Fgr are expressed in monocytes (Bolen *et al.*, 1992). Identification of their role in activation-induced intracellular events is one goal of future research.

To study signal transduction mechanisms used by other bacterial activators, pathways similar to those described have been investigated. Signals transduced by phagocytes via FcRγ seem to be associated with activation of certain protein kinases (Brozna *et al.*, 1988; Connelly *et al.*, 1991; Ting *et al.*, 1991), enhanced phosphatidylinositol turnover (Imamichi and Koyama, 1991), and an increase in $[Ca^{2+}]_i$ (Brunkhorst *et al.*, 1991; Imamichi *et al.*, 1990; Young *et al.*, 1984). Activation of phospholipase C is essential for IgG-stimulated intracellular killing of *S. aureus* (Zheng *et al.*, 1993). Signal transduction pathways utilized by the chemotactic factor *N*-formylmethionyl-leucyl-phenylalanine (fMLP) consist of activation of a pertussis toxin-sensitive G-protein ($G_{i\alpha2}$), activation of PLC leading to the formation of Ins P_3, and DAG. Cytosolic Ca^{2+} concentrations increase, and Ca^{2+}-dependent protein kinases, among them PKC, are activated (Verghese and Snyderman, 1989).

The bacterial lipopeptides, potent activators of macrophages (Hauschildt *et al.*, 1990a), fail to induce PKC activation, phosphoinositide degradation, or the generation of the second messengers cAMP and cGMP (Steffens *et al.*, 1989; Hauschildt *et al.*, 1990b). Associated with lipopeptide- and LPS-induced activation in macrophages is the change of the ADP-ribosylation state of a 32-kDa cytosolic protein (Hauschildt *et al.,* 1994). Like phosphorylation, ADP-ribosylation constitutes a covalent modification by which cells regulate protein functions. Whether this reversible protein modification is related to the functional activation of macrophages is not yet clear.

IV. Macrophages as Effector Cells in Bacteria–Macrophage Interactions

A. General Considerations

Macrophages provide a defense system against microbial invasion. They move toward microbial particles along a concentration gradient of chemotactic molecules. After opsonization, bacteria are phagocytosed, phagocytosis being a dynamic process in which bacteria are attached to the macrophage membrane in preparation for ingestion. The ability of the ingested particles or invading microorganisms to activate macrophages for antigen presentation is an important factor in the initiation of the immune response. It involves activation of T cells mediated by the interaction of the T cell with the antigen–MHC complex together with other surface or secreted molecules of the macrophage. Once ingested, bacteria are subject to killing by both oxidative and nonoxidative mechanisms. The optimum antibacterial activity depends on the assistance of a diverse range of substances, including cytokines, multifunctional substances that influence the ability of the host to respond to the bacteria.

B. Antigen Processing and Presentation

Macrophages belong to a group of cells, including B cells and dendritic cells, that are capable of taking up antigen and processing it before returning fragments to their surface. Fragments can also be presented that are derived from endogenous antigen. The fragments are recognized by effector cells [T cells, natural killer (NK) cells] in association with MHC class I or class II molecules. MHC class I molecules are found on all nucleated cells, MHC class II molecules on dendritic cells, B cells, monocytes and macrophages, and some activated human T cells. Both molecules are recognized by $\alpha\beta$- and possibly by $\gamma\delta$-T cells (Kronenberg, 1994). Recognition by the effector cells serves at least two purposes: defense against intracellular pathogens and feedback regulation of immune responses by elimination of stimulatory antigen-presenting cells.

MHC class I molecules consist of a heavy chain (H) and a light chain, known as β-2 microglobulin (β_2M) (Fig. 15). They are synthesized in the endoplasmic reticulum (ER), where they bind antigens in the form of peptides of restricted length (8–10 amino acids) (Falk *et al.*, 1991; Van Bleek and Nathenson, 1990). The newly formed class I antigen heterotrimer is then transported through the Golgi to the cell surface. The vast majority of peptides presented by class I molecules are derived from normal cellular proteins located in the nucleus and cytoplasm. In some cases also peptides derived from foreign material have been found to be associated with class I molecules (Pamer *et al.*, 1991). Cytosolic and nuclear proteins are degraded by an ATP-dependent large proteolytic complex, the proteasome (Goldberg and Rock, 1992). The peptides generated in the cytosol are then translocated to the lumen of the ER, probably by the peptide transporters TAP1 and

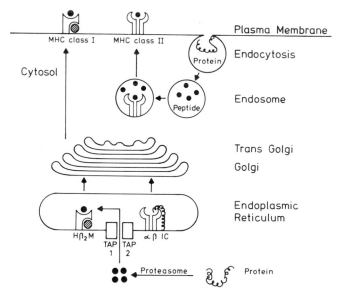

FIG. 15   A simplified view of antigen processing and presentation by MHC class I and II molecules. Processing and presentation by MHC class I molecules: Proteins are degraded in the cytosol by proteasomes to peptides, which are translocated by TAP1 and TAP2 to the ER. In the ER the peptides bind to newly synthesized MHC class I molecules, consisting of a heavy chain (H) β-2 microglobulin (β_2M) heterodimer. The loaded MHC molecules leave the ER, pass the Golgi, and reach the surface of the cell. Processing and presentation by MHC class II molecules: Proteins are taken up by endocytosis and degraded to peptides in endosomes. In late endocytic compartments with lysosomal characteristics, peptides are loaded to MHC class II molecules. These molecules, consisting of $\alpha\beta$ chains and the invariant chains (IC), leave the ER, pass the Golgi and trans-Golgi, and when they take up peptides consist only of the $\alpha\beta$ chain. The loaded molecules leave the endocytic route and reach the surface of the cell.

TAP2 (Powis *et al.*, 1992; Levy *et al.*, 1991; Koppelman *et al.*, 1992) (for transporter associated with antigen presentation). TAPs are essential for proper antigen processing; their exact function, however, is not known.

Once the MHC peptide heterotrimer reaches the cell surface, it can interact with cytotoxic class I-restricted CD8$^+$ α/β T cells. Upon interaction, CD8$^+$ cytotoxic T cells are triggered to release products such as perforin, enzymes, and different lymphokines, including IFN-γ, which can induce lethal processes in target cells (Apasov *et al.*, 1993).

The finding of class I-restricted CD8$^+$ T cells with reactivity for intracellular parasites (Rodrigues *et al.*, 1991; Kurlander *et al.*, 1992; Pamer *et al.*, 1992) raises the question of how such antigens are introduced into the class I pathway. Until recently it was thought that class II molecules predominantly present peptides from exogeneous antigens, whereas MHC class I molecules present peptides derived from intracellular proteins. However, this concept does not seem to be valid any longer. A number of class II ligands are known (Rudensky *et al.*, 1991; Nelson *et al.*, 1992) which are of endogenous origin, and exogenous protein can be presented by class I molecules (Moore *et al.*, 1988). Identification of class I-restricted CD8$^+$ T cells with reactivity for *Mycobacterium tuberculosis*, for example, suggests that *Mycobacterium tuberculosis*, which replicates inside the endosomal compartment, must have left the compartment as a result of endosomal damage. Leakage of mycobacterial proteins into the cytoplasm then leads to their processing and presentation in association with MHC class I molecules. Macrophages and other cells expressing the antigen in the context of class I molecules may be targeted by class I-restricted cytotoxic T cells and destroyed. Released bacteria from these cells are then taken up by monocytes.

MHC class II molecules are assembled in the ER from an α and β chain and the invariant chain (IC). Functions of the invariant chain may be to circumvent premature peptide binding in the ER (Roche and Cresswell, 1990; Teyton *et al.*, 1990) and to induce efficient transport of class II molecules from the ER (Anderson and Miller, 1992). A nonamer consisting of three invariant chain molecules and three α,β-dimers is transported through the Golgi to the trans-Golgi reticulum. From there MHC class II molecules enter the endocytic pathway. The invariant chain is degraded by proteases (Blum and Cresswell, 1988), thereby restoring the peptide binding capacity of class II molecules (Neefjes and Ploegh, 1992).

It is not yet clear where degradation of proteins and association of the resulting peptides with class II molecules takes place. Several data suggest that lysosomes may be the prime site for generation of peptides and their binding to class II molecules (Peters *et al.*, 1991; Neefjes and Ploegh, 1992). Peptides presented by class II molecules are derived from proteins that are present or directed to endosomal/lysosomal compartments. These peptides do not necessarily have to be generated from antigens taken up by antigen-presenting cells. Peptides bound to class II molecules are recognized by α/β CD4$^+$ T-helper (T$_H$) cells. Stimulated by the antigen and by interactions between other cell surface molecules, T cells proliferate and produce cytokines such as IL-2.

Using functional criteria and lymphokine profiles, one can distinguish two major subsets of T$_H$ cells, namely T$_H$1 and T$_H$2 cells. T$_H$1 cells, which produce IL-2 and IFN-γ, are primarily involved in helping inflammatory-type immune responses, whereas T$_H$2 cells, producing IL-4, IL-5, and IL-10, mainly mediate humoral immunity (Mosmann and Coffman, 1989a; Torbett et al., 1990; Firestein et al., 1989; Mosmann and Coffman, 1989b).

The type of T$_H$ cells presented at the site of an immune response to infectious agents will influence the response of the surrounding lymphocytes. For example, T$_H$2 cells can induce activation and differentiation of B cells, and T$_H$1 cells can stimulate macrophages. Macrophages, on the other hand, by interaction with bacteria, can produce cytokines which provide preferential stimulation of T$_H$1 cells (Gajewski et al., 1991). Future research is needed to determine how far certain cytokine patterns found after bacterial infections can be correlated with the clinical outcome of the infection. In leprosy patients, cytokine patterns involving IFN-γ, IL-2, and TNF-α tend to be associated with the resistance of patients to the disease, whereas IL-4, IL-5, and IL-10 tend to be associated with increased susceptibility (Uyemura et al., 1992).

In addition to α/β T cells, T lymphocytes that express the γ/δ T-cell antigen receptor also seem to play an important role in bacterial infection. These cells, predominantly CD4$^-$ and CD8$^-$, represent a minority of human blood cells. The nature of the antigen and the rules that govern antigen recognition by these cells are mainly unknown. It seems that antigen recognition by some γ/δ T cells does not necessarily require the participation of MHC class I and class II molecules. Antigens such as surface proteins, inducible self-proteins, proteins secreted by bacteria, or stress response proteins may be recognized by these cells in an MHC-unrestricted way. It remains to be clarified how far macrophages play a role in activating these cells.

C. Phagocytosis

Macrophages provide a major defense line against microbial invasion. They take up large particles, such as bacteria, and destroy them. They move toward the microbial particles guided by a gradient of chemotactic molecules. Once the particles are bound to the surface of the macrophage, the plasma membrane expands along the surface of the particles and engulfs them. This process requires the active participation of actin-containing microfilaments present under the cell surface. Attachment of the particles to the cell surface is mediated by specific macrophage receptors and is dependent upon the nature of the bacterial surface. Often bacteria become coated with "opsonins," such as antibodies (Ig) and fragments of the third component of complement (C3), thus preparing them for phagocytosis by macrophages that have specific receptors for several isotypes of C3 and for the Fc region of various antibodies. Complement components can be generated by macrophages as well as by bacteria that are capable of activating the complement cascade via the classical or alternative pathway. Phagocytosis of IgG and complement opsonized targets seems to be a property of activated cells.

Although unstimulated phagocytes moderately ingest IgG-opsonized targets, the uptake is enhanced in the presence of stimulating agents (Brown, 1991). This enhancement has been related to the generation of the oxidative burst (Gresham et al., 1990) and to the engagement of receptors other than CR3 and CR4 (Gresham et al., 1989) that recognize the sequence Arg-Gly-Asp. Activation of macrophages also seems to be necessary for phagocytosis via CR1, the receptor of C3b-containing complexes (Brown, 1989). The mechanism underlying activation of CR1 may involve protein kinase C-mediated phosphorylation of CR1 (Fallman et al., 1989; Bussolino et al., 1989). For IgG-Fc receptor-mediated or CR1-mediated phagocytosis, it has been suggested that CR3 may be important in organizing cytoskeletal events required for ingestion (Graham et al., 1989). This would imply a function of CR3 quite distinct from its ability to act as a receptor for iC3b.

Opsonized particles are engulfed by the plasma membrane of the macrophage by a sequential and circumferential interaction between ligand and receptor. This interaction, linking the two surfaces, is like that of a zipper closing and thus is called the "zipper interaction." Complete engulfment only occurs if the ligand is distributed uniformly over the whole particle surface. Whereas the "zipper interaction" or conventional phagocytosis applies to the mode of entry of most extracellular pathogens, other pathogens such as *Legionella pneumophila* enter macrophages by a process called "coiling phagocytosis" in which long phagocyte pseudopods coil around the organism as it is internalized (Horwitz, 1984). Whether this mode of entry can be related to its intracellular fate is not clear. Ingestion of bacteria also occurs in the absence of opsonins. It depends very much on the charge, hydrophobicity, and chemical composition of the bacterial surface and the complementary macrophage receptors. Thus, pili of *Pseudomonas aeruginosa* enhance their susceptibility to phagocytosis (Speert et al., 1986), whereas the carbohydrate surface structures (capsules) of *Streptococcus pneumoniae* inhibit their being bound to and ingested by macrophages.

Once the bacterial particles are taken up by macrophages, the vesicles containing the particles (phagosomes) fuse with primary lysosomes, forming secondary lysosomes (phagolysosomes) (Nichols *et al.*, 1971). Degradation of the ingested particles takes place within the lysosome, which contains hydrolases and other enzymes that generate substances which are essential for killing the bacteria.

Some bacteria are taken up but not killed by macrophages. *Mycobacterium tuberculosis* prevents fusion of the phagosome with the lysosome, thereby avoiding killing by lysosomal enzymes (Shurin and Stossel, 1978). Other bacteria, such as *Mycobacterium leprae,* resist degradation of lysosomal enzymes (Nathan *et al.*, 1980), and others not only resist, but even require an acidified phagolysosome for growth (Hackstadt and Williams, 1981).

D. Chemotaxis

Chemotaxis is the directed movement of cells in response to a gradient of a chemical substance. The chemical substance, also termed chemotactic factor or chemoattractant, is released at sites of inflammation, immune reactions, and tissue damage. It attracts phagocytic cells, such as polymorphonuclear leukocytes (PMN), mononuclear phagocytes, and tissue macrophages, but also lympocytes to the inflammatory focus. By attracting and activating the cells of the immune system to move toward the inflammation, chemotactic substances play an important role in helping to fight off pathogens.

At sites of inflammation, the substances diffuse to the adjoining capillaries, causing phagocytes to adhere to the vascular endothelium. Endothelial cell adhesiveness for neutrophils can be markedly enhanced by pretreatment with LPS, IL-1, and phorbolester (Schleimer and Rutledge, 1986) possibly by increasing the CD11/18 complex on neutrophils and ICAM-1 on the endothelium surface (Dustin *et al.*, 1986). After attachment to endothelium (pavement), the cells insert pseudopodia between the endothelial cells, dissolve the basement membrane (diapedesis), leave the blood vessel, and migrate along increasing concentrations of inflammatory mediators (Fig. 16).

There are numerous chemotactically active substances, including N-formylated peptides, which are present in *Escherichia coli* supernatants, the complement fragment C5a, leukotriene LTB$_4$, and IL-8. The latter belongs to a family of proinflammatory molecules called chemokines (Schall, 1991; Oppenheim *et al.*, 1991; Horuk, 1994). Among the IL-8 family one can distinguish between the two subfamilies, the cxc and cc chemokines. The distinction is based on whether the first two cysteine residues in a conserved

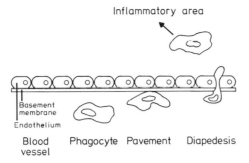

Inflammatory area

Basement
membrane
Endothelium

Blood   Phagocyte Pavement   Diapedesis
vessel

FIG. 16 Chemotaxis. Chemotactic substances released at sites of inflammation attract and activate phagocytes. Phagocytes adhere to the basement membrane of the blood vessel (pavement), move across the endothelial cell layer (diapedesis), leave the blood vessel, and move toward the inflammatory focus.

motif are separated by an intervening residue (Schall, 1991). The cxc subfamily comprises chemokines such as IL-8, neutrophil-activating peptide 2 (NAP-2), and melanoma growth-stimulating activity (MGSA); the cc subfamily contains, for example, the monocyte chemotactic protein 1 (MIP-1), macrophage inflammatory protein 1 (MPI-1), and RANTES. Whereas the former group is a chemoattractant for neutrophils but not monocytes, the opposite is true for the second group (Oppenheim et al., 1991).

The phagocytes bear receptors for the chemoattractant factors, and exposure to any of these factors results in desensitization to a second exposure to the same stimulus, but not to other unrelated chemotactic factors. Once receptors are occupied, an immediate increase of the intracellular second messengers Ca^{2+} and cAMP can be observed. Ca^{2+} and cAMP activate respective kinases and, as supposed by Snyderman et al. (1986), activation of Ca^{2+}-dependent protein kinases leads to cellular activation, whereas cAMP-dependent protein kinases inactivate the system. Chemoattractants also increase cellular cAMP levels by a calcium-mediated inhibition of cAMP degradation. This may serve as an autoregulatory action in chemoattractant-induced leukocyte activation (Gallin et al., 1978). Chemotactic factors, including fMLP, C5a, LTB_4, and PAF, activate phospholipase C, which catalyzes the formation of inositol trisphosphate and 1,2-diacylglycerol from phosphatidylinositol 4,5-bisphosphate. DAG is known to activate protein kinase C, and the activation is potentiated by increased cytosolic Ca^{2+} concentrations (Bell, 1986).

There is evidence that the receptors are coupled to polyphosphoinositide hydrolysis via G-proteins (Verghese and Snyderman, 1989). These signal transduction events are followed by shape changes and by changes in physiological function, such as degranulation and superoxide production (Snyderman and Uhing, 1988). *Bordetella pertussis,* by producing pertussis toxin which ADP-ribosylates G-proteins, impairs macrophage chemotaxis (Meade *et al.,* 1984). This can lead to diminished macrophage bacteriocidal capacity, resulting in dramatic host defense problems.

E. Bactericidal Mechanisms

1. Oxygen-Dependent Bactericidal Mechanisms

The professional phagocytic cells, neutrophils, eosinophils, monocytes, and macrophages exposed to soluble or particulate stimuli rapidly consume a large amount of oxygen. This cyanide-insensitive oxygen consumption, termed respiratory burst, is important for killing ingested bacteria (Babior *et al.,* 1973; Drath and Karnovsky, 1975). Oxygen is consumed in a reaction involving reduction of oxygen to superoxide:

$$2\ O_2 + NADPH \rightarrow 2\ O_2^- + NADP^+ + H^+$$

The reaction is catalyzed by the NADPH oxidase, an enzyme complex which has both membrane and cytoplasmic structural units. Cytoplasmic components of the NADPH oxidase included a 47-kDa and a 67-kDa protein. Activation of the enzyme is associated with phosphorylation of the 47-kDa protein and translocation of both the 47-kDa and 67-kDa protein to the plasma membrane (Dusi *et al.,* 1993). Activation of the enzyme also seems to involve a small GTP-binding protein (Abo *et al.,* 1991).

O_2^- produced by the respiratory burst is the precursor of other bactericidal reactive oxygen intermediates (ROI), such as hydrogen peroxide (H_2O_2), the hydroxyl radical ($HO^{\cdot}$) (Fridovich, 1986), and perhaps singlet oxygen, which is oxygen with one electron in a high energy state. Among these bactericidal products, the hydroxyl radical is the strongest oxidizing agent. It is formed from superoxide in a reaction which seems to be catalyzed by iron, thus making iron availability in a crucial step in the generation of $HO^{\cdot}$.

$$O_2^- + O_2^- + 2H^+ \rightarrow H_2O_2 + O_2$$
$$Fe^{2+} + H_2O_2 \rightarrow Fe^{3+} + OH^{\cdot} + OH^-$$
$$Fe^{3+} + O_2^- \rightarrow Fe^{2+} + O_2$$
$$\text{Net } O_2^- + H_2O_2 \rightarrow OH^{\cdot} + OH^- + O_2$$

The bactericidal ROI are generated in the absence of the lysosomal myeloperoxidase (MPO) and therefore may not require fusion of the lysosome with the phagosome.

NADPH oxidase is located in the plasma membrane. Upon activation, which is triggered by membrane receptors, O_2^- is released into the extracellular space, where it can damage surrounding tissues (target cells). The degree of the damage depends on the concentration of molecules in the extracellular fluid that can inactivate the ROI. During phagocytosis ROIs are confined to the phagocytic vacuole. Here they serve as highly toxic agents to the internalized microbial agent.

ROI also contribute to akalinization of phagocytic vacuoles. The rise in pH results from an increased pumping of electrons (as O_2^-) unaccompanied by protons into the vacuole. Moreover ROIs, such as superoxide and peroxide, are anionic and consume protons as they disproportionate to the protonated form (Segal, 1989). This elevation of pH to about pH 7.8–8.0 leads to activation of neutral proteases and cationic proteins that help to kill and digest bacteria (Segal, 1989). After initial alkalinization, the pH within the phagocytic vacuoles gradually declines again to about pH 6.0. The increase in pH is not seen in phagocytic vacuoles in chronic granulomatous disease (CGD). These vacuoles, which show abnormal activity, are filled with undigested debris (Klebanoff, 1975), possibly owing to the inability of neutral enzymes to digest the particles.

Other highly reactive substances, such as hypochlorite, may also contribute to bactericidal mechanisms. Myeloperoxidase, which is present in lysosomes, catalyzes the oxidation of chloride to hypochlorite in the presence of hydrogen peroxide.

$$H_2O_2 + Cl^- + H^+ \rightarrow H_2O + HOCl$$
$$HOCl + H_2O_2 \rightarrow Cl^- + H_2O + {}^1O_2 + H^+$$

Although hypochlorite is a potent oxidizing agent, it efficiently oxidizes amines to chloramines. The role of MPO in defense against bacteria is not clear. The fact that a hereditary lack of this enzyme is not associated with an increased risk for serious infection (Kitahara *et al.,* 1981), and that macrophages in contrast to monocytes hardly contain any MPO, argues against a major role for MPO in the microbicidal activity of monocytes and macrophages. MPO may play an important part in degrading hydrogen peroxide to water.

Some bacteria have developed means to resist the oxidative burst to which they are exposed once within the phagolysosomes. Superoxide dismutase from *Mycobacterium leprae* has the potential to eliminate superoxide (Wheeler and Gregory, 1980). Glycolipids from *Mycobacterium leprae* (Neill and Klebanoff, 1988) and mucoid exopolysaccharide from *Pseudomonas aeruginosa* (Learn *et al.,* 1987) can function as scavengers of hypochlorite, a potent oxidizing and halogenating agent.

2. Oxygen-Independent Bactericidal Mechanisms

There is evidence that phagocytes can kill bacteria in the absence of an oxidative burst (Catterall *et al.,* 1987). They are capable of secreting granule-associated proteins that are involved in antimicrobial activity. Among these proteins are neutral proteases, lysosomal hydrolases, lysozyme, and cationic proteins. The group of neutral proteases includes digestive enzymes such as collagenase, elastase, and plasminogen activator (Nathan, 1987). While unstimulated macrophages are poor secretors of most proteases, activated macrophages release substantial amounts of these enzymes. Their activity is greatest at pH 7.0 to 8.0, values that are reached in phagocytic vacuoles prior to acidification. There are bacteria that seem capable of interfering with the acidification step (Horwitz and Maxfield, 1984), thereby increasing their chance of survival. Phagosomes containing live *Legionella pneumophila* have a mean pH of 6.1, which is 0.8 pH units higher than phagosomes containing killed *Legionella pneumophila*. *Legionella pneumophila* inhibits phagosome-lysosome fusion (Horwitz, 1983) so that the bacteria escape degradation by acid hydrolases, such as phosphatases, glycosidases, sulfatases, deoxyribonucleases, and lipases. Except for acid hydrolases, macrophages elaborate a wide range of other enzymes, including lysozyme, a cationic protein.

Lysozyme, which is present in macrophages in relatively high concentrations, mediates digestion of bacterial cell walls. It has the ability to cleave the β 1-4 linkage between *N*-acetylglucosamine and *N*-acetylmuramic acid residues, which form the carbohydrate component of cell-wall peptidoglycan (Ganz *et al.*, 1985). Unlike lysosomal hydrolases, which are only released in response to stimulation, lysozyme is secreted continuously (Gordon *et al.*, 1974). Owing to its cationic nature, it may be involved in antimicrobial activity.

Cationic proteins with microbicidal activity have been purified from various cells, including rabbit alveolar macrophages (macrophage protein 1 and 2) (Ganz *et al.*, 1986). Two of the proteins, lysozyme and bacterial/permeability-increasing protein (BPI), differ in their structure from the remaining cationic proteins which either belong to a family of related mammalian antibiotics named "defensins" or to a family named "serprocidins" (Gabay and Almeida, 1993; Ganz *et al.*, 1990). All defensins are basic peptides that comprise 29–43 amino acids. Human and animal defensins are active against a variety of bacteria, including pathogens such as *Mycobacterium avium* and *Mycobacterium intracellulare* (Ogata *et al.*, 1992).

Members of the "serprocidin" family are cathepsin G, elastase, proteinase 3, and azurocidin (Gabay *et al.*, 1989; Elsbach and Weiss, 1992). Azurocidin is a potent antibiotic showing selective activity against Gram-negative bacteria (Spitznagel, 1990). It is present in monocytes and almost disappears as monocytes mature into macrophages. Killing by azurocidin seems to involve binding to LPS, followed by a lethal injury at the inner membrane of Gram-negative bacteria (Gabay and Almeida, 1993).

Similar to azurocidin, another cationic protein, the BPI, also interacts with LPS. Its cytotoxic activity is confined to Gram-negative bacteria. Once bound to LPS at the outer membrane of the bacteria, a number of intracellular events, such as growth arrest, changes in outer membrane permeability, and activation of enzymes degrading phospholipids and peptidoglycan, occur (Mannion *et al.*, 1990), leading to bacterial damage. BPI also binds to isolated LPS, suggesting a role for BPI as an extracellular bactericidal and endotoxin-neutralizing agent (Elsbach and Weiss, 1992).

Some bacteria, lile *Salmonella typhimurium*, are resistant to degradation by lysosomal enzymes. In contrast to killed bacteria, the live bacteria survive in the phagolysosome and do not get degraded (Carrol *et al.*, 1979). Resistance to the effects of defensins may be the reason for their survival (Groisman and Saier, 1990).

Recently, a great deal of interest has been focused on the role of nitric oxide (NO) in macrophage antimicrobial activity (Nathal and Hibbs, 1991). NO is regarded as a cytotoxic effector molecule generated from the guanidino nitrogen atom of L-arginine by the inducible isoform of NO synthase (iNOS) (Hibbs *et al.*, 1990; Stuehr and Nathan, 1989; Marletta, 1989). Expression of the NO synthase occurs several hours after activation of macrophages (Hauschildt *et al.*, 1990c) and, once expressed, it synthesizes large amounts of NO (nanomoles) for long periods. Activation requires a priming agent, such as IFN-γ (Weinberg *et al.*, 1978; Adams and Hamilton, 1991), followed by a second signal, such as IL-2 (Cox *et al.*, 1992), TNF-α (Ding *et al.*, 1988; Drapier *et al.*, 1988), bacterial lipopolysaccharide (Ding *et al.*, 1988), bacterial synthetic lipopeptide (Hauschildt *et al.*, 1990c), muramyl dipeptide (Drapier *et al.*, 1988), lipoteichoic acid (Auguet *et al.*, 1992; Keller *et al.*, 1992), and picolinic acid (Melillo *et al.*, 1993).

NO produced by activated macrophages has been shown to be cytostatic or cytotoxic for a variety of invading microorganisms, such as *Mycobacterium leprae* (Adams *et al.*, 1991) and *Mycobacterium tuberculosis* (Chan *et al.*, 1992; Denis, 1991). Little is known about the mechanisms underlying NO-induced cytotoxicity for intracellular microorganisms. Presumably they are similar to those observed in mammalian cells. The effector functions of NO have been shown to involve iron-dependent reactions (Drapier and Hibbs, 1986, 1988; Drapier *et al.*, 1991). NO inhibits the activity of enzymes that contain catalytically active nonheme iron coordinated with sulfur atoms (Hibbs *et al.*, 1990). The enzymes affected are ribonucleotide reductase, cis-aconitase in the tricarboxylic acid cycle, and complex 1 and complex 2 in the mitochondrial electron transport chain. The inhibition of enzyme activity is accompanied by a release of intracellular iron from the target cells (Hibbs *et al.*, 1984) and by the appearance of paramagnetic dinitrosyl–iron–sulfur complexes in target and effector cells (Drapier *et al.*, 1991; Peliat *et al.*, 1990; Vanin *et al.*, 1993). It has been suggested that the iron in these paramagnetic complexes stems from the iron of the enzymes mentioned here (Hibbs *et al.*, 1990; Stuehr and Nathan, 1989; Drapier *et al.*, 1991). However, it cannot be excluded that iron from ferritin, the main iron storage pool in cells, also contributes to the detected paramagnetic signals. The release of iron may be causually related to cytotoxicity because iron is vital to many cell functions, including those depending on iron-dependent enzymes.

Apart from targeting intracellular iron and thereby inhibiting enzyme functions, NO may react with the superoxide anion to form peroxinitrite (Beckman *et al.*, 1990).

$$NO^{\cdot} + O_2^{\cdot} \rightarrow ONOO^-$$
$$ONOOH \rightarrow HO^{\cdot} + NO_2^{\cdot}$$

Peroxynitrous acid can decompose to form the highly reactive hydroxyl radical and nitrogen dioxide (Hogg *et al.*, 1992). These radicals can produce severe damage to both mirobes and host cells (Halliwell and Gutterage, 1990; Perutz *et al.*, 1985). Zhu *et al.* (1992) showed the bactericidal activity of peroxynitrite on *Escherichia coli*, and Radi *et al.* (1993) demonstrated the ability of peroxynitrite to kill *Trypanosoma cruzi*.

The interaction of NO with ROI in macrophages is not very likely because the formation of these two highly reactive substances does not occur simultaneously. However, since ROI formation precedes NO formation, the sequential exposure of ROI followed by NO may be an efficient means to induce killing of bacteria.

Although human monocytes and macrophages express iNOS (Reiling *et al.*, 1994), it has been difficult to demonstrate NO production by these cells. Whether the well-established role of iNOS in murine macrophages as an antimicrobial system is conserved in human macrophages remains to be clarified.

3. Cytokine Production

When exposed to biologically active agents, including those of bacterial origin, macrophages respond by secreting a wide variety of substances. Among these substances are cytokines which can interfere with the ability of the host to respond to infections (Bendtzen, 1988). One group of the cytokine family consists of the interleukins.

a. Interleukin-1   Interleukin-1 is produced by monocytes and macrophages after stimulation by a variety of infectious agents (Dinarello, 1984), including LPSs (Dinarello, 1991). IL-1 possesses a wide spectrum of biological activities and affects numerous cell systems (Nathan, 1987). The best-studied property of IL-1 is its ability to cause fever by inducing hypothalamic prostaglandin E_2 synthesis (Dinarello *et al.*, 1986). A rise in temperature is an important defense mechanism, since it enhances several immune responses, such as T-cell proliferation, induction of cytotoxic T cells, antibody secretion, and leads to a diminished microbial replication (Sisson and Dinarello, 1989). Furthermore, IL-1 helps to protect the host against infections by inducing the synthesis of hepatic acute phase proteins, such as serum amyloid A, C-reactive proteins, complement components, fibrinogen, and antiproteases, such as α-1-antitrypsin and α-2-macroglobulin. In addition, metalloproteins are synthesized that bind serum iron and zinc, thereby decreasing the availability of the elements essential for growth of bacteria. IL-1 participates in the mechanism of T- and B-cell activation. In the presence of antigens or mitogens, it amplifies T-cell activation by inducing IL-2 and IL-2 receptor gene expression, and in B cells it costimulates proliferation and antibody production. Besides other cytokines, such as TNF-α and IFN-γ, IL-1 can induce production of prostaglandins, complement, and its own synthesis by an autoregulatory cycle. IL-1 was found to increase resistance to infection with bacteria, i.e., *Staphylococcus pneumoniae, Staphylococcus aureus,* and *Pseudomonas aeruginosa* (Ozaki *et al.,* 1987; Minami *et al.,* 1988). It can replace bacterial products in the induction of this resistance and, like bacterial substances, the agent is most effective when administered 24 hr prior to the challenge (Dinarello, 1991).

b. Interleukin-6   Interleukin-6 is a multifunctional interleukin produced by a variety of nonlymphoid and lymphoid cells, including monocytes and macrophages (see Hirano, 1991). It acts on B cells activated with *Staphylococcus aureus* Cowan 1 or pokeweed mitogen to induce antibody production, but not on resting B cells (Muraguchi *et al.,* 1988). Similarly to IL-1, IL-6 promotes the proliferation of human T cells stimulated with PHA (Lotz *et al.,* 1988). Some effects of IL-6, such as induction of thymocyte proliferation (Le *et al.,* 1988), enhancement of phagocytosis, and expression of FcγR, are synergistic with IL-1. IL-6 can induce the biosynthesis of a variety of acute phase proteins by hepatocytes, and it inhibits the production of IL-1 and TNF-α, both of which are potent inducers of IL-6 (Hirano, 1991). Its participation in acute phase reactions and in immune responses suggests that it is a particularly important mediator of infection, inflammation, and tissue injury.

Further interleukins produced by monocytes and macrophages include IL-8, IL-10, and IL-12. IL-8 exhibits chemotactic activity for neutrophils, T cells, and basophils. It activates neutrophils to release lysosomal enzymes and to adhere to endothelial cells (Mukaida *et al.,* 1992; Oppenheim *et al.,* 1991). Similar to IL-6, IL-8 is efficiently produced by peripheral blood monocytes after bacterial infection. IL-10 is a potent immunosuppressant of macrophage functions (Moore *et al.,* 1993; Rennick *et al.,* 1992). It depresses their antigen-presenting capacity. IL-10 inhibits the production of

IL-2 and IFN-γ by T_H1 cells and enhances B-cell proliferation and antibody secretion (Rousset *et al.*, 1992).

IL-12 is produced by phagocytic cells in response to bacteria, bacterial products, and intracellular parasites. Its main targets are T- and natural killer cells. These cells respond to IL-12 with increased cytokine production, primarily IFN-γ, and enhanced generation of cytotoxic cells. IFN-γ acts as a potent enhancer of the bactericidal activity of phagocytic cells. Thus, through its ability to induce IFN-γ, IL-12 together with other cytokines, such as IL-1 and TNF-α, can activate some of the inflammatory mechanisms that represent the first innate resistance against infection. In addition, IL-12 facilitates the generation of T_H1 cells and suppression of T_H2 cell responses (Trincheri, 1993). Thus, IL-12 seems to affect both antigen-independent and antigen-specific immune responses (Trincheri, 1994).

c. Tumor Necrosis Factor-α   Like most interleukins, the cytokine TNF-α exhibits pleiotropic biological activities on various cell types (Wong and Goeddel, 1989). Most of the activities can be attributed to its ability to activate a great variety of genes in the target cells. It triggers a cascade of mediators which interact to effect and amplify additional cellular and systemic changes. TNF-α is not produced constitutively; it requires bacteria, viruses, or their products to induce its synthesis. The best-known cellular source of TNF-α is the activated macrophage (Carswell *et al.*, 1975), and the most potent biological inducer of TNF-α production by macrophages is LPS (Beutler *et al.*, 1986). TNF-α acts in an autocrine fashion, thereby activating macrophages and enhancing their cytotoxic potential (Philip and Epstein, 1986; Hoffman and Weinberg, 1987).

Treatment with TNF-α can cause a minor induction of MHC class I (Martin *et al.*, 1989; Collins *et al.*, 1986) and class II antigens (Lapierre *et al.*, 1988). TNF-α synergizes strongly with IFN-γ in this regard. TNF-α acts as a chemoattractant for monocytes and polymorphonuclear leukocytes (PMN) (Ming *et al.*, 1987), and it enhances their adherence to endothelial cells. TNF-α can play a beneficial role as immunostimulant and important mediator of host resistance to many infectious agents. It can protect mice against bacterial infection (Parant, 1987), and it enhances the killing of *Mycobacterium avium* by macrophages (Bermudez and Young, 1988). The beneficial effects of TNF-α—mild fever and recruitment of both microbe-specific and less specialized immune components—help to protect against microbial infection. On the other hand, an overproduction of TNF-α during infection can lead to severe systemic toxicity and life-threatening shock. It has been shown to play a major endocrine role in the pathogenesis of shock in Gram-negative bacterial septicemia (Tracey *et al.*, 1988) and in experimental murine listeriosis. The severity of the infection increased with increasing TNF-α production (Havell, 1987).

It has been suggested that the direct cytotoxicity of TNF-α is related to the generation of intracellular reactive oxygen intermediates, which activate preexisting potentially toxic molecules (Wong and Goeddel, 1989). Not only as a direct effector, but also as an inducer of cytokines and other mediators, TNF-α can cause extensive tissue damage and multiple organ failure.

In addition to the interleukins and TNF-α, macrophages secrete numerous other cytokines, including those of the transforming growth factor, colony-stimulating factor, and interferon family. Not only do macrophages secrete cytokines, they also serve as targets for many cytokines, such as IFN-γ, IL-4, IL-2, and granulocyte-macrophage CSF (Nacy and Meltzer, 1991). For example, IFN-γ increases the expression of MHC class II antigens, a crucial step in the initiation of an immune response and a step which can be inhibited by IFN-α/β (Ling et al., 1985). IFN-γ enhances the production of toxic oxygen intermediates and the ability of macrophages to kill target cells. It "primes" macrophages which, when exposed to a second signal such as LPS, acquire cytolytic competence. GM–CSF cooperates with IFN-γ in inducing macrophage resistance to infection with *Leishmania major* (Belosević et al., 1988) and cytostasis of *Leishmania mexicana amazonensis* in infected human monocytes (Ho et al., 1990). A combination of TNF-α and GM–CSF enhances the intracellular destruction of *Mycobacterium avium* (Bermudez and Young, 1990). Except for their ability to display synergistic and antagonistic interactions, another characteristic feature of cytokines is their ability to stimulate or inhibit the production of other cytokines (Vilcek and Le, 1991).

Thus, IL-1 has been reported to stimulate the production of IL-6 (Content et al., 1985), GM–CSF (Zucali et al., 1986), and IL-8 (Matsushima et al., 1988). IL-1 as well as TNF-α stimulate each other as well as their own production (Neta et al., 1991). One example of an inhibitory action of cytokines on cytokine production is the immunosuppressive effect of the cytokine transforming growth factor-β (TGF-β). TGF-β, which is produced by activated macrophages (Wahl et al., 1990), inhibits intracellular killing when given prior to IFN-γ treatment and infection. It has been suggested that TGF-β acts by preventing TNF-α production by activated macrophages, which as an autostimulatory cytokine is responsible for intracellular killing (Green et al., 1990).

Cytokines can have diverse activities, and many cytokines have overlapping effects. Many of them are secreted by macrophages when encountered with bacteria. Not only do cytokines play a role in the regulation of macrophages themselves, they also regulate T- and B-cell growth and differentiation. Thus, they serve as powerful tools in assisting in the control and fine tuning of the immune response in bacterial infection.

V. Conclusions and Perspectives

Bacteria have the capacity to enter, survive, and multiply within macro-phages. In order to secure their survival and replication, they try to outwit the host cells, which in turn have developed means to control bacterial invasion and replication. Whereas the macrophage takes a strategy that leads to killing of bacteria, death of the macrophage may not necessarily be the aim of bacteria. Each step leading to bacterial infection, adherence, entry, persistance, and multiplication is crucial for the survival of both macrophage and bacteria and is subject to manipulation by either host or invader. By defining the molecular and genetic basis underlying each step, it will be possible to interfere with the option to cause damage to the bacteria. Furthermore, the use of mutant strains will help to clarify the significance of each step for bacterial infection.

Little is known about the existence of subpopulations of macrophages that may be susceptible to infection by distinct bacterial strains. If such subpopulations can be identified, it will be important to search for stimuli that enhance their ability to resist bacterial infection.

Clearance of intracellular bacteria requires T lymphocytes which recognize bacterial peptides associated with MHC antigens present on the surface of infected macrophages. Upon interacting with infected macrophages, the T cells produce a variety of cytokines. The nature of the cytokines depends on the subset of T cells that have been stimulated. These cytokines influence macrophage functions and induce macrophages to secrete other cytokines, which can have synergistic, antagonistic, or autostimulatory effects. In identifying the effects of these cytokines, cultured cell models have been very useful; however, they may not provide a realistic picture of their mode of action in the complex intact organism, in which infected macrophages are also exposed to numerous other biologically active agents. The identification of possible bacteria-induced manipulations of host–cell signal transduction mechanisms and manipulation of antigen processing and presentation pathways will increase our understanding of immune responses infected bacteria and provide a means to impair the mechanisms of microbial virulence.

Acknowledgment

The secretarial assistance of Mrs. Irene Hinz is gratefully acknowledged.

References

Abo, A., Pick, E., Hall, A., Totty, N., Teahan, C. G., and Segal, A. W. (1991). Activation of the NADPH oxidase involves the small GTP-binding protein p21[racl]. *Nature* (*London*) **353,** 668–670.

Adams, D. O., and Hamilton, T. A. (1984). The cell biology of macrophage activation. *Annu. Rev. Immunol.* **2**, 283–318.

Adams, D. O., and Hamilton, T. A. (1991). Molecular basis of macrophage activation: Diversity and its origins. *In* "The Macrophage" (C. E. Lewis and J. O. D. McGee, eds.), pp. 77–117. IRL Press, Oxford and New York.

Adams, L. B., Franzblau, S. G., Vavrin, Z., Hibbs, J. B., Jr., and Krahenbuhl, J. L. (1991). L-Arginine-dependent macrophage effector functions inhibit metabolic activity of *Mycobacterium leprae. J. Immunol.* **147**, 1642–1646.

Aderem, A. A., Keum, M. M., Pure, E., and Cohn, Z. A. (1986). Bacterial lipopolysaccharides, phorbol myristate acetate and zymosan induce the myristoylation of specific macrophage proteins. *Proc. Natl. Acad. Sci. U.S.A.* **83**, 5817–5821.

Aderem, A. A., Albert, K. A., Keum, M. M., Wang, J. K. T., Greengard, P., and Cohn, Z. A. (1988). Stimulus-dependent myristoylation of a major substrate of protein kinase C. *Nature (London)* **332**, 362–364.

Anderson, C. L. (1982). Isolation of the receptor for human IgG from human monocyte cell line (U937) and from human peripheral blood monocytes. *J. Exp. Med.* **156**, 1794–1806.

Anderson, M. S., and Miller, J. (1992). Invariant chain can function as a chaperone protein for class II major histocompatibility complex molecules. *Proc. Natl. Acad. Sci. U.S.A.* **89**, 2282–2286.

Apasov, S., Redegeld, F., and Sitkovsky, M. (1993). Cell-mediated cytotoxicity: Contact and secreted factors. *Curr. Opin. Immunol.* **5**, 404–410.

Armstrong, J. A., and D'Arcy Hart, P. (1971). Response of cultured macrophages to *Mycobacteria tuberculosis* with observations on fusion of lysosomes with phagosomes. *J. Exp. Med.* **134**, 7113–7140.

Armstrong, J. A., and D'Arcy Hart, P. (1975). Phagosome lysosome interactions in cultured macrophages infected with virulent tubercule bacilli. *J. Exp. Med.* **142**, 1–16.

Asano, T., and Kleinerman, E. S. (1993). Liposome-encapsulated MTP-PE: A novel biologic agent for cancer therapy. *J. Immunother.* **14**, 286–292.

Asano, T., McWatters, A., An, T., Matsushima, K., and Kleinerman, E. S. (1994). Liposomal muramyl tripeptide upregulates interleukin-1 alpha, interleukin-1 beta, tumor necrosis factor-alpha, interleukin-6 and interleukin-8 gene expression in human monocytes. *J. Pharmacol. Exp. Ther.* **268**, 1032–1039.

Auguet, M., Lonchampt, M.-O., Delaflotte, S., Goulin-Schulz, J., Chabrier, P. E., and Braquet, P. (1992). Induction of nitric oxide synthase by lipoteichoic acid from *Staphylococcus aureus* in vascular smooth muscle cells. *FEBS Lett.* **297**, 183–185.

Babior, B., Kipnes, R., and Curnutte, J. T. (1973). Biological defense mechanisms. The production of superoxide, a potential bactericidal agent. *J. Clin. Invest.* **52**, 741–744.

Bazil, V., Horejsi, V., Baudys, M., Kristopova, H., Strominger, J., Kostra, W., and Hilgert, J. (1986). Biochemical characterization of a soluble form of the 53 kDa monocyte surface antigen. *Eur. J. Immunol.* **16**, 1583–1589.

Beck, G., Benach, J. L., and Habicht, G. S. (1990). Isolation, preliminary chemical characterization, and biological activity of *Borrelia burgdorferi* peptidoglycan. *Biochem. Biophys. Res. Commun.* **167**, 89–95.

Becker, E. L., Kermode, J. C., Naccache, P. H., Yassin, R., Munoz, J. J., Marsh, M. L., Huang, C. K., and Shaafi, R. (1986). Pertussin toxin as a probe of neutrophil activation. *Fed. Proc., Fed. Am. Soc. Exp. Biol.* **45**(2), 151.

Beckman, J. S., Beckman, T. W., Chen, J., Marschall, P. A. and Freeman, B. A. (1990). Apparent hydroxyl radical production by peroxynitrite: Implications for endothelial injury from nitric oxide and superoxide. *Proc. Natl. Acad. Sci. U.S.A.* **87**, 1620–1624.

Behr, T., Fischer, W., Peter-Katalinic, J., and Egge, H. (1992). The structure of pneumococcal lipoteichoic acid. Improved preparation, chemical and mass spectrometric studies. *Eur. J. Biochem.* **207**, 1063–1075.

Bell, R. M. (1986). Protein kinase C activation by diacylglycerol second messengers. *Cell* (*Cambridge, Mass.*) **45,** 631–632.

Belosević, M., Davis, C. E., Meltzer, M. S., and Nacy, C. A. (1988). Regulation of activated macrophage antimicrobial activities. Identification of lymphokines that cooperate with IFN-γ for induction of resistance to infection. *J. Immunol.* **141,** 890–896.

Bendtzen, K. (1988). Interleukin 1, interleukin 6 and tumor necrosis factor in infection, inflammation and immunity. *Immunol. Lett.* **19,** 183–192.

Benz, R., and Bauer, K. (1988). Permeation of hydrophilic molecules through the outer membrane of Gram-negative bacteria: Review on bacterial porins. *Eur. J. Biochem.* **176,** 1–19.

Bermudez, L. E., and Young, L. S. (1988). Tumor necrosis factor, alone or in combination with IL-2, but not IFN-gamma, is associated with macrophage killing of *Mycobacterium avium* complex. *J. Immunol.* **140,** 3006–3013.

Bermudez, E. L., and Young, L. S. (1990). Recombinant granulocyte-macrophage colony-stimulating factor activates human macrophages to inhibit growth or kill *Mycobacterium avium* complex. *J. Leukocyte Biol.* **48,** 67–73.

Bernardini, M. L., Sanna, M. G., Fontaine, A., and Sansonetti, P. J. (1993). OmpC is involved in invasion of epithelial cells by *Shigella flexneri. Infect. Immun.* **61,** 3625–3635.

Berridge, M. J., and Irvine, R. F. (1984). Inositol trisphosphate, a novel second messenger in cellular signal transduction. *Nature* (*London*) **312,** 315–321.

Bessler, W., Cox, M., Lex, A., Suhr, B., Wiesmüller, K.-H., and Jung, G. (1985). Synthetic lipopeptide analogues of bacterial lipoprotein are potent polyclonal activators for murine B-lymphocytes. *J. Immunol.* **135,** 1900–1905.

Bessler, W. G., Resch, K., Hancock, E., and Hantke, K. (1977). Induction of lymphocyte proliferation and membrane changes by lipopeptide derivatives of the lipoprotein from the outer membrane of *E. coli. Z. Immunitaetsforsch.* **153,** 11–23.

Beutler, B., Krochin, N., Milsark, I. W., Luedke, C., and Cerami, A. (1986). Control of cachectin (tumor necrosis factor) synthesis: Mechanisms for endotoxin. *Science* **232,** 977–980.

Bhakdi, S., Klonisch, T., Nuber, P., and Fischer, W. (1991). Stimulation of monokine production by lipoteichoic acids. *Infect. Immun.* **59,** 4614–4620.

Blum, J. S., and Cresswell, P. (1988). Role for intracellular proteases in the processing and transport of class II HLA antigens. *Proc. Natl. Acad. Sci. U.S.A.* **85,** 3975–3979.

Bolen, J. B., Rowley, R. B., Spana, C., and Tsygankov, A. Y. (1992). The Src family of tyrosine protein kinases in hemopoietic signal transduction. *FASEB J.* **6,** 3403–3409.

Braun, V., and Bosch, V. (1972). Sequence of the murein lipoprotein and the attachment site of the lipid. *Eur. J. Biochem.* **28,** 51–69.

Braun, V., and Bosch, V. (1973). In vivo biosynthesis of murein-lipoprotein of the outer membrane of *E. coli. FEBS Lett.* **34,** 302–306.

Braun, V., and Hantke, K. (1974). Biochemistry of bacterial cell envelopes. *Annu. Rev. Biochem.* **43,** 89–121.

Braun, V., and Hantke, K. (1975). Characterization of the free form of murein-lipoprotein from the outer membrane of *Escherichia coli* B/r. *FEBS Lett.* **60,** 26–28.

Braun, V., and Rehn, K. (1969). Chemical characterization, spatial distribution and function of a lipoprotein (murein-lipoprotein) of the *E. coli* cell wall. The specific effect of trypsin on the membrane structure. *Eur. J. Biochem.* **10,** 426–438.

Brown, E. J. (1989). Internalization of C3b oligomers by human polymorphonuclear leukocytes. *J. Biol. Chem.* **264,** 6196–6201.

Brown, E. J. (1991). Complement receptor and phagocytosis. *Curr. Opin. Immunol.* **3,** 76–82.

Brozna, J. P., Hauff, N. F., Phillips, W. A., and Johnston, R. B., Jr. (1988). Activation of the respiratory burst in macrophages. Phosphorylation specifically associated with Fc receptor-mediated stimulation. *J. Immunol.* **141,** 1642–1647.

Brunkhorst, B. A., Lazzari, K. G., Strohmeier, G., Weil, G., and Simons, E. R. (1991). Calcium changes in immune complex-stimulated human neutrophils. Simultaneous measurement of receptor occupancy and activation reveals full population stimulus binding but subpopulation activation. J. Biol. Chem. 266, 13035–13043.

Bussolino, F., Fischer, E., Turrini, F., Kazatchkine, M. D., and Arese, P. (1989). Platelet-activating factor enhances complement-dependent phagocytosis of diamide-treated erythrocytes by human monocytes through activation of protein kinase C and phosphorylation of complement receptor type one (CR-1). J. Biol. Chem. 264, 21711–21719.

Cady, A. B., Riveau, G., Chédid, L., Dinarello, C. A., Johannsen, L., and Krueger, J. M. (1989). Interleukin-1-induced sleep and febrile responses differentially altered by a muramyl dipeptide derivative. Int. J. Immunopharmacol. 11, 887–893.

Carrol, M. E. W., Jackett, P. S., Aber, V. R., and Lowrie, D. B. (1979). Phagolysosome formation, cyclic adenosine 3'5'-monophosphate and the fate of Salmonella typhimurium within mouse peritoneal macrophages. J. Gen. Microbiol. 110, 421–429.

Carswell, E. A., Old, L. J., Kassel, R. L., Green, S., Fiore, N., and Williamson, B. (1975). An endotoxin-induced serum factor that causes necrosis of tumors. Proc. Natl. Acad. Sci. U.S.A. 72, 3666–3670.

Catterall, J. R., Black, C. M., Leventhal, J. P., Rizk, N. W., Wachtel, J. S., and Remington, J. S. (1987). Nonoxidative microbicidal activity in normal human alveolar and peritoneal macrophages. Infect. Immun. 55, 1635–1640.

Chan, J., Xing, Y., Magliozzo, R. S., and Bloom, B. R. (1992). Killing of virulent Mycobacterium tuberculosis by reactive nitrogen intermediates produced by activated murine macrophages. J. Exp. Med. 175, 1111–1122.

Chang, Z. L., Novotney, M., and Suzuki, T. (1990). Phospholipase C and A_2 in tumoricidal activation of murine macrophage-like cell lines. FASEB J. 4, A1753.

Chatila, T., Scholl, P., Spertini, F., Ramesh, N., Trede, N., Fuleihan, R., and Geha, R. S. (1991). Toxic shock syndrome toxin-1, toxic shock, and the immune system. Curr. Top. Microbiol. Immunol. 174, 63–79.

Chédid, L. A., Parant, M. A., Audibert, F. M., Riveau, G. J., Parant, F. J., Lederer, E., Choay, J. P., and Lefrancier, P. L. (1982). Biological activity of a new synthetic muramyl peptide adjuvant devoid of pyrogenicity. Infect. Immun. 35, 417–424.

Chen, T. Y., Lei, M. G., Suzuki, T., and Morrison, D. C. (1992). Lipopolysaccharide receptors and signal transduction pathways in mononuclear phagocytes. Curr. Top. Microbiol. Immunol. 181, 169–188.

Clarkson, S. B., and Ory, P. A. (1988). CD16. Developmentally regulated IgG Fc receptors on cultured human monocytes. J. Exp. Med. 167, 408–420.

Cobb, M. H., Boulton, T. G., and Robbins, D. J. (1991). Extracellular signal-regulated kinases. ERKs in progress. Cell Regul. 2, 965–978.

Cockcroft, S., and Gomperts, B. D. (1985). Role of guanine nucleotide binding protein in the activation of polyphosphoinositide phosphodiesterase. Nature (London) 314, 534–536.

Coley, J., Tarelli, E., Archibald, A. R., and Baddiley, J. (1978). The linkage between teichoic acid and peptidoglycan in bacterial cell walls. FEBS Lett. 88, 1–9.

Collier, R. J. (1990). Corynebacteria. In "Microbiology" (B. D. Davis, R. Dulbecco, H. N. Eisen, and H. S. Ginsberg, eds.), pp. 507–514. Lippincott, Philadelphia.

Collins, T., Lapierre, L. A., Fiers, W., Strominger, J. L., and Pober, J. S. (1986). Recombinant human tumor necrosis factor increases mRNA levels and surface expression of HLA-A,B antigens in vascular endothelial cells and dermal fibroblasts in vitro. Proc. Natl. Acad. Sci. U.S.A. 83, 446–450.

Conlan, J. W., and North, R. J. (1992). Roles of Listeria monocytogenes virulence factors in survival: Virulence factors distinct from listeriolysin are needed for the organism to survive an early neutrophil-mediated host defense mechanism. Infect. Immun. 60, 951–957.

Connelly, P. A., Farrell, C. A., Merenda, J. M., Conklyn, M. J., and Showell, H. S. (1991). Tyrosine phosphorylation is an early signalling event common to Fc receptor crosslinking in human neutrophils and rat basophilic leukemia cells (RBC-2H3). *Biochem. Biophys. Res. Commun.* **177,** 192–201.

Content, J., DeWit, L., Poupart, P., Opdenakkar, G., Van Damme, J., and Billiau, A. (1985). Induction of a 26-kDa-protein in RNA in human cells treated with an interleukin-1-related, leukoccyte-derived factor. *Eur. J. Biochem.* **152,** 253–257.

Cox, G. W., Melillo, G., Chattopadhyay, U., Mullet, D., Fertel, R. H., and Varesio, L. (1992). Tumor necrosis factor-α dependent production of reactive nitrogen intermediates mediates IFN-γ plus IL-2-induced murine macrophage tumoricidal activity. *J. Immunol.* **149,** 3290–3296.

Cross, A. S. (1990). The biological significance of bacterial encapsulation. *Curr. Top. Microbiol. Immunol.* **150,** 87–95.

Cunningham, M. W., Antone, S. M., Gulizia, J. M., McManus, B. M., Fischetti, V. A., and Gauntt, C. J. (1992). Cytotoxic and viral neutralizing antibodies crossreact with streptococcal M protein, enteroviruses, and human cardiac myosin. *Proc. Natl. Acad. Sci. U.S.A.* **89,** 1320–1324.

Daemen, T., Veninga, A., Roerdink, F. H., and Scherphof, G. L. (1989). Conditions controlling tumor cytotoxicity of rat liver macrophages mediated by liposomal muramyl dipeptide. *Biochim. Biophys. Acta* **991,** 145–151.

Davis, B. D. (1990). Bacterial architecture. *In* "Microbiology" (B. D. Davis, R. Dulbecco, H. N. Eisen, and H. S. Ginsberg, eds.), pp. 21–50. Lippincott, Philadelphia.

Dekker, L., and Parker, P. (1994). Protein kinase C, a question of specificity. *Trends Biochem. Sci.* **19,** 73–77.

Denis, M. (1991). Interferon-gamma-treated murine macrophages inhibit growth of tubercle bacilli via the generation of reactive nitrogen intermediates. *Cell. Immunol.* **132,** 150–157.

Dinarello, C. A. (1984). Interleukin-1 and the pathogenesis of the acute-phase response. *N. Engl. J. Med.* **311,** 1413–1418.

Dinarello, C. A. (1991). Interleukin 1. *In* "The Cytokine Handbook" (A. Thompson, ed.), pp. 47–82. Academic Press, London.

Dinarello, C. A., Goldin, N. P., and Wolff, S. M. (1974). Demonstration and characterization of two distinct human leukocytic pyrogens. *J. Exp. Med.* **139,** 1369.

Dinarello, C. A., Carnnon, J. G., Wolff, S. M., Bernheim, H. A., Beutler, B., Cerami, A., Figari, I. S., Palladiner, M. A., Jr., and O'Connor, J. V. (1986). Tumor necrosis factor (cachectin) is an endogenous pyrogen and induces production of interleukin 1. *J. Exp. Med.* **163,** 1433–1450.

Ding, A. H., Nathan, C. F., and Stuehr, D. J. (1988). Release of reactive nitrogen intermediates and reactive oxygen intermediates from mouse peritoneal macrophages. Comparison of activating cytokines and evidence for independent production. *J. Immunol.* **141,** 2407–2412.

Drapier, J.-C., and Hibbs, J. B., Jr. (1986). Murine cytotoxic activated macrophages inhibit aconitase in tumor cells. Inhibition involves the iron-sulfur-prosthetic group and is reversible. *J. Clin. Invest.* **78,** 790–797.

Drapier, J.-C., and Hibbs, J. B., Jr. (1988). Differentiation of murine macrophages to express nonspecific cytotoxicity for tumor cells results in L-arginine-dependent inhibition of mitochondrial iron-sulfur enzymes in the macrophage effector cells. *J. Immunol.* **140,** 2829–2838.

Drapier, J.-C., Wietzerbin, J., and Hibbs, J. B., Jr. (1988). Interferon-γ and tumor necrosis factor induce the L-arginine-dependent cytotoxic effector mechanism in murine macrophages. *Eur. J. Immunol.* **18,** 1587–1592.

Drapier, J.-C., Pellat, C., and Yann, H. (1991). Generation of ERR-detectable nitrosyl-iron complexes in tumor target cells cocultured with activated macrophages. *J. Biol. Chem.* **266,** 10162–10167.

Drath, D. B., and Karnovsky, M. L. (1975). Superoxide production by phagocytic leukocytes. *J. Exp. Med.* **141**, 257–260.

Drevets, D. A., Leenen, P. J. M., and Campbell, P. A. (1993). Complement receptor type 3 (CD11b/CD18) involvement is essential for killing of *Listeria monocytogenes* by mouse macrophages. *J. Immunol.* **151**, 5431–5439.

Drysdale, P. E., Yapundich, R. A., Shin, M. L., and Shin, H. S. (1987). Lipopolysaccharide-mediated macrophage activation: The role of calcium in the generation of tumoricidal activity. *J. Immunol.* **139**, 951–956.

Dusi, S., Della-Bianca, V., Grzeskowiak, M., and Rossi, F. (1993). Relationship between phosphorylation and translocation to the plasma membrane of p47 phox and p67 phox and activation of NADPH oxidase in normal and Ca^{2+}-depleted human neutrophils. *Biochem. J.* **290**, 173–178.

Dustin, M. L., Rothlein, R., Bhan, A. K., Dinarello, C. A., and Springer, T. A. (1986). Induction of IL-1 and interferon-γ: Tissue distribution, biochemistry, and function of a natural adherence molecule (ICAM-1). *J. Immunol.* **137**, 245–254.

Edelson, P. J. (1982). Intracellular parasites and phagocytic cells: Cell biology and pathophysiology. *Rev. Infect. Dis.* **4**, 124–135.

Elsbach, P., and Weiss, J. (1992). Oxygen-independent antimicrobial systems of phagocytes. *In* "Inflammation Basic Principles and Medical Correlation" (J. J. Gallin, I. M. Goldstein, and R. Snyderman, eds.), 2nd ed., pp. 603–636. Raven Press, New York.

Endl, J., Seidl, H. P., Fiedler, F., and Schleifer, K. H. (1983). Chemical composition and structure of cell wall teichoic acids of staphylococci. *Arch. Microbiol.* **135**, 215–223.

Ezekowitz, R. A. B., and Stahl, P. D. (1988). The structure and function of vertebrate mannose lectin-like proteins. *J. Cell Sci., Suppl.* **9**, 121–133.

Falk, K., Rötzschke, O., Stevanović, S., Jung, G., and Rammensee, H. G. (1991). Allele-specific motifs revealed by sequencing of self-peptides eluted from MHC molecules. *Nature (London)* **351**, 290–296.

Falkow, S., and Mekalanos, J. (1990). Enteric bacilli and vibrios. *In* "Microbiology" (B. D. Davis, R. Dulbecco, H. N. Eisen, and H. S. Ginsberg, eds.), pp. 561–587. Lippincott, Philadelphia.

Falkow, S., Isberg, R. R., and Portnoy, D. A. (1992). The interaction of bacteria with mammalian cells. *Annu. Rev. Cell Biol.* **8**, 333–363.

Fallman, M., Lew, D. P., Stendahl, O., and Andersson, T. (1989). Receptor-mediated phagocytosis in human neutrophils is associated with increased formation of inositol phosphates and diacylglycerol: Elevation in cytosolic free calcium and formation of inositol phosphates can be dissociated from accumulation of diacylglycerol. *J. Clin. Invest.* **84**, 886–891.

Fidler, I. J., Nii, A., Utsugi, T., Brown, D., Bakouche, O., and Kleinerman, E. S. (1990). Differential release of TNF-alpha, IL 1, and PGE2 by human blood monocytes subsequent to interaction with different bacterial derived agents. *Lymphokine Res.* **9**, 449–463.

Firestein, G. S., Roeder, W. D., Laxer, J. A., Townsend, K. S., Weaver, C. T., Hom, J. T., Linton, J., Torbett, B. E., and Glasebrook, A. L. (1989). A new murine CD4+ T cell subset with an unrestricted cytokine profile. *J. Immunol.* **143**, 518–525.

Fischer, W., Mannsfeld, T., and Hagen, G. (1990). On the basic structure of poly(glycerophosphate) lipoteichoic acids. *Biochem. Cell Biol.* **68**, 33–43.

Fischer, W., Behr, T., Hartmann, R., Peter-Katalinic, J., and Egge, H. (1993). Teichoic acid and lipoteichoic acid of *Streptococcus pneumoniae* possess identical chain structures. A reinvestigation of teichoic acid (C polysaccharide). *Eur. J. Biochem.* **215**, 851–857.

Fischetti, V. A. (1989). Streptococcal M protein: Molecular design and biological behavior. *Clin. Microbiol. Rev.* **2**, 285–314.

Fleischer, B. (1991). The human T cell response to mitogenic microbial endotoxins. *Curr. Top. Microbiol. Immunol.* **174**, 53–62.

Frehel, C., and Rostogi, N. (1987). *Mycobacterium leprae* surface components intervene in the early phagosome-lysosome fusion inhibition event. *Infect. Immun.* **55,** 2916–2921.

Fridovich, I. (1986). Biological effects of superoxide radical. *Arch. Biochem. Biophys.* **247,** 1–11.

Gabay, J. E., and Almeida, R. P. (1993). Antibiotic peptides and serine homologs in human polymorphonuclear leukocytes: Defensins and azurocidin. *Curr. Opin. Immunol.* **5,** 97–102.

Gabay, J. E., Scott, R. W., Campanelli, D., Griffith, J., Wilde, C., Marra, M. N., Seeger, M., and Nathan, C. F. (1989). Antibiotic proteins of human polymorphonuclear leukocytes. *Proc. Natl. Acad. Sci. U.S.A.* **86,** 5610–5614.

Gajewski, T. F., Pinnas, M., Wong, T., and Fitch, F. W. (1991). Murine T_H1 and T_H2 clones proliferate optimally in response to distinct antigen presenting cell population. *J. Immunol.* **196,** 1750–1758.

Galanos, C., Lüderitz, O., Rietschel, E. T., and Westphal, O. (1977). Newer aspects of the chemistry and biology of bacterial lipopolysaccharides, with special reference to their lipid A component. *Int. Rev. Biochem., Biochem. Lipids II* **14,** 239–335.

Gallin, J. I., Sandler, J. A., Clyman, R. J., Manganiello, V. C., and Vaughan, M. (1978). Agents that increase cyclic AMP inhibit accumulation of cGMP and depress human monocyte locomotion. *J. Immunol.* **120,** 492–496.

Ganz, T., Selsted, M. E., Szklarek, D., Harwig, S. S., Daher, K., Bainton, D. F., and Lehrer, R. I. (1985). Defensins. Natural peptide antibiotics of human neutrophils. *J. Clin. Invest.* **76,** 1427–1435.

Ganz, T., Selsted, M. E., and Lehrer, R. K. (1986). Antimicrobial activity of phagocyte granule proteins. *Semin. Respir. Infect.* **1,** 107–117.

Ganz, T., Selsted, M. E., and Lehrer, R. I. (1990). Defensins. *Eur. J. Haematol.* **44,** 1–8.

Geyer, R., Galanos, C., Westphal, O., and Golecki, J. R. (1979). A lipopolysaccharide-binding cell-surface protein from *Salmonella minnesota.* Isolation, partial characterization and occurrence in different *Enterobacteriaceae. Eur. J. Biochem.* **98,** 27–38.

Gilman, A. G. (1987). G proteins: Transducers of receptor-generated signals. *Annu. Rev. Biochem.* **56,** 614–649.

Goldberg, A. L., and Rock, K. L. (1992). Proteolysis, proteasomes and antigen presentation. *Nature (London)* **357,** 875–879.

Goldman, R. C., White, D., and Leive, L. (1981). Identification of outer membrane proteins including known lymphocyte mitogens as the endotoxin proteins of *E. coli* O111th. *J. Immunol.* **127,** 1290–1295.

Goodman, G. W., and Sultzer, B. M. (1979). Characterization of the chemical and physical properties of a novel B-lymphocyte activator, endotoxin protein. *Infect. Immun.* **24,** 685–696.

Gordon, S., and Mokoena, T. (1989). Receptors for mannosyl structures on mononuclear phagocytes. *In* "Human Monocytes" (M. Zembala and G. L. Asherson, eds.), pp. 141–149. Academic Press, London.

Gordon, S., Todd, J., and Cohn, Z. A. (1974). In vitro synthesis and secretion of lysozyme by mononuclear phagocytes. *J. Exp. Med.* **139,** 1228–1248.

Gordon, S., Perry, V. H., Rabinowitz, S., Chung, L. P., and Rosen, H. (1988). Plasma membrane receptors of the mononuclear phagocyte system. *J. Cell Sci., Suppl.* **9,** 1–26.

Goren, M. B., D'Arcy Hart, P., Young, M. R., and Armstrong, J. A. (1976). Prevention of phagosome-lysosome fusion in cultured macrophages by sulfatides of *Mycobacterium tuberculosis. Proc. Natl. Acad. Sci. U.S.A.* **73,** 2510–2514.

Graham, I. L., and Brown, E. J. (1991). Extracellular calcium binding results in a conformational change in Mac-1 (CD11b/CD18) on neutrophils. Differentiation of adhesion and phagocytosis functions of Mac-1. *J. Immunol.* **146,** 685–691.

Graham, I. L., Gresham, H. D., and Brown, E. J. (1989). An immobile subset of plasma membrane CD11b/CD18 (Mac-1) is involved in phagocytosis of targets recognized by multiple receptors. *J. Immunol.* **142,** 2352–2358.

Green, S. J., Crawford, R. M., Meltzer, M. S., Hibbs, J. B., and Nacy, C. A. (1990). Leishmania provide a second signal for nitric oxide production in interferon-γ-treated macrophages by stimulation with TNF-α. *J. Immunol.* **144,** 4290–4297.

Gresham, H. D., Goodwin, J. L., Allen, P. M., Anderson, D. C., and Brown, E. J. (1989). A novel member of the integrin receptor family mediates Arg-Gly-Asp-stimulated neutrophil phagocytosis. *J. Cell Biol.* **108,** 1935–1943.

Gresham, H. D., Zheleznyak, A., Mormol, J. S., and Brown, E. J. (1990). Studies on the molecular mechanisms of human neutrophil Fc-receptor-mediated phagocytosis: Evidence that a distinct pathway for activation of the respiratory burst results in reactive oxygen metabolite-dependent amplification of ingestion. *J. Biol. Chem.* **265,** 7819–7826.

Groisman, E. A., and Saier, M. H., Jr. (1990). Salmonella virulence: New clues to intramacrophage survival. *Trends Biol. Sci.* **15,** 30–33.

Hackstadt, T., and Williams, J. C. (1981). Biochemical stratagun for obligate parasitism of eukaryotic cells by *Coxiella burnetti. Proc. Natl. Acad. Sci. U.S.A.* **78,** 3240–3244.

Halliwell, B., and Gutterage, J. M. C. (1990). Role of free radicals and catalytic metal ions in human disease: An overview. *In* "Methods in Enzymology" (L. Packer and A. Glazer, eds.), Vol. 186, pp. 1–85. Academic Press, San Diego, CA.

Han, J., Lee, J. D., Tobias, P. S., and Ulevitch, R. J. (1993). Bacterial lipopolysaccharide induces rapid protein tyrosine phosphorylation of 70Z/3 cells expressing CD14. *J. Biol. Chem.* **268,** 25009–25014.

Harrop, P. J., O'Grady, R. L., Knox, K. W., and Wicken, A. J. (1980). Stimulation of lysosomal enzyme release from macrophages by lipoteichoic acid. *J. Periodontal Res.* **15,** 492–501.

Hauschildt, S., Hoffmann, P., Beuscher, H. U., Dufhues, G., Heinrich, P., Wiesmüller, K. H., Jung, G., and Bessler, W. G. (1990a). Activation of bone marrow-derived mouse macrophages by bacterial lipopeptide: Cytokine production, phagocytosis and Ia expression. *Eur. J. Immunol.* **20,** 63–68.

Hauschildt, S., Wolf, B., Lückhoff, A., and Bessler, W. G. (1990b). Determination of second messengers and protein kinase C in bone marrow-derived macrophages stimulated with a bacterial lipopeptide. *Mol. Immunol.* **27,** 473–479.

Hauschildt, S., Bassenge, E., Bessler, W., Busse, R., and Mülsch, A. (1990c). L-arginine-dependent nitric oxide formation and nitrite release in bone marrow-derived macrophages stimulated with bacterial lipopeptide and lipopolysaccharide. *Immunology* **70,** 332–337.

Hauschildt, S., Scheipers, P., and Bessler, W. G. (1994). Lipopolysaccharide-induced change of ADP-ribosylation of a cytosolic protein in bone marrow-derived macrophages. *Biochem. J.* **297,** 17–20.

Havell, E. A. (1987). Production of tumor necrosis factor during murine listeriosis. *J. Immunol.* **139,** 4225–4231.

Hibbs, J. B., Taintor, R. R., and Vavrin, Z. (1984). Iron depletion: Possible cause of tumor cell cytotoxicity induced by activated macrophages. *Biochem. Biophys. Res. Commun.* **123,** 716–723.

Hibbs, J. B., Taintor, R. R., Vavrin, Z., Granger, D. L., Drapier, J.-C., Amber, I. J., and Lancaster, J. R. (1990). Synthesis of nitric oxide from a terminal guanidinonitrogen atom of L-arginine: a molecular mechanism regulating cellular proliferation that targets intracellular iron. *In* "Nitric Oxide from L-Arginine: A Bioregulatory System" (S. Moncada and E. A. Higgs, eds.), pp. 189–224. Excerpta Medica, Amsterdam.

Hirano, T. (1991). Interleukin 6. *In* "The Cytokine Handbook" (A. Thomson, ed.), pp. 169–190. Academic Press, London.

Ho, J. L., Reed, S. G., Wick, F. A., and Giordano, M. (1990). Granulocyte-macrophage and macrophage colony-stimulating factors activate intramacrophage killing of *Leishmania mexicana amazonensis. J. Infect. Dis.* **162,** 224–230.

Hoffman, M., and Weinberg, J. B. (1987). Tumor necrosis factor-α induces increased hydrogen peroxide production and Fc receptor expression but not increased Ia antigen expression by peritoneal macrophages. *J. Leukocyte Biol.* **47,** 704–707.

Hoffmann, P., Heinle, S., Schade, U. F., Loppnow, L., Ulmer, A. J., Flad, H. D., and Bessler, W. G. (1988). Stimulation of human and murine adherent cells by bacterial lipoprotein and synthetic lipopeptide analogues. *Immunobiology* **177,** 158–170.

Hoffmann, P., Wiesmüller, K.-H., Metzger, J., Jung, G., and Bessler, W. (1989). Induction of tumorcytoxicity in murine bone marrow-derived macrophages by two synthetic lipopeptide analogues. *Biol. Chem. Hoppe-Seyler* **370**, 575–582.

Hogg, N., Darley-Usmar, V. M., Wilson, M. T., and Moncada, S. (1992). Production of hydroxyl radicals from the simultaneous generation of superoxide and nitric oxide. *Biochem. J.* **281**, 419–424.

Holst, O., and Brade, H. (1992). Chemical structure of the core region of lipopolysaccharides. *In* "Bacterial Endotoxic Lipopolysaccharides" (D. C. Morrison and J. L. Ryan, eds.), Vol. I, pp. 135–170. CRC Press, Boca Raton, FL.

Holzer, T. J., Nelson, K. E., Schauf, V., Crispen, R. G., and Anderson, B. R. (1986). *Mycobacterium leprae* fails to stimulate phagocytic cell superoxide anion generation. *Infect. Immun.* **51**, 514–520.

Hooijkaas, H., van der Linde-Preesman, A. A., Bitter, M. W., Benner, R., Pleasants, R. J., and Wostmann, S. B. (1985). Frequency analysis of functional immunoglobulin C- and V-gene expression by mitogen-reactive B cells in germfree mice fed chemically defined ultrafiltered "antigen-free" diet. *J. Immunol.* **134**, 2223–2227.

Horuk, R. (1994). The interleukin-8 receptor family: From chemokine to malaria. *Immunol. Today* **15**, 169–174.

Horwitz, M. A. (1983). The legionnaires' disease bacterium (*Legionella pneumophila*) inhibits phagosome-lysosome fusion in human monocytes. *J. Exp. Med.* **158**, 2108–2126.

Horwitz, M. A. (1984). Phagocytosis of the legionnaires' disease bacterium (*Legionella pneumophila*) occurs by a novel mechanism: Engulfment within a pseudopod coil. *Cell (Cambridge, Mass.)* **36**, 33–37.

Horwitz, M. A. (1992). Interaction between macrophages and *Legionella pneumophila. Curr. Top. Microbiol. Immunol.* **181**, 265–282.

Horwitz, M. A., and Maxfield, F. R. (1984). *Legionella pneumophila* inhibits acidification of its phagosome in human monocytes. *J. Cell Biol.* **99**, 1936–1943.

Hurme, M., and Serkkola, E. (1991). Differential activation signals are required for the expression of interleukin 1-α and -β genes in human monocytes. *Scand. J. Immunol.* **33**, 713–718.

Hurme, M., Viherluoto, J., and Nordström, T. (1992). The effect of calcium mobilization on LPS-induced IL-1β production depends on the differentiation stage of the monocytes/macrophages. *Scand. J. Immunol.* **36**, 507–511.

Ikuta, K., Takami, M., Kim, C. W., Honyo, T., Miyoshi, T., Tagaya, Y., Kawabe, T., and Yodoi, J. (1987). Human lymphocyte Fc receptor for IgE: Sequence homology of its cloned cDNA with animal lectins. *Proc. Natl. Acad. Sci. U.S.A.* **84**, 819–823.

Imamichi, T., and Koyama, J. (1991). Functions of two types of Fc gamma receptor on guinea pig macrophages in immune complex-induced arachidonic acid release. *Mol. Immunol.* **28**, 359–365.

Imamichi, T., Sato, H., Iwaki, S., Nakamura, T., and Koyama, J. (1990). Different abilities of two types of Fc-gamma receptor on guinea pig macrophages to trigger the intracellular Ca^{2+} mobilization and O_2^- generation. *Mol. Immunol.* **27**, 829–838.

Ishibashi, Y., and Arai, T. (1990). Roles of the complement receptor type 1 (CR1) and type 3 (CR3) on phagocytosis and subsequent phagosome-lysosome fusion in Salmonella-infected murine macrophages. *FEMS Microbiol. Immunol.* **64**, 89–96.

Jacobs, R. F., Locksley, R. M., Wilson, C. B., Haas, J. E., and Klebanoff, S. J. (1984). Interaction of primate alveolar macrophages and *Legionella pneumophila. J. Clin. Invest.* **73**, 1515–1523.

Jakway, J. P., and DeFranco, A. L. (1986). Pertussis toxin inhibition of B cell and macrophage responses to bacterial lipopolysaccharide. *Science* **234**, 734–746.

Jann, K., and Jann, B. (1984). Structure and biosynthesis of O-antigens. *In* "Handbook of Endotoxin" (E. T. Rietschel, ed.), pp. 138–186. Elsevier, Amsterdam.

Jann, K., and Jann, B., eds. (1990). Bacterial capsules. *Curr. Top. Microbiol. Immunol.* **150**, 1–157.

Johannsen, L., Wecke, J., Obal, F., Jr., and Krueger, J. M. (1991). Macrophages produce somnogenic and pyrogenic muramyl peptides during degestion of staphylococci. *Am. J. Physiol.* **260**, R126–R133.

Joiner, K. A., Fuhrman, S. A., Miettinen, H. M., Kasper, L. H., and Mellman, I. (1990). *Toxoplasma gondii:* Fusion competence of parasitophorous vacuoles in Fc-receptor-transfected fibroblasts. *Science* **249**, 641–646.

Joy, C., and Maher, D. M., (1981). The quantitation of electron energy-loss spectra. *J. Microsc. (Oxford)* **124**, 37–48.

Kang, Y. H., Dwivedi, R. S., and Lee, C. H. (1990). Ultrastructural and immunocytochemical study of the uptake and distribution of bacterial lipopolysaccharide in human monocytes. *J. Leukocyte Biol.* **48**, 316–332.

Kaplan, G. (1993). Recent advances in cytokine therapy in leprosy. *J. Infect. Dis.* **167**, Suppl. 1, S18–S22.

Kaya, S., Araki, Y., and Ito, E. (1985). Characterization of a novel linkage unit between ribitol teichoic acid and peptidoglycan in *Listeria monocytogenes* cell walls. *Eur. J. Biochem.* **146**, 517–522.

Keller, R., Fischer, W., Kleist, R., and Bassetti, S. (1992). Macrophage response to bacteria: Induction of marked secretory and cellular activities of lipoteichoic acids. *Infect. Immun.* **60**, 3664–3672.

Kishimoto, A., Mikawa, K., Hashimoto, K., Yasuda, D., Tanaka, S., Tominaga, M., Kurada, T., and Nishizuka, Y. (1989). Limited proteolysis of protein kinase C subspecies by calcium-dependent neutral protease (calpain). *J. Biol. Chem.* **264**, 4088–4092.

Kitahara, M., Eyre, II. J., Simonian, Y., Atkin, C. L., and Hasstedt, S. J. (1981). Hereditary myeloperoxidase deficiency. *Blood* **57**, 888–893.

Klebanoff, S. J. (1975). Antimicrobial mechanisms in neutrophilic polymorphonuclear leukocytes. *Semin. Hematol.* **12**, 117–142.

Kleine, B., Rapp, W., Wiesmüller, K.-H., Edinger, M., Beck, W., Metzger, J., Ataullakhanov, R., Jung, G., and Bessler, W. G. (1994). Lipopeptide-polyoxyethylene-conjugates as mitogens and adjuvants. *Immunobiology* **190**, 53–66.

Kleinerman, E. S., Maeda, M., and Jaffe, N. (1993). Liposome-encapsulated muramyl tripeptide: A new biologic response modifier for the treatment of osteosarcoma. *Cancer Treat Res.* **62**, 101–107.

Koppelman, B., Zimmerman, D. L., Walter, P., and Brodsky, F. M. (1992). Evidence for peptide transport across microsomal membranes. *Proc. Natl. Acad. Sci. U.S.A.* **89**, 3908–3912.

Kovacs, E. J., Radzioch, D., Young, H. A., and Varesio, L. (1988). Differential inhibition of IL-1 and TNF-α mRNA expression by agents which block second messenger pathways in murine macrophages. *J. Immunol.* **141**, 30101–30105.

Kronenberg, M. (1994). Antigens recognized by γST cells. *Curr. Opin. Immunol.* **6**, 64–71.

Kuhn, H.-M., Meier-Dieter, U., and Mayer, H. (1988). ECA, the enterobacterial common antigen. *Microbiol. Rev.* **54**, 195–222.

Kurlander, R. J., Shawar, S. M., Brown, M. L., and Rich, R. R. (1992). Specialized role for a murine class Ib MHC molecule in prokaryotic host defenses. *Science* **257**, 678–679.

Kurzman, I. D., Shi, F., and MacEwen, E. G. (1993). In vitro and in vivo canine mononuclear cell production of tumor necrosis factor induced by muramyl peptides and lipopolysaccharide. *Vet. Immunol. Immunopathol.* **38**, 45–56.

Kuwano, K., Akashi, A., Matsu-ura, I., Nishimoto, M., and Arai, S. (1993). Induction of macrophage-mediated production of tumor necrosis factor alpha by an L-form derived from *Staphylococcus aureus. Infect. Immun.* **61**, 1700–1706.

Lapierre, L. A., Fiers, W., and Pober, J. S. (1988). Three distinct classes of regulatory cytokines control endothelial cell MHC antigen expression. Interactions with immune gamma interferon differentiate the effects of tumor necrosis factor and lymphotoxin from those of leukocyte alpha and fibroblast beta interferons. *J. Exp. Med.* **167**, 794–804.

Latsch, M., Mollerfeld, J., Ringsdorf, H., and Loos, M. (1990). Studies on the interaction of C1q, a subcomponent of the first component of complement, with porins from *Salmonella minnesota* incorporated into artificial membranes. *FEBS Lett.* **276**, 201–204.

Le, J., Fredrickson, G., Reis, L. F. L., Diamantstein, T., Hirano, T., Kishimoto, T., and Vilcek, J. (1988). Interleukin 2-dependent and interleukin 2-independent pathways of regulation of thymocyte function by interleukin 6. *Proc. Natl. Acad. Sci. U.S.A.* **85**, 8643–8647.

Learn, D. B., Brestel, E. P., and Seetharama, S. (1987). Hydrochlorite scavenging by *Pseudomonas aeruginosa* alginate. *Infect. Immun.* **55**, 1813–1818.

Lee, J. D., Kravchenko, V., Kirkland, T. N., Han, J., Mackman, N., Moriarty, A., Leturcq, D., Tobias, P. S., and Ulevitch, R. J. (1993). GPI-anchored or integral membrane forms of CD14 mediate identical cellular responses to endotoxin. *Proc. Natl. Acad. Sci. U.S.A.* **90**, 9930–9934.

Lee, P. K., and Schlievert, P. M. (1991). Molecular genetics of pyrogenic exotoxin "superantigens" of group A streptococci and *Staphylococcus aureus*. *Curr. Top. Microbiol. Immunol.* **174**, 1–19.

Levy, F., Gabathuler, R., Larsson, R., and Krist, S. (1991). ATP is required for in vitro assembly of MHC class I antigens but not for transfer of peptides across the ER membrane. *Cell (Cambridge, Mass.)* **67**, 265–274.

Levy, R., Kotb, M., Nagauker, O., Majumdar, G., Alkan, M., Ofek, I., and Beachey, E. H. (1990). Stimulation of oxidative burst in human monocytes by lipoteichoic acids. *Infect. Immun.* **58**, 566–568.

Lex, A., Wiesmüller, K.-H., Jung, G., and Bessler, W. (1986). A synthetic analogue of *Escherichia coli* lipoprotein, tripalmitoylpentapeptide, constitutes a potent immune adjuvant. *J. Immunol.* **137**, 2676–2681.

Ling, P. D., Warren, M. K., and Vogel, S. N. (1985). Antagonistic effect of interferon-β on the interferon-γ-induced expression of Ia antigen in murine macrophages. *J. Immunol.* **135**, 1857–1863.

Litosch, J., Wallis, C., and Fain, J. (1985). 5-hydroxytryptamine stimulates inositol phosphate production in a cell-free system from blowfly salivary glands. *J. Biol. Chem.* **260**, 5464–5471.

Loleit, M., Deres, K., Wiesmüller, K.-H., Jung, G., Eckert, M., and Bessler, W. G. (1994). Biological activity of the *Escherichia coli* lipoprotein: Detection of novel lymphocytes activating peptide segments of the molecule and their conformational characterization. *Biol. Chem. Hoppe-Seyler* **375**, 407–412.

Lory, S. (1990). Pseudomonas and other nonfermenting bacilli. *In* "Microbiology" (B. D. Davis, R. Dulbecco, H. N. Eisen, and H. S. Ginsberg, eds.), pp. 595–600. Lippincott, Philadelphia.

Lotz, M., Jirik, F., Kabouridis, R., Tsoukas, C., Hirano, T., Kishimoto, T., and Carson, D. A. (1988). B cell stimulating factor 2/interleukin 6 is a costimulant for human thymocytes and T lymphocytes. *J. Exp. Med.* **167**, 1253–1258.

Lüderitz, O., Freudenberg, M. A., Galanos, C., Lehmann, V., Rietschel, E. T., and Shaw, D. H. (1982). Lipopolysaccharides of Gram negative bacteria. *Curr. Top. Membr. Transp.* **17**, 79–151.

Manjula, B. N., and Fischetti, V. A. (1986). Sequence homology of group A streptococcal Pep M5 protein with other coiled-coil proteins. *Biochem. Biophys. Res. Commun.* **140**, 684–690.

Mannion, B. A., Weiss, J., and Elsbach, P. (1990). Separation of sublethal and lethal effects of the bactericidal/permeability-increasing protein of *Escherichia coli*. *J. Clin. Invest.* **85**, 853–860.

Marletta, M. A. (1989). Nitric oxide: Biosynthesis and biological significance. *Trends Biol. Sci.* **14**, 488–492.

Martin, M., Schwinzer, R., Schellekens, H., and Resch, K. (1989). Glomerular mesangial cells in local inflammation. Induction of the expression of MHC class antigens by interferon-gamma. *J. Immunol.* **142**, 1887–1894.

Matsushima, K., Morishita, K., Yashimura, T., Lavu, S., Kobayashi, Y., Lew, W., Appella, E., Kung, H. F., Leonard, E. J., and Oppenheim, J. J. (1988). Molecular cloning of a human monocyte-derived neutrophil chemotactic factor (MDNCF) and the induction of MDNCF mRNA by interleukin 1 and tumor necrosis factor. *J. Exp. Med.* **167**, 1883–1893.

Mattsson, E., Verhage, L., Rollof, J., Fleer, A., Verhoef, J., and van Dijk, H. (1993). Peptidoglycan and teichoic acid from *Staphylococcus epidermidis* stimulate human monocytes to release tumour necrosis factor-alpha, interleukin-1 beta and interleukin-6. *FEMS Immunol. Med. Microbiol.* **7**, 281–287.

McCarty, M. (1990). Streptococci. *In* "Microbiology" (B. D. Davis, R. Dulbecco, H. N. Eisen, and H. S. Ginsberg, eds.), pp. 525–538. Lippincott, Philadelphia.

McCoy, D. M., Brown, G. L., Ausobsky, J. R., and Polk, H. C., Jr. (1985). Muramyl dipeptide enhances in vitro peritoneal macrophage phagocytic activity. *Am. Surg.* **51**, 634–636.

Meade, B. D., Kind, P. D., Ewell, J. B., McGrath, P. P., and Manclark, C. R. (1984). In vitro inhibition of murine macrophage migration by *Bordetella pertussis* lymphocytosis-promoting factor. *Infect. Immun.* **45,** 718–725.

Mekalanos, J. J. (1985). Cholera toxin: Genetic analysis, regulation and role in pathogenesis. *Curr. Top. Microbiol. Immunol.* **118,** 97–118.

Melillo, G., Cox, G. W., Radzioch, D., and Varesio, L. (1993). Picolinic acid, a catabolite of L-tryptophan, is a costimulus for the induction of reactive nitrogen intermediate production in murine macrophages. *J. Immunol.* **150,** 4031–4040.

Metzger, J., Jung, G., Bessler, W. G., Hoffmann, P., Strecker, M., Lieberknecht, A., and Schmidt, U. (1991). Lipopeptides containing 2-(Palmitoylamino)-6,7-bis(palmitoyloxy)heptanoic acid: Synthesis, stereospecific stimulation of B-lymphocytes and macrophages, and adjuvanticity in vivo and in vitro. *J. Med. Chem.* **34,** 1969–1974.

Minami, A., Fujimoto, K., Ozaki, Y., and Nakamura, S. (1988). Augmentation of host resistance to microbial infections by recombinant human interleukin-1 alpha. *Infect. Immun.* **56,** 3116–3120.

Ming, W. J., Bersam, L., and Mantovani, A. (1987). Tumor necrosis factor is chemotactic for monocytes and polymorphonuclear leukocytes. *J. Immunol.* **138,** 1469–1477.

Moll, H., Fuchs, H., Blank, C., and Röllinghoff, M. (1993). Langerhans cells transport *Leishmania major* from the infected skin to the draining lymph node for presentation to antigen-specific T cells. *Eur. J. Immunol.* **23,** 1595–1601.

Moore, K., O'Garra, A., de Waal Malefyt, R., Vieira, P., and Mosmann, T. R. (1993). Interleukin-10. *Annu. Rev. Immunol.* **11,** 165–190.

Moore, M. W., Carbone, F. R., and Bevan, M. J. (1988). Introduction of soluble protein into the class I pathway of antigen processing and presentation. *Cell (Cambridge, Mass.)* **54,** 777–785.

Mosmann, T. R., and Coffman, R. L. (1989a). Different patterns of lymphokine secretion lead to different functional properties. *Annu. Rev. Immunol.* **7,** 145–173.

Mosmann, T. R., and Coffman, R. L. (1989b). Heterogeneity of cytokine secretion patterns and functions of helper T cells. *Adv. Immunol.* **46,** 111–147.

Mosser, D. M., and Edelson, P. J. (1985). The mouse macrophage receptor for CRbi (cR3) is a major mechanism in the phagocytosis of *Leishmania* promastigotes. *J. Immunol.* **135,** 2785–2789.

Moxon, E. R., and Kroll, J. S. (1990). The role of bacterial polysaccharide capsules as virulence factors. *Curr. Top. Microbiol. Immunol.* **150,** 65–85.

Mukaida, N., Harada, A., Yasumoto, K., and Matsushima, K. (1992). Properties of pro-inflammatory cell type-specific leukocyte chemotactic cytokines, interleukin 8 (IL-8) and monocyte chemotactic and activating factor (MCAF). *Microbiol. Immunol.* **36,** 773–789.

Muraguchi, A., Hirano, T., Tang, B., Matsuda, T., Horii, Y., Nakajima, K., and Kishimoto, T. (1988). The essential role of B cell stimulatory factor 2 (BSF-2/IL-6) for the terminal differentiation of B cells. *J. Exp. Med.* **67,** 332–344.

Myones, B. L., Dalzell, J. G., Hogg, N., and Ross, G. D. (1988). Neutrophil and monocyte cell surface p 150,95 has iC3b-receptor (CR4) activity resembling CR3. *J. Clin. Invest.* **82,** 640–651.

Nacy, C. A., and Meltzer, M. S. (1991). T-cell-mediated activation of macrophages. *Curr. Opin. Immunol.* **3,** 330–335.

Nagao, S., Akagawa, K. S., Okada, F., Harada, Y., Yagawa, K., Kato, K., and Tanigawa, Y. (1992). Species dependency of in vitro macrophage activation by bacterial peptidoglycans. *Microbiol. Immunol.* **36,** 1155–1171.

Nathan, C. F. (1987). Secretory products of macrophages. *J. Clin. Invest.* **79,** 319–326.

Nathan, C. F., and Hibbs, J. B., Jr. (1991). Role of nitric oxide synthesis in macrophage antimicrobial activity. *Curr. Opin. Immunol.* **3,** 65–70.

Nathan, C. F., Murray, H. W., and Cohn, L. A. (1980). The macrophage as an effector cell. *N. Engl. J. Med.* **303,** 622–626.

Neefjes, J. J., and Ploegh, H. L. (1992). Inhibition of endosomal proteolytic activity by leupeptin blocks surface expression of MHC class II molecules and their conversion to SDS resistant $\alpha\beta$ heterodimers in endosomes. *EMBO J.* **11,** 411–416.

Neill, M. A., and Klebanoff, S. J. (1988). The effect of phenolic glycolipid-1 from *Mycobacterium leprae* on the antimicrobial activity of human macrophages. *J. Exp. Med.* **167,** 30–42.

Nelson, C. A., Roof, R. W., McCourt, D. W., and Unanue, E. R. (1992). Identification of the naturally processed form of hen egg white lysozyme bound to the murine major histocompatibility complex class II molecule I-A^k. *Proc. Natl. Acad. Sci. U.S.A.* **89,** 7380–7383.

Nesmeyanov, V. A., Khaidukov, S. V., Komaleva, R. L., Andronova, T. M., and Ivanov, V. T. (1990). Muramylpeptides augment expression of Ia-antigens on mouse macrophages. *Biomed. Sci.* **1,** 151–154.

Neta, R., Sayers, T., and Oppenheim, J. J. (1991). Relationship of TNF to interleukins. *In* "Tumor Necrosis Factor: Structure, Function and Mechanism of Action" (B. B. Aggarwal and J. Vilcek, eds.), pp. 499–566. Dekker, New York.

Nichols, B. A., Bainton, D. F., and Farquhar, M. G. (1971). Differentiation of monocytes. Origin, nature, fate of their azurophil granules. *J. Cell Biol.* **50,** 498–515.

Nikaido, H. (1979). Nonspecific transport through the outer membrane. *In* "Bacterial Outer Membranes, Biogenesis and Functions" (M. Inouye, ed.), pp. 361–407. Wiley, New York.

Nishizuka, Y. (1984). The role of protein kinase C in cell surface signal transduction and tumor promotion. *Nature (London)* **308,** 693–698.

Nishizuka, Y. (1986). Studies and perspectives of protein kinase C. *Science* **233,** 305–312.

Nixdorff, K., Weber, G., Kaniecki, K., Ruiner, W., and Schell, S. (1992). Bacterial protein-LPS complexes and immunomodulation. *In* "Microbial Infections: Role of Biological Response Modifiers" (H. Friedman and T. W. Klein, eds.), pp. 49–61. Plenum, New York.

Novick, R. P. (1990). Staphylococci. *In* "Microbiology" (B. D. Davis, R. Dulbecco, H. N. Eisen, and H. S. Ginsberg, eds.), pp. 539–550. Lippincott, Philadelphia.

Novotney, M., Chang, Z. L., Uchiyama, H., and Suzuki, T. (1991). Protein kinase C in tumoricidal activation of murine macrophage-like cell lines. *Biochemistry* **30,** 5597–5604.

O'Brien, A. D., Tesh, V. L., Donohue-Rolfe, A., Jackson, M. P., Olsnes, S., Sandvig, K., Lindberg, A. A., and Keusch, G. T. (1992). Shiga toxin: Biochemistry, genetics, mode of action, and role in pathogenesis. *Curr. Top. Microbiol. Immunol.* **180,** 65–94.

Ofek, J., and Sharon, N. (1988). Lectinophagocytosis: A molecular mechanism of recognition between cell surface sugar and lectins in the phagocytosis of bacteria. *Infect. Immun.* **56,** 539–547.

Ogata, K., Linzer, B. A., Zuberi, R. I., Ganz, T., Lehrer, R. I., and Catanzaro, A. (1992). Activity of defensins from human neutrophilic granulocytes against *Mycobacterium avium–Mycobacterium intracellulare*. *Infect. Immun.* **60,** 4720–4725.

Ohshima, Y., Beuth, J., Yassin, A., Ko, H. L., and Pulverer, G. (1988). Stimulation of human monocyte chemiluminescence by staphylococcal lipoteichoic acid. *Med. Microbiol. Immunol.* **177,** 115–121.

Oppenheim, J. J., Zachariae, C. O., Mukaida, N., and Matsushima, K. (1991). Properties of the novel proinflammatory supergene "intercrine" cytokine family. *Annu. Rev. Immunol.* **9,** 617–648.

Ozaki, Y., Ohashi, F., Minami, A., and Nakamura, S. (1987). Enhanced resistance of mice to bacterial infection induced by recombinant human interleukin-1 alpha. *Infect. Immun.* **55,** 1436–1440.

Pabst, N. J., Hedegaard, H. B., and Johnston, R. B., Jr. (1982). Cultured human monocytes require exposure to bacterial products to maintain an optimal oxygen radical response. *J. Immunol.* **128,** 123–128.

Pabst, M. J., Gross, J. M., Brozna, J. P., and Goren, M. B. (1988). Inhibition of macrophage priming by sulfatide from *Mycobacterium tuberculosis*. *J. Immunol.* **140,** 634–640.

Pamer, E. G., Harty, J. T., and Bevan, M. J. (1991). Precise prediction of a dominant class I MHC-restricted epitope of *Listeria monocytogenes*. *Nature (London)* **353,** 852–855.

Pamer, E. G., Wang, C. R., Flaherty, I., Fischer Lindahl, K., and Bevan, M. J. (1992). H-2M3 presents a *Listeria monocytogenes* peptide to cytotoxic T lymphocytes. *Cell (Cambridge, Mass.)* **70,** 215–223.

Parant, M. (1987). TNF as a mediator of LPS- and MDP-induced enhancement of the mouse nonspecific immunity. *J. Leukocyte Biol.* **42,** 576.

Parant, M., Riveau, G., Parant, F., and Chédid, L. (1984). Inhibition of endogenous pyrogen-induced fever by a muramyl dipeptide derivative. *Am. J. Physiol.* **247**, C169–174.

Peliat, C., Henry, Y., and Drapier, J. C. (1990). IFN-γ-activated macrophages: detection by electron paramagnetic resonance of complexes between L-arginine-derived nitric oxide and nonheme iron proteins. *Biochem. Biophys. Res. Commun.* **166**, 119–125.

Perutz, W., Monig, H., Butler, J., and Land, E. (1985). Reactions of nitrogen dioxide in aqueous model systems: Oxidation of tyrosine units in peptides and proteins. *Arch. Biochem. Biophys.* **243**, 125–134.

Peters, P. J., Neefjes, J. J., Vorschort, V., Ploegh, H. L., and Geuze, H. J. (1991). Segregation of MHC class II molecules from MHC class I molecules in the Golgi complex for transport to lysosomal compartments. *Nature (London)* **349**, 669–676.

Philip, R., and Epstein, L. B. (1986). Tumor necrosis factor as immunomodulator and mediator of monocyte cytotoxicity induced by itself, γ-interferon and interleukin-1. *Nature (London)* **323**, 86–89.

Potekhina, N. V., Naumova, I. B., Shashkov, A. S., and Terekhova, L. P. (1991). Structural features of cell wall teichoic acid and peptidoglycan of *Actinomadura cremea* INA 292. *Eur. J. Biochem.* **199**, 313–316.

Potekhina, N. V., Tul'skaya, E. M., Naumova, I. B., Shashkov, A. S., and Evtushenko, L. I. (1993). Erythritolteichoic acid in the cell wall of *Glycomyces tenuis* VKM Ac-1250. *Eur. J. Biochem.* **218**, 371–375.

Powis. S. J., Deverson, E. V., Coadwell, W. J., Ciruela, A., Huskisson, N. S., Smith, H., Butcher, G. W., and Howard, J. C. (1992). Effect of polymorphism of an MHC-linked transporter on the peptides assembled in a class I molecule. *Nature (London)* **357**, 211–215.

Prpic, V., Weiel, J. E., Somers, S. D., Diguiseppi, J., Gomias, S., Pizzo, S. V., Hamilton, T. A., Herrman, B., and Adam, D. O. (1987). Effect of bacterial lipopolysaccharide on the hydrolysis of phosphatidylinositol-4,5-bisphosphate in murine macrophages. *J. Immunol.* **139**, 526–533.

Radi, R., Cosgrove, T. P., Beckman, J. S., and Freeman, B. A. (1993). Peroxynitrite-induced luminol chemiluminescence. *Biochem. J.* **290**, 52–57.

Reiling, N., Ulmer, A. J., Duchrow, M., Ernst, M., Flad, H.-D., and Hauschildt, S. (1994). Nitric oxide synthase: mRNA expression of different isoforms in human monocytes/macrophages. *Eur. J. Immunol.* **24**, 1941–1944.

Reiner, N. E. (1994). Altered cell signaling and mononuclear phagocyte deactivation during intracellular infection. *Immunol. Today* **15**, 374–381.

Relman, D., Tuomanen, E., Falkow, S., Golenbock, D. T., Sankkonen, K., and Wright, S. D. (1990). Recognition of a bacterial adhesion by an integrin: Macrophage CR3 ($\alpha_M\beta_2$, CD11b/CD18) binds filamentous hemagglutinin of *Bordetella pertussis*. *Cell (Cambridge, Mass.)* **61**, 1375–1382.

Rennick, D., Berg, D., and Holland, G. (1992). Interleukin 10: An overview. *Prog. Growth Factor Res.* **4**, 207–227.

Rietschel, E. T., and Brade, H. (1992). Bacterial endotoxins. *Sci. Am.* **267**, 54–61.

Rietschel, E. T., Brade, L., Lindner, B., and Zähringer, U. (1992). Biochemistry of lipopolysaccharides. *In* "Bacterial Endotoxic Lipopolysaccharides" (D. C. Morrison and J. L. Ryan, eds.), Vol. I, pp. 3–41. CRC Press, Boca Raton, FL.

Riveau, G., Parant, M., Damais, C., Parant, F., and Chédid, L. (1986). Dissociation between muramyl dipeptide-induced fever and changes in plasma metal levels. *Am. J. Physiol.* **250**, C572–577.

Robbins, J. B., and Pittman, M. (1990). Bordetella. *In* "Microbiology" (B. D. Davis, R. Dulbecco, H. N. Eisen, and H. S. Ginsberg, eds.), pp. 621–624. Lippincott, Philadelphia.

Roche, P. A., and Cresswell, P. (1990). Invariant chain association with HLA-DR molecules inhibits immunogenic peptide binding. *Nature (London)* **345**, 615–618.

Rodrigues, M. M., Cordey, A. S., Arreaza, G., Corradin, G., Romero, P., Maryanski, J. L., Nussenzweig, R. S., and Zavala, F. (1991). CD8$^+$ cytolytic T-cell clones derived against the *Plasmodium-yoelii* circumsporozoite protein protects against malaria. *Int. Immunol.* **3**, 579–585.

Rosenfeld, S. I., Looney, R. J., Leddy, J. P., Abraham, D. L., and Anderson, C. L. (1985). Human platelet Fc receptors for immunoglobulin G. Identification as a 40000 molecular weight membrane protein shared by monocytes. *J. Clin. Invest.* **76**, 2317–2322.

Ross, G., Walport, M. J., and Hogg, N. (1989). Receptor for IgG Fc and fixed C3. *In* "Human Monocytes" (M. Zembala and G. L. Asherson, eds.), pp. 37–47. Academic Press, London.

Rousset, F., Garcia, E., Defrance, T., Peronne, C., Vizzio, N., Hsu, D. H., Kastelein, R., Moore, K. W., and Banchereau, J. (1992). Interleukin 10 is a potent growth and differentiation factor for activated human B-lymphocytes. *Proc. Natl. Acad. Sci. U.S.A.* **89**, 1890–1893.

Rudensky, A. Y., Preston-Hurlburt, P., Hong, S.-C., Barlow, A., and Janeway, C. A. (1991). Sequence analysis of peptides bound to MHC class II molecules. *Nature (London)* **353**, 622–627.

Salgame, P., Yamamura, M., Bloom, B. R., and Modlin, R. L. (1992). Evidence for functional subsets of CD4+ and CD8+ T cells in human disease: Lymphokine patterns in leprosy. *Chem. Immunol.* **54**, 44–59.

Schall, T. J. (1991). Biology of the RANTES/SIS cytokine family. *Cytokine* **3**, 165–183.

Schleimer, R., and Rutledge, B. (1986). Cultured human vascular endothelial cells acquire adhesiveness for neutrophils after stimulation with interleukin 1, endotoxin, and tumor-promoting phorbol diesters. *J. Immunol.* **136**, 649–654.

Schlesinger, L. S., and Horwitz, M. A. (1990). Phagocytosis of leprosy bacilli is mediated by complement receptors CR1 and CR3 on human monocytes and complement components C3 in serum. *J. Clin. Invest.* **85**, 1304–1314.

Schlesinger, L. S., Bellinger-Kawahara, C. G., Payne, N. R., and Horwitz, M. A. (1990). Phagocytosis of *Mycobacterium tuberculosis* is mediated by human monocyte complement receptors and complement component C3. *J. Immunol.* **144**, 2771–2780.

Sebaldt, R. J., Prpić, V., Hollenbach, P., Adams, D. O., and Uhing, R. J. (1990). Interferon-gamma potentiates the accumulation of diacylglycerol in murine macrophages. *J. Immunol.* **145**, 684–689.

Segal, A. W. (1989). The electron transport chain of the microbicidal oxodase of phagocytic cells and involvement in the molecular pathology of chronic granulomatous disease. *J. Clin. Invest.* **83**, 1785–1793.

Sharon, N. (1984). Carbohydrates as recognition determinants in phagocytosis and in lectin-mediated killing of target cells. *Biol. Cell.* **51**, 239–246.

Shurin, S. B., and Stossel, T. P. (1978). Complement (C3) activated phagocytosis by lung macrophages. *J. Immunol.* **120**, 1305–1312.

Sisson, S. D., and Dinarello, C. A. (1989). Interleukin 1 in human monocytes. *In* "Human Monocytes" (M. Zembala and G. L. Asherson, eds.), pp. 187–193. Academic Press, London.

Sjögren, H. O. (1991). T cell activation by superantigens—dependence on MHC class II molecules. *Curr. Top. Microbiol. Immunol.* **174**, 39–51.

Snyderman, R., and Uhing, R. J. (1988). Chemoattractant stimulus-response coupling. *In* "Inflammation: Basic Principles and Clinical Correlates" (J. J. Gallin, I. M. Goldstein, and R. Snyderman, eds.), pp. 421–439. Raven Press, New York.

Snyderman, R., Smith, C. D., and Verghese, M. W. (1986). A chemoattractant receptor on macrophages exists in two affinity states regulated by guanine nucleotides. *J. Leukocyte Biol.* **40**, 785–800.

Speert, D. P., Lorth, B. A., Cabral, D. A., and Solit, I. E. (1986). Nonopsonic phagocytosis of nonmucoid *Pseudomonas aeruginosa* by human neutrophils and monocyte-derived macrophages is correlated with bacterial piliation and hydophobicity. *Infect. Immun.* **53**, 207–212.

Speert, D. P., Wright, S. D., Silverstein, S. C., and Mah, B. (1988). Functional observation of macrophage receptors for in vitro phagocytosis of unopsonized *Pseudomonas aeruginosa. J. Clin. Invest.* **82**, 872–879.

Spiegel, A. M. (1987). Signal transduction by guanine nucleotide binding proteins. *Mol. Cell. Edocrinol.* **49**, 1–16.

Spitznagel, J. (1990). Antibiotic proteins of human neutrophils. *J. Clin. Invest.* **86**, 1381–1386.

Standiford, T. J., Arenberg, D. A., Danforth, J. M., Kunkel, S. L., VanOtteren, G. M., and Strieter, R. M. (1994). Lipoteichoic acid induces secretion of interleukin-8 from human blood monocytes: A cellular and molecular analysis. *Infect. Immun.* **62**, 119–125.

Steffens, U., Bessler, W. G., and Hauschildt, S. (1989). B cell activation by synthetic lipopeptide analogues of bacterial lipoprotein by passing phosphatidylinoitol metabolism and protein kinase C translocation. *Mol. Immunol.* **26**, 897–904.

Sternweis, P. C., and Smrca, A. V. (1992). Regulation of phospholipase C by G proteins. *Trends Biol. Sci.* **17**, 502–507.

Straub, R. E., and Gershengron, M. C. (1986). Thyrotropin-releasing hormone and GTP activate inositol trisphosphate formation in membranes isolated from rat pituitary cells. *J. Biol. Chem.* **261**, 2712–2717.

Stuehr, D. J., and Nathan, C. F. (1989). Nitric oxide. A macrophage product responsible for cytostasis and respiratory inhibition in tumor target cells. *J. Exp. Med.* **169**, 1543–1555.

Swartz, M. N. (1990a). Aerobic spore-forming bacilli. *In* "Microbiology" (B. D. Davis, R. Dulbecco, H. N. Eisen, and H. S. Ginsberg, eds.), pp. 625–631. Lippincott, Philadelphia.

Swartz, M. N. (1990b). Anaerobic spore-forming bacilli: The Clostridia. *In* "Microbiology" (B. D. Davis, R. Dulbecco, H. N. Eisen, and H. S. Ginsberg, eds.), pp. 633–646. Lippincott, Philadelphia.

Takayama, K., and Qureshi, N. (1992). Chemical structure of lipid A. *In* "Bacterial Endotoxic Lipopolysaccharides" (D. C. Morrison and J. L. Ryan, eds.), Vol. I, pp. 43–65. CRC Press, Boca Raton, FL.

Tarsi-Tsuk, D., and Levy, R. (1990). Stimulation of the respiratory burst in peripheral blood monocytes by lipoteichoic acid. The involvement of calcium ions and phospholipase A2. *J. Immunol.* **144**, 2665–2670.

Tenu, J.-P., Adam, A., Souvannavong, V., Barratt, G., Yapo, A., Petit, J.-F., Level, M., Clemance, M., and Douglas, K. (1987). A novel muramyl peptide derivative stimulates tumoricidal activity of macrophages and antibody production by B cells. *FEBS Lett.* **220**, 93–97.

Teyton, L., O'Sulivan, D., Dickson, P. W., Lotteau, V., Selk, A., Fink, P., and Peterson, P. A. (1990). Invariant chain distinguishes between the exogenous and endogenous antigen presentation pathway. *Nature (London)* **348**, 39–44.

Thomas, G. (1992). MAP kinase by another name smells just as sweet. *Cell (Cambridge, Mass.)* **68**, 3–6.

Ting, A. T., Einspahr, K. J., Abraham, R. T., and Leibson, P. J. (1991). Fc gamma receptor signal transduction in natural killer cells. Coupling to phospholipase C via a G protein-independent, but tyrosine kinase-dependent pathway. *J. Immunol.* **147**, 3122–3127.

Tobias, P. S., Soldau, K., and Ulevitch, R. (1986). Isolation of a lipopolysaccharide-binding acute phase reactant from rabbit serum. *J. Exp. Med.* **164**, 777–793.

Tobias, P. S., Soldau, K., Kline, L., Lee, J. D., Kato, K., Martin, T. P., and Ulevitch, R. J. (1993). Crosslinking of lipopolysaccahride to CD14 on THP-1 cells mediated by lipopolysaccharide binding protein. *J. Immunol.* **150**, 3011–3021.

Torbett, B. E., Laxer, J. A., and Glasebrook, A. L. (1990). Frequencies of T cells secreting IL-2 and/or IL-4 among unprimed CD4$^+$ populations. Evidence that clones secreting IL-2 and IL-4 give rise to clones which secrete only IL-4. *Immunol. Lett.* **23**, 227–233.

Tracey, K. J., Lowry, S. F., and Cerami, A. (1988). Cachectin: A hormone that triggers acute shock and chronic cachexia. *J. Infect. Dis.* **157**, 413–420.

Trincheri, G. (1993). Interleukin-12 and its role in the generation of TH1 cells. *Immunol. Today* **14**, 335–338.

Trincheri, G. (1994). The missing link between natural and adaptive immunity. *In* "Intensive Care Medicine" (E. Faist, A. E. Bane, and L. Thijs, eds.), Suppl. 1, p. 54. Harwood Academic Publishers Chu, Switzerland.

Tul'skaya, E. M., Streshinskaya, G. M., Naumova, I. B., Shashkov, A. S., and Terekhova, L. P. (1993). A new structural type of teichoic acid and some chemotaxonomic criteria of two species *Nocardiopsis dassonvillei* and *Nocardiopsis antarcticus. Arch. Microbiol.* **160**, 299–305.

Uchikawa, K., Sekikawa, I., and Azuma, I. (1986). Structural studies on teichoic acids in cell walls of several serotypes of *Listeria monocytogenes. J. Biochem. (Tokyo)* **99**, 315–327.

Uhing, R. J., Prpic, V., Jiang, H., and Exton, J. H. (1986). Hormone-stimulated polyphosphoinositide breakdown in rat liver plasma membranes. *J. Biol. Chem.* **261,** 2140–2146.

Uhl, B., Wolf, B., Schwinde, A., Metzger, J., Jung, G., Bessler, W. G., and Hauschildt, S. (1991). Intracellular localization of a lipopeptide macrophage activator: Immunocytochemical investigations and EELS analysis on ultrathin cryosections of bone marrow-derived macrophages. *J. Leukocyte Biol.* **50,** 10–18.

Uhl, B., Speth, V., Wolf, B., Jung, G., Bessler, W. G., and Hauschildt, S. (1992). Rapid alterations in the plasma membrane structure of macrophages stimulated with bacterial lipopeptides. *Eur. J. Cell Biol.* **58,** 90–98.

Unkeless, J. C., Fleit, H., and Mellman, I. S. (1981). Structural aspects and heterogeneity of immunoglobin Fc receptors. *Adv. Immunol.* **31,** 247–270.

Unkeless, J. C., Scigliono, E., and Freedman, V. H. (1988). Structure and function of human and murine receptors for IgG. *Annu. Rev. Immunol.* **6,** 251–281.

Uyemura, K., Ho, C. T., Ohmen, J., Rea, T. H., Bloom, B. R., and Modlin, R. L. (1992). Cytokine patterns of immunologically-mediated tissue damage. *J. Immunol.* **149,** 1470–1475.

van Alphen, L., Verkleij, A., Leunissen-Bijvelt, J., and Lugtenberg, B. (1978). Architecture of the outer membrane of *Escherichia coli.* III. Protein-lipopolysaccharide complexes in intramembraneous particles. *J. Bacteriol.* **134,** 1089–1098.

Van Bleek, G. M., and Nathenson, S. G. (1990). Isolation of an endogeneously processed immunodominant viral peptide from the class I H-2K^b molecule. *Nature (London)* **348,** 213–216.

Vanin, A. F., Mordvintcev, P. J., Hauschildt, S., and Mülsch, A. (1993). The relationship between L-arginine-dependent nitric oxide synthesis, nitrite release and dinitrosyl-iron complex formation by activated macrophages. *Biochim. Biophys. Acta* **1177,** 37–42.

Verghese, M. W., and Snyderman, R. (1989). Signal transduction and intracellular messengers. *In* "Human Monocytes" (M. Zembala and G. L. Asherson, eds.), pp. 101–112. Academic Press, London.

Vilcek, J., and Le, J. (1991). Immunology of cytokines: An introduction. *In* "The Cytokine Handbook" (A. Thomson, ed.). pp. 1–17. Academic Press, London.

Vordermeier, H.-M., and Bessler, W. G. (1987). Polyclonal activation of murine B-lymphocytes in vitro by *Salmonella typhimurium* porins. *Immunobiology* **175,** 245–251.

Vordermeier, H.-M., Drexler, H., and Bessler, W. G. (1987). Polyclonal activation of human peripheral blood lymphocytes by bacterial porins and defined porin fragments. *Immunol. Lett.* **15,** 121–126.

Vordermeier, H.-M., Hoffmann, P., Gombert, F. O., Jung, G., and Bessler, W. G. (1990). Synthetic peptide segments from the *Escherichia coli* porin OmpF constitute leukocyte activators. *Infect. Immun.* **58,** 2719–2724.

Wahl, S. M., McCartney-Francis, N., Allen, J. B., Dougherty, F. B., and Dougherty, S. F. (1990). Macrophage production of TGF-beta and regulation by TGF-beta. *Ann. N.Y. Acad. Sci.* **593,** 188–196.

Wallace, M. A., and Fain, J. N. (1985). Guanosine 5'-O-thiotriphosphate stimulates phospholipase C activity in plasma membranes of rat hepatocytes. *J. Biol. Chem.* **260,** 9527–9530.

Weber, G., Heck, D., Bartlett, R. R., and Nixdorff, K. (1992). Modulation of effects of lipopolysaccharide on macrophages by a major outer membrane protein of *Proteus mirabilis* as measured in a chemiluminescence assay. *Infect. Immun.* **60,** 1069–1075.

Weber, G., Link, F., Ferber, E., Munder, P. G., Zeitter, D., Bartlett, R. R., and Nixdorff, K. (1993). Differential modulation of the effects of lipopolysaccharide on macrophages by a major outer membrane protein of *Proteus mirabilis. J. Immunol.* **151,** 415–424.

Weinberg, J. B., Chapman, H. A., Jr., and Hibbs, J. B., Jr. (1978). Characterization of the effects of endotoxin on macrophage tumor cell killing. *J. Immunol.* **178,** 72–80.

Weinstein, S. L., Gold, M. R., and DeFranco, A. L. (1991). Bacterial lipopolysaccharide stimulates protein tyrosine phosphorylation in macrophages. *Proc. Natl. Acad. Sci. U.S.A.* **88,** 4148–4152.

Weinstein, S. L., Sanghera, J. S., Lemke, K., DeFranco, A. L., and Pelech, S. L. (1992). Bacterial lipolysaccharide induces tyrosine phosphorylation and activation of mitogen-activated protein kinase in macrophages. *J. Biol. Chem.* **267,** 14955–14962.

Weinstein, S. L., June, C. H., and DeFranco, A. L. (1993). Lipopolysaccharide-induced protein tyrosine phosphorylation in human macrophages is mediated by CD14. *J. Immunol.* **151,** 3829–3838.

Wheeler, K., Pound, J. D., Gordon, J., and Jefferis, R. (1993). Engagement of CD40 lowers the threshold for activation of resting B cells via antigen receptor. *Eur. J. Immunol.* **23,** 1165–1168.

Wheeler, P. R., and Gregory, D. (1980). Superoxide dismutase, peroxidatic activity and catalase in *Mycobacterium leprae* purified from armadillo liver. *J. Gen. Microbiol.* **121,** 457–464.

Wiesmüller, K.-H., Hess, G., Bessler, W. G., and Jung, G. (1990). Diastereomers of tripalmitoyl-S-glyceryl-L-cysteinyl carrier adjuvant systems induce different mitogenic and protective immune responses. *In* "Chirality and Biological Activity, Proceedings of the International Symposium, Tübingen 1988" (B. Holmstedt, H. Frank, and B. Testa, eds.), pp. 267–272. Alan R. Liss, New York.

Wolf, B., Hauschildt, S., Uhl, B., Metzger, J., Jung, G., and Bessler, W. G. (1989). Localization of the cell activator lipopeptide in bone marrow-derived macrophages by electron energy loss spectroscopy (EELS). *Immunol. Lett.* **20,** 121–126.

Wolinski, E. (1990). Mycobacteria. *In* "Microbiology" (B. D. Davis, R. Dulbecco, H. N. Eisen, and H. S. Ginsberg, eds.), pp. 648–664. Lippincott, Philadelphia.

Wong, G. H. W., and Goeddel, D. V. (1989). Tumor necrosis factor. *In* "Tumor Necrosis Factor in Human Monocytes" (M. Zembala and G. L. Asherson, eds.), pp. 195–215. Academic Press, London.

Wright, S. D., and Griffin, F. M., Jr. (1985). Activation of phagocytic cells' C3 receptors for phagocytosis. *J. Leukocyte Biol.* **38,** 327–329.

Wright, S. D., and Silverstein, S. C. (1982). Tumor-promoting phorbol esters stimulate C3b and C3b' receptor-mediated phagocytosis in cultured human monocates. *J. Exp. Med.* **156,** 1149–1164.

Wright, S. D., and Silverstein, S. C. (1983). Receptors for C3b and C3bi promote phagocytosis but not the release of toxic oxygen from human phagocytes. *J. Exp. Med.* **158,** 2016–2023.

Yamamoto, K., and Johnston, R. B., Jr. (1984). Dissociation of phagocytosis from stimulation of the oxidative metabolic burst in macrophages. *J. Exp. Med.* **159,** 405–416.

Young, J. D. E., Ko, S. S., and Cohn, Z. A. (1984). The increase in intracellular free calcium associated with IgG γ2b/γ1 Fc-receptor-ligand interactions: Role in phagocytosis. *Proc. Natl. Acad. Sci. U.S.A.* **81,** 5430–5434.

Zheng, L., Nibbering, P. H., and van Furth, R. (1993). Stimulation of the intracellular killing of *Staphylococcus aureus* by human monocytes mediated by Fc-gamma receptors I and II. *Eur. J. Immunol.* **23,** 2826–2833.

Zhu, L., Gunn, C., and Beckman, J. S. (1992). Bactericidal activity of peroxynitrite. *Arch. Biochem. Biophys.* **298,** 452–457.

Zucali, J. R., Dinarello, C. A., Oblon, D. J., Gross, M. A., Anderson, L., and Weiner, R. S. (1986). Interleukin 1 stimulates fibroblasts to produce granulocyte-macrophage colony-stimulating activity and prostaglandin E$_2$. *J. Clin. Invest.* **11,** 1857–1863.

Index